NANOPHOTONICS WITH SURFACE PLASMONS

Advances in
NANO-OPTICS AND NANO-PHOTONICS

Series Editors

Satoshi Kawata
Department of Applied Physics
Osaka University, Japan

Vladimir M. Shalaev
Purdue University
School of Electrical and Computer Engineering
West Lafayette, IN, USA

NANOPHOTONICS WITH SURFACE PLASMONS

Edited by

V.M. SHALAEV
Purdue University
School of Electrical & Computer Engineering
Indiana, USA

S. KAWATA
Department of Applied Physics
Osaka University, Japan

ELSEVIER

AMSTERDAM • BOSTON • HEIDELBERG • LONDON NEW YORK • OXFORD
PARIS • SAN DIEGO • SAN FRANCISCO • SINGAPORE • SYDNEY • TOKYO

Elsevier
Radarweg 29, PO Box 211, 1000 AE Amsterdam, The Netherlands
The Boulevard, Langford Lane, Kidlington, Oxford OX5 1GB, UK

First edition 2007

Library of Congress Cataloging in Publication Data
A catalogue record for this book is available from the Library of Congress.

British Library Cataloguing in Publication Data
A catalogue record for this book is available from the British Library.

ISBN-13: 978-0-444-52838-4

ISSN: 1871-0018

Transferred to digital print 2008

Printed and bound in Great Britain by CPI Antony Rowe, Eastbourne

Preface

There is an undeniable and ever-increasing need for faster information processing and transport. Many believe that the current electronic techniques are running out of steam due to issues with RC-delay times, meaning that fundamentally new approaches are needed to increase data processing operating speeds to THz and higher frequencies. The photon is the ultimate unit of information because it packages data in a signal of zero mass and has unmatched speed. The power of light is driving the photonic revolution, and information technologies, which were formerly entirely electronic, are increasingly enlisting light to communicate and provide intelligent control. Today we are at a crossroads in this technology. Recent advances in this emerging area now enable us to mount a systematic approach toward the goal of full system-level integration.

The mission that researchers are currently trying to accomplish is to fully integrate photonics with nanotechnology and to develop novel photonic devices for manipulating light on the nanoscale, including molecule sensing, biomedical imaging, and processing information with unparalleled operating speeds. To enable the mission one can use the unique property of metal nanostructures to "focus" light on the nanoscale. Metal nanostructures supporting collective electron oscillations – plasmons – are referred to as plasmonic nanostructures, which act as optical nanoantennae by concentrating large electromagnetic energy on the nanoscale.

There is ample evidence that photonic devices can be reduced to the nanoscale using optical phenomena in the near field, but there is also a scale mismatch between light *at the microscale and devices and processes at the nanoscale* that must first be addressed. Plasmonic nanostructures can serve as optical couplers across the nano–micro interface. They also have the unique ability to enhance local electromagnetic fields for a number of ultra-compact, subwavelength photonic devices. Nanophotonics is not only about very small photonic circuits and chips, but also about new ways of sculpting the flow of light with nanostructures and nanoparticles exhibiting fascinating optical properties never seen in macro-world.

Plasmonic nanostructures utilizing surface plasmons (SPs) have been extensively investigated during the last decade and show a plethora of amazing effects and fascinating phenomena, such as extraordinary light transmission, giant field enhancement, SP nano-guides, and recently emerged metamaterials that are often based on plamonic nanostructures. Nanoplasmonics-based metamaterials are expected to open a new gateway to unprecedented electromagnetic properties and functionality unattainable from naturally occurring materials. The structural units of metamaterials can be tailored in shape and size, their composition and morphology can be artificially tuned, and inclusions can be designed and placed at desired locations to achieve new functionality.

As the Editors of this volume we are deeply grateful to all contributing authors, leading experts in various areas of nanoplasmoincs, for their effort and their willingness to share recent results within the framework of this volume.

Vladimir M. Shalaev and Satoshi Kawata

Contents

List of Contributors

Sergey I. Bozhevolnyi	Department of Physics and Nanotechnology, Aalborg University, Aalborg Øst, Denmark
Mark Brongersma	Geballe Laboratory for Advanced Materials, Stanford University, Stanford, CA, USA
A. Dereux	Laboratoire de Physique de l'Université de Bourgogne, Optique Submicronique, Dijon, France
Vladimir P. Drachev	School of Electrical and Computer Engineering and Birck Nanotechnology Center, Purdue University, West Lafayette, IN, USA
Alexander V. Kildishev	School of Electrical and Computer Engineering and Birck Nanotechnology Center, Purdue University, IN, USA
Thomas A. Klar	School of Electrical and Computer Engineering and Birck Nanotechnology Center, Purdue University, West Lafayette, IN, USA
	Physics Department and CeNS, Ludwig-Maximilians-Universität, Amalienstr. 54 München, Germany
Alexey V. Krasavin	EPSRC Nanophotonics Portfolio Centre, School of Physics and Astronomy, University of Southampton, Highfield, Southampton, UK
J. R. Krenn	Institute of Physics and Erwin Schrödinger Institute for Nanoscale Research, Karl–Franzens University, Graz, Austria

Kevin F. MacDonald	EPSRC Nanophotonics Portfolio Centre, School of Physics and Astronomy, University of Southampton, Highfield, Southampton, UK
M. A. Noginov	Center for Materials Research, Norfolk State University, Norfolk, VA, USA
Vladimir M. Shalaev	School of Electrical and Computer Engineering and Birck Nanotechnology Center, Purdue University, West Lafayette, IN, USA
Igor I. Smolyaninov	Department of Electrical and Computer Engineering, University of Maryland, College Park, MD, USA
Junji Tominaga	National Institute of Advanced Industrial Science and Technology, AIST, Center for Applied Near-Field Optics Research, Tsukuba, Japan
J.-C. Weeber	Laboratoire de Physique de l'Université de Bourgogne, Optique Submicronique, Dijon, France
Anatoly V. Zayats	Centre for Nanostructured Media, IRCEP, The Queen's University of Belfast, Belfast, UK
Nikolay I. Zheludev	EPSRC Nanophotonics Portfolio Centre, School of Physics and Astronomy, University of Southampton, Highfield, Southampton, UK
G. Zhu	Center for Materials Research, Norfolk State University, Norfolk, VA, USA
Rashid Zia	Brown University, Division of Engineering, Box D, Providence, RI 02912

Chapter 1

Dynamic components utilizing long-range surface plasmon polaritons

by

Sergey I. Bozhevolnyi

Department of Physics and Nanotechnology, Aalborg University, Skjernvej 4A, DK-9220 Aalborg Øst, Denmark

Nanophotonics with Surface Plasmons
Advances in Nano-Optics and Nano-Photonics
ISSN: 1871-0018

V.M. Shalaev & S. Kawata (Editors)
© 2007 Published by Elsevier B.V.
DOI: 10.1016/S1871-0018(06)02001-2

Contents

§ 1. Introduction

Integrated optical devices and circuits are being increasingly used for light routing and switching in the rapidly developing area of broadband optical communications. Such devices are traditionally based on guiding of light in a dielectric waveguide consisting of a core and a cladding, with the refractive index of the former being larger than that of the latter (Marcuse, 1974). Electromagnetic radiation propagating in and confined to the core (by virtue of total internal reflection) in the form of waveguide modes can be controlled with externally applied electrical signals via, for example, electro-, magneto-, and thermo-optic effects, depending on the dielectric properties and electrode configuration (Hunsperger, 1995). The necessity of introducing controlling electrodes, which are usually metallic, close to waveguides bring about a problem associated with the incurrence of additional loss of radiation due to its absorption. The effect of absorption can be minimized with increasing the electrode–waveguide separation, but that would decrease the aforementioned (useful) effects as well, a circumstance that makes the positioning of electrodes in conventional waveguide modulators and switches a challenging design problem. Ideally, one would like to send the light and electrical signals along the same channel facilitating the information transfer from electronic to optical circuits.

We have recently demonstrated that the aforementioned problem can be circumvented by using thin metal stripes surrounded by dielectric for *both* guiding of radiation in the form of plasmon–polariton modes and control, i.e., modulation and switching, of its propagation (Nikolajsen et al., 2004). Surface plasmon polaritons (SPPs) are light waves that are coupled to oscillations of free electrons in a conductor, usually a metal, and propagating along the metal–dielectric interface (Raether, 1988). For a sufficiently thin metal film embedded in dielectric, the SPPs associated with the upper and lower interfaces couple and form a symmetric mode, a long-range SPP (LRSPP), whose propagation loss decreases with the decrease of the film thickness (Burke et al., 1981). Furthermore, a thin metal stripe surrounded by dielectric supports the propagation of an LRSPP stripe mode, whose field distribution can be adjusted (by varying the stripe thickness and width) close to that of a single-mode fiber (Berini, 2000; Charbonneau et al., 2000; Nikolajsen et al., 2003). Thus, efficient LRSPP excitation and guiding (at telecom wavelengths) along 10-nm-thin gold stripes embedded in polymer (fig. 1) was realized demonstrating the coupling loss of ~0.5 dB and propagation loss of ~6–8 dB/cm (Nikolajsen et al., 2003).

Low propagation and coupling loss attainable with LRSPPs have stimulated experimental studies of LRSPP-based integrated optics, and different passive components including straight and bent waveguides, Y-splitters, multimode interference devices and directional couplers have been recently demonstrated (Boltasseva et al., 2005b; Charbonneau et al., 2005). As an alternative approach for making photonic circuits, LRSPP stripe waveguides have a unique feature – the possibility of using the same metal stripe circuitry for both guiding optical radiation and transmitting electrical signals that control its guidance. Lately, efficient LRSPP-based dynamic devices with low power consumption, including various modulators and switches, have been realized utilizing the thermo-optic effect in the polymer cladding and demonstrating thereby first examples of electrically controlled plasmonic components (Nikolajsen et al., 2004,

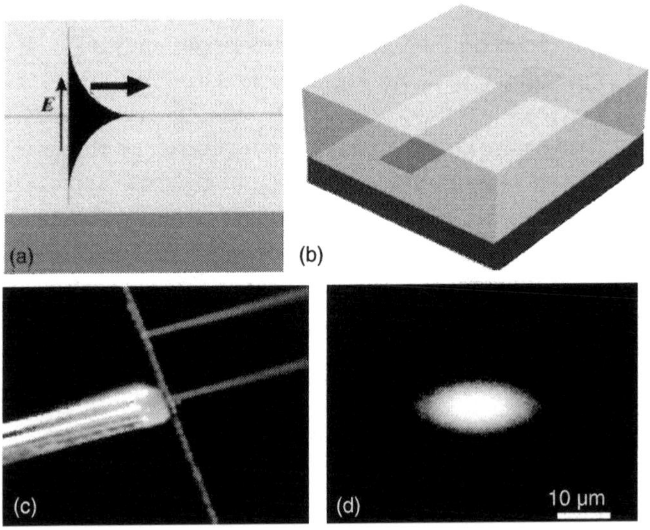

Fig. 1. (a) Schematic representation of the LRSPP field distribution near a thin metal film embedded in dielectric along with the orientation of the dominant electric field component. (b) Schematic layout of an LRSPP stripe waveguide. (c) Optical microscope image of the end-fire in/out coupling arrangement showing a cleaved single-mode fiber and a fabricated sample with stripe waveguides. (d) Optical microscope image of the intensity distribution of fundamental LRSPP mode at the output facet of the stripe waveguide excited at the wavelength of 1.55 μm.

2005). It has also been shown that essentially the same metal stripes, which constitute the heart of LRSPP-based modulators and switches, can be used to monitor the transmitted LRSPP power by means of measuring variations in the stripe resistance (Bozhevolnyi et al., 2005b). In addition, together with different passive and active LRSPP-based components for integrated optics, two different approaches for making Bragg gratings based on LRSPP-supporting configurations, i.e., by varying widths (Jetté-Charbonneau et al., 2005) and thickness (Bozhevolnyi et al., 2005a) of the metal stripe, have been recently reported where a very broad range of LRSPP-based grating performance (from weak narrow-band gratings up to very strong and broad-band gratings) has been experimentally demonstrated. Furthermore, LRSPP gratings (with variable metal thickness) tilted with respect to the stripe direction have been used to realize a compact and efficient Z-add-drop filter (Boltasseva et al., 2005a). Overall, recent investigations demonstrate convincingly that LRSPP-based components constitute quite a promising alternative for integrated photonic circuits meeting low-cost, simplicity of fabrication, flexibility as well as performance requirements.

Here, first examples of thermo-optic LRSPP-based components, i.e., a Mach-Zehnder interferometric modulator (MZIM), directional-coupler switch (DCS), in-line extinction modulator (ILEM) and integrated power monitor, whose operation utilizes thin gold stripes embedded in polymer and transmitting both LRSPPs and electrical signal currents, are reviewed. This chapter is organized as follows. Fundamentals of the LRSPP planar and stripe waveguides, including the influence of asymmetry in the refractive index distribution, are considered in Section 2. Section 3 is devoted to basic LRSPP stripe waveguide fabrication and characterization. Realization and investigations of thermo-optic MZIMs and DCSs are described in Section 4. Design, fabrication and characterization of ILEMs and power monitors are presented in Sections 5 and 6, respectively. The chapter terminates with the outlook in Section 7.

§ 2. Fundamentals of long-range surface plasmon polaritons

It has been long known that any interface between two media having dielectric susceptibilities with opposite signs of their real parts can support propagation of surface waves (polaritons), whose fields decrease exponentially into both neighbor media. Negative values of the dielectric function are achieved due to the resonant material response, e.g., at the long-wavelength side of plasmon resonance in metals (i.e., the resonance in free electron oscillations) with surface polaritons being conveniently termed SPPs (Raether, 1988). The corresponding (SPP) propagation constant β can be found from matching the tangential electric and magnetic field components across the interface:

$$\beta = \frac{\omega}{c}\sqrt{\frac{\varepsilon_{\mathrm{d}}\varepsilon_{\mathrm{m}}}{\varepsilon_{\mathrm{d}} + \varepsilon_{\mathrm{m}}}}, \tag{1.1}$$

where ω and c are the frequency and speed of electromagnetic waves in vacuum, ε_{d} and ε_{m} are the dielectric susceptibilities of dielectric and metal, respectively. Assuming that $\mathrm{Re}\{\varepsilon_{\mathrm{d}}\}>0$ and $\mathrm{Re}\{\varepsilon_{\mathrm{m}}\}<0$, it is seen that the condition of SPP existence is in fact the following unequality: $\mathrm{Re}\{\varepsilon_{\mathrm{d}}\}<-\mathrm{Re}\{\varepsilon_{\mathrm{m}}\}$.

The metal susceptibility is a complex number containing an imaginary part related to the absorption of radiation by the metal (ohmic loss). Consequently, the SPP propagation constant β is also complex number, with the real part determining the SPP wavelength $\lambda_{\mathrm{SPP}} = 2\pi/\mathrm{Re}\beta < \lambda = 2\pi c/\omega$ and the imaginary part – the SPP propagation length $L_{\mathrm{SPP}} = (2\mathrm{Im}\beta)^{-1}$. Due to the relatively small propagation length (~30 μm in visible and ~300 μm in the near-infrared wavelength range for a silver–air interface (Raether, 1988)), SPPs are considered to be somewhat limited in their applications. However, by changing a metal–dielectric interface to a symmetrical structure of a thin metal film embedded in dielectric, one can significantly decrease the SPP propagation loss (Sarid, 1981). In this symmetrical structure, two identical SPPs associated with the two (upper and lower) metal–dielectric interfaces become coupled, forming symmetrical and asymmetrical (with respect to the orientation of the main electric field component) modes whose propagation constants can be found from the implicit dispersion relation (Burke et al., 1986):

$$\tanh(S_{\mathrm{m}}t) = -\frac{2\varepsilon_{\mathrm{d}}S_{\mathrm{d}}\varepsilon_{\mathrm{m}}S_{\mathrm{m}}}{\varepsilon_{\mathrm{d}}^2S_{\mathrm{m}}^2 + \varepsilon_{\mathrm{m}}^2S_{\mathrm{d}}^2}, \quad S_{\mathrm{d}} = \sqrt{\beta^2 - \varepsilon_{\mathrm{d}}k_0^2}, \quad S_{\mathrm{m}} = \sqrt{\beta^2 - \varepsilon_{\mathrm{m}}k_0^2}, \tag{1.2}$$

where t is the metal film thickness and $k_0 = \omega/c$ is the light wave number in vacuum.

2.1. Long-range surface plasmon polaritons

It turns out that, of two modes described by the above dispersion relation (1.2), the symmetrical mode, called LRSPP (fig. 1(a)), extends progressively into the dielectric cladding (up to several micrometers) and becomes only weakly attached to the metal for thinner metal films. Consequently, the part of mode field within the metal becomes also progressively small, decreasing drastically the mode absorption and propagation loss. Due to an increased field penetration in the dielectric cladding, a thin metal stripe (surrounded by dielectric) supports the propagation of an LRSPP stripe mode, whose field distribution can be adjusted (by varying the stripe thickness and width) rather close to that of a single-mode fiber (fig. 1(b)–(d)). An accurate theoretical description of the LRSPP dispersion and mode field profiles in the case of finite-width and finite-thickness metal stripes is rather complicated, and requires elaborate numerical modeling (Berini, 2000; Al-Bader, 2004). Here, a simple approach based on the effective index approximation is used (Boltasseva et al., 2005b).

As a first step, we considered planar (symmetrical) geometry shown in fig. 2(a). A metal film of variable thickness t is surrounded by two identical dielectric layers characterized by the refractive index $n = 1.535$, corresponding to the refractive index of BCB (benzocyclobutene) polymer at the light wavelength of 1.55 μm, and variable thickness d. The structure is placed on a silicon substrate with a refractive index of 3.47. The metal in

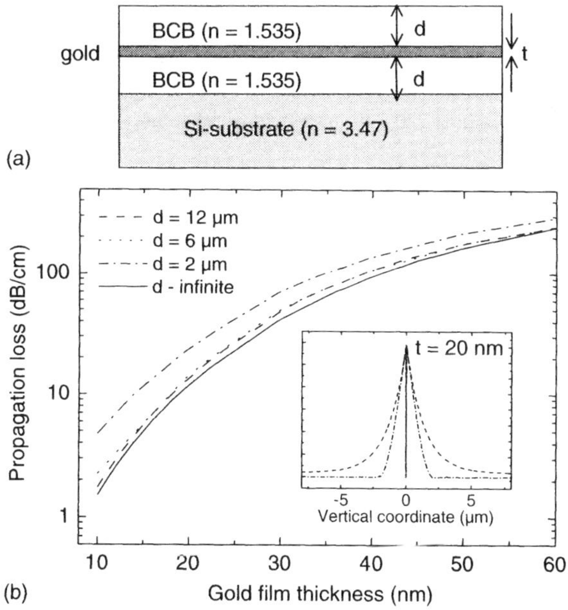

Fig. 2. (a) Symmetrical geometry of an infinitely wide metal film of variable thickness t surrounded by two identical polymer ($n = 1.535$) layers of variable thickness d. The structure is placed on a silicon substrate ($n = 3.47$). (b) Dependence of the LRSPP propagation loss on the gold film thickness at the wavelength of 1550 nm for different thickness of polymer cladding layers. The vertical mode profiles for the 20-nm-thick gold film are shown in the inset for two different cladding thicknesses. (This figure is taken from Boltasseva et al., 2005b.)

our analysis is gold with the complex refractive index $n = 0.55 + 11.5i$ (this value is in fact also close to that of silver at 1.55 μm).

We analyzed the LRSPP propagation loss at the wavelength of 1.55 μm for different thicknesses of metal film and BCB cladding (fig. 2(b)). For infinite polymer cladding the propagation loss was found to increase monotonically when increasing film thickness from ~1.5 dB/cm (for a 10-nm-thick gold film) to ~250 dB/cm (for the film thickness of 60 nm). It should be emphasized that in order to support LRSPP propagation one should ensure a symmetrical structure. This means that two polymer layers should have the same refractive index and be thick enough, so that the LRSPP field is located inside the polymer and does not penetrate into the silicon substrate or air. The LRSPP mode profile in depth (perpendicular to the sample surface) is mainly determined by the metal thickness and reflects how tight the LRSPP is bound to the metal. Here we should mention that, in turn, the cladding (polymer) thickness can be used to tune the LRSPP depth profile (Nikolajsen et al., 2003), as demonstrated in the inset of fig. 2(b). For the gold thickness of 20 nm, the breadth of the LRSPP depth profile changes from ~10 μm for a 12-μm-thick cladding to ~4 μm for the polymer thickness of 2 μm. However, besides the control of the LRSPP depth profile, the decrease in the cladding thickness increases the propagation loss. For example, reducing polymer thickness to 2 μm will change, for a 10-nm-thick metal film, the LRSPP propagation loss from ~1.5 to ~5 dB/cm (fig. 2(b)).

To study the influence of asymmetry in the cladding indexes on LRSPP properties we analyzed the same geometry as in fig. 2(a) for the cladding thickness of 12 μm but with a variable refractive index of the top cladding (fig. 3(a)). The dependence of the LRSPP propagation loss on the refractive index difference between top and bottom cladding layers is shown in fig. 3(b) for gold thicknesses of 10 and 15 nm. For example, for a 10-nm-thick film the LRSPP mode was found to have the propagation loss increasing from 1.7 dB/cm (for the symmetrical structure) to ~4 dB/cm (for the refractive index difference of ±0.006). The increase in the propagation loss with the increasing asymmetry is accompanied with the change from a symmetrical LRSPP mode depth profile to an asymmetrical one (inset of fig. 3(b)). Further increase of the refractive index difference (more than ±0.006) will create a conventional slab waveguide formed by a polymer layer with a higher refractive index surrounded by two media with lower refractive indexes, resulting in the propagating mode of the slab waveguide instead of the LRSPP mode.

The dependence of the LRSPP normalized effective refractive index b on the gold film thickness is presented in fig. 4 with the normalized index

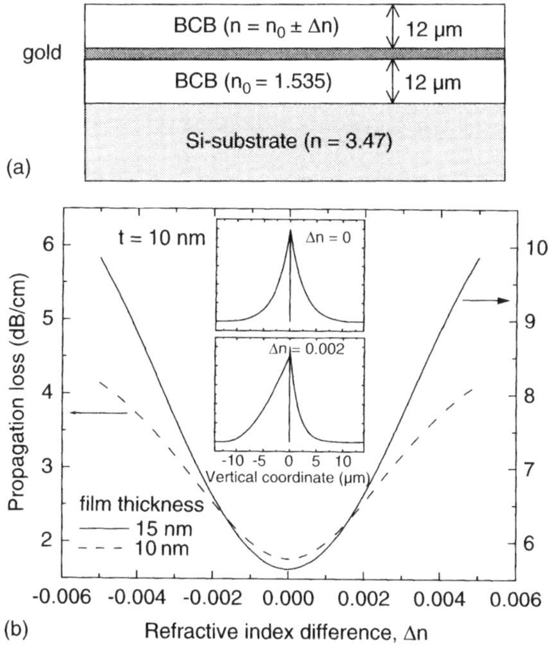

Fig. 3. (a) Same geometry as in fig. 2(a) for a polymer cladding thickness of 12 μm only with the variable refractive index of the top polymer cladding. (b) Dependence of the LRSPP propagation loss on the refractive index difference between two polymer claddings at the wavelength of 1550 nm for 10- and 15-nm-thick gold films. The vertical mode profiles for the 10-nm-thick gold film are shown in the inset for 0 and 0.002 differences between cladding indices. (This figure is taken from Boltasseva et al., 2005b.)

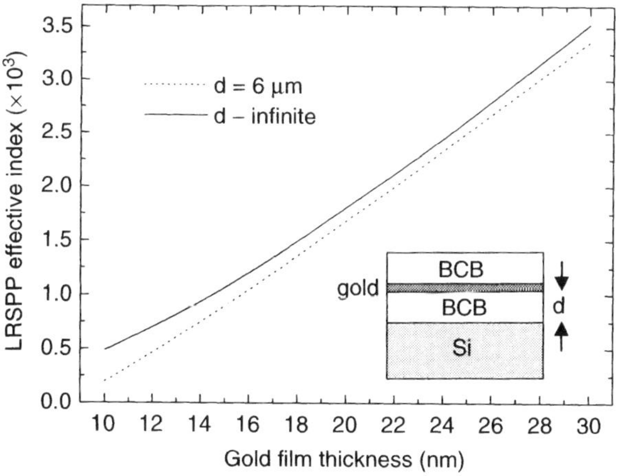

Fig. 4. The dependence of the LRSPP effective refractive index on the gold film thickness for the infinite and 6-μm-thick polymer cladding. (This figure is taken from Boltasseva et al., 2005b.)

b being conveniently determined as

$$b = \frac{\beta - k_0 n_{\text{cl}}}{k_0 n_{\text{cl}}} = \frac{N_{\text{eff}} - n_{\text{cl}}}{c_{\text{cl}}}, \tag{1.3}$$

where $\beta = (2\pi/\lambda)$, N_{eff} is the LRSPP propagation constant, λ is the light wavelength (1.55 μm), n_{cl} is the refractive index of the cladding (1.535) and N_{eff} is the LRSPP mode effective refractive index. It should be noted that the normalized index depends very weakly on the cladding refractive index, allowing one to use the dependencies shown in fig. 4 for determination of the LRSPP propagation constant for the configurations with different cladding materials.

2.2. *LRSPP stripe modes*

The properties of LRSPP modes guided by a waveguide structure composed of a thin lossy metal film of finite width, surrounded by dielectric, were for the first time considered theoretically by Berini (2000). In our simple qualitative analysis, the characteristics of the LRSPP mode propagating in a stripe metal waveguide of finite width were found by using the effective refractive index method, which is considered to be reasonably accurate for waveguide modes being far from cutoff (Kogelnik, 1979) and found to give fairly good predictions for the behavior of LRSPP stripe waveguides. The geometry that we considered is shown in fig. 5 (a). A metal strip of variable thickness t and width w is surrounded by polymer characterized by the refractive index n, and the whole structure is placed on a silicon substrate.

In the first step, the structure with an infinitely wide metal film is analyzed resulting in the vertical LRSPP mode profile and the effective index, which is used in the second step as the refractive index of a core in the slab waveguide configuration (the core thickness is considered equal to the stripe width). The waveguide analysis at the second step provides us with the lateral mode profile (parallel to the sample surface) as well as the corrected value for the mode effective refractive index and propagation loss. The lateral LRSPP mode field diameter (MFD) is shown in fig. 5(b) as a function of the stripe width for gold film thicknesses of 10 and 14 nm. A typical behavior of the lateral LRSPP MFD was found first to decrease following the decrease in the stripe width and then to increase again demonstrating a poor light confinement by narrow stripes (Berini, 2000).

The LRSPP mode effective index together with the propagation loss as a function of the waveguide width for a 10-nm-thick stripe is shown in

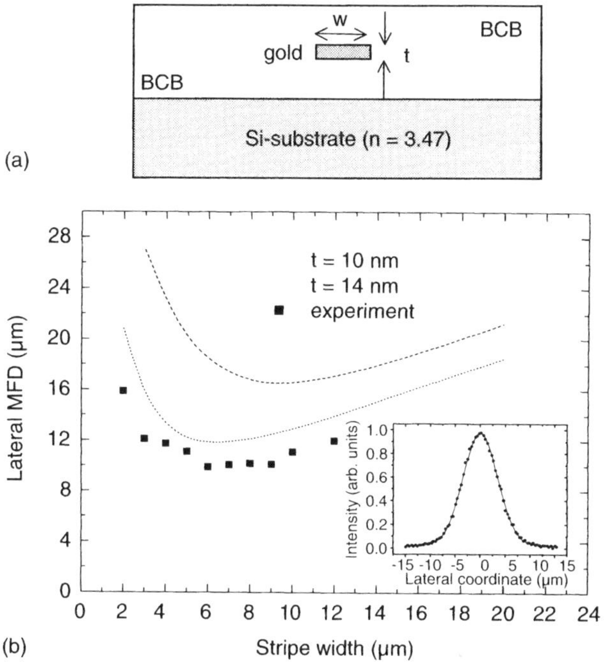

Fig. 5. (a) The geometry of a metal stripe of variable thickness t and width w surrounded by polymer ($n = 1.535$) layers. The structure is placed on a silicon substrate ($n = 3.47$). (b) The dependence of the lateral LRSPP MFD on the stripe width for gold film thicknesses of 10 and 14 nm. Modeling performed using the effective index approach. Dots represent the values measured for 15-nm-thick stripes sandwiched between 15-μm-thick polymer cladding layers. The inset shows an example of the lateral intensity profile fitted to a Gaussian distribution. (This figure is taken from Boltasseva et al., 2005b.)

fig. 6. The simulations indicate that, for the stripe thickness of 10 nm, the multimode regime sets in for stripes wider than 20 μm (fig. 6(a)). This feature was used to design multimode-interference (MMI) waveguide structures (Boltasseva et al., 2005b). The propagation loss was found to decrease with the stripe width (a similar trend was also predicted in (Berini, 2000)) implying the possibility to reach very low propagation loss. For example, the propagation loss below 1 dB/cm can be achieved for a 10-nm-thick stripe by reducing its widths below 5 μm (fig. 6(b)). Finally, one can also analyze the influence of asymmetry of the refractive index cladding distribution on the LRSPP stripe modes similarly to what has been done for the planar geometry (fig. 3). Elaborate modeling of this problem based on the normal mode analysis using a fully vectorial formulation has also been recently reported (Breukelaar et al., 2006).

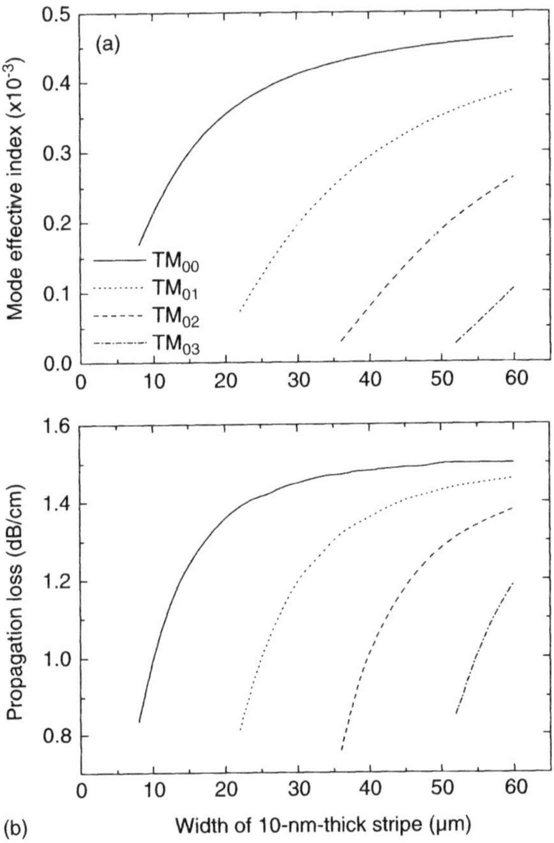

Fig. 6. The LRSPP mode effective index (a) together with the propagation loss (b) as a function of the waveguide width for a 10-nm-thick stripe. Modeling is performed using the effective index approach. (This figure is taken from Boltasseva et al., 2005b.)

§ 3. Basic waveguide fabrication and characterization

Fabrication of LRSPP stripe waveguides involved spin coating of a silicon substrate (4″ or 6″) with a layer of polymer BCB having a thickness of 13–15 μm and then with a layer of UV resist material. Straight stripe waveguides and various waveguide structures were patterned using standard UV lithography, gold deposition and liftoff. As a final fabrication step the spin coating with the top cladding, comprising another 13- to 15-μm-thick BCB layer, was performed. Since the symmetry of the structure is very important for the LRSPP properties (propagation loss, MFD) we controlled carefully that the cladding layers had the same refractive index and were thick enough to accommodate the EM field of the

LRSPP. This was guaranteed by applying the same polymer for the top and bottom claddings and using identical spinning and curing conditions. After the final polymer curing the wafer was cut into individual samples.

For optical characterization of the LRSPP stripe waveguides and waveguide devices standard transmission measurements were performed. In order to excite the LRSPP mode end-fire coupling of light was performed using a tunable laser (1550 nm or 1570 nm) or a broadband light source (two multiplexed EE-LED diodes – 1310 nm and 1550 nm) together with a polarization controller, as a source. Light polarized perpendicular to the waveguide plane was launched into the LRSPP waveguide via butt coupling from a polarization-maintaining (PM) fiber with a MFD of 10.8 μm. To ensure that the polarization of light was orthogonal to the waveguide layer angular adjustments of the PM fiber were performed. Coiled standard single-mode fiber of 1 km was used as out-coupling fiber in order to strip off all light coupled into the fiber cladding. Index-matching gel was used to decrease the reflection at the sample edges. The output signal was detected by a power meter (for measurements performed with the laser) or optical spectrum analyzer (for broadband measurements). The final adjustment of the in- and out-coupling fibers with respect to a stripe waveguide was accomplished by maximizing the amount of light transmitted through the waveguide.

The propagation loss measurements were performed for 8-μm-wide straight stripe waveguides of different thicknesses (thickness of the deposited gold layer) from approximately 8.5 to 35 nm. At a particular wavelength the propagation loss was found as the slope of the linear fit to the experimental values of loss obtained for different lengths of the LRSPP waveguide (4, 8, 14 mm for waveguide thicknesses up to 15 nm and 2, 3, 4 mm for thicknesses up to 35 nm) (cutback method). This linear fitting technique allowed us to estimate the coupling loss from the intersection point on the loss axis corresponding to zero length of the waveguide. The value of the coupling loss for a 15-nm-thick stripe waveguide varied from approximately 0.5 dB per facet for a 10-μm-wide waveguide to ~1.5 dB for a 4-μm-wide stripe. Figure 7 shows the experimental results for the propagation loss at 1550 nm together with the LRSPP propagation loss curve calculated for infinitely wide stripes. Good agreement between experimental and calculated values, observed for waveguide thicknesses higher than 15 nm, clearly indicates that, for thick stripes, the internal damping in metal (ohmic loss) is dominating. For thin stripes, higher values of experimentally obtained propagation loss compared to the calculated values can be explained by the presence of other loss mechanisms such as the scattering by inhomogeneities in the gold

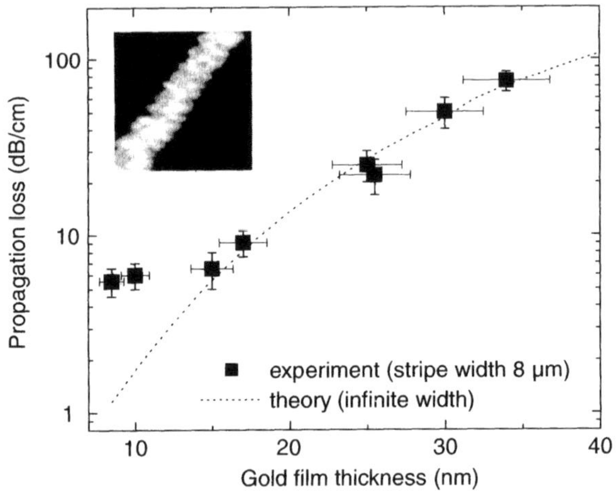

Fig. 7. Experimental measurements of the propagation loss dependence on the thickness of the 8-μm-wide stripe at the wavelength of 1550 nm together with the propagation loss curve calculated for infinitely wide stripes. The inset shows a typical near-field optical image ($69 \times 69\,\mu m^2$) obtained with a 5-μm-wide 10-nm-thick gold stripe. (This figure is adapted from Boltasseva et al., 2005b and Nikolajsen et al., 2003.)

structure, at the waveguide edges, and scattering and absorption in the polymer. By eliminating the described loss mechanisms one should achieve the loss limit set by the internal damping in metal, which is ~1.5 dB/cm for a 10-nm-thick infinitely wide stripe and decreases with the stripe width (Berini, 2000). Further reduction of the stripe thickness (< 10 nm) will hardly lead to a significant decrease in the propagation loss in practice due to fabrication difficulties in creating a very thin *homogeneous* metal layer. Since the flatness of a nanometer-thin film can be strongly influenced by that of a substrate surface, it is a rough polymer surface that sets, in our case, a 10–15 nm limit on the thickness of a film exhibiting thickness variations on the scale much smaller than the thickness itself.

In order to study the LRSPP mode profile the output intensity distribution from a stripe waveguide was monitored with a microscope arrangement imaging the waveguide output on an infrared vidicon camera with 200 × magnification. The PM fiber output with the known MFD was used for the calibration of the mode profile measurement system. The output intensity distribution at the output of the 15-nm-thick stripe waveguide for three different waveguide widths (4, 8 and 12 μm) is shown in fig. 8. The mode depth profile consists of two exponential decays with the decaying parameters, which are primarily determined by the metal thickness. However, for narrow stripes (less than 6 μm wide) the depth

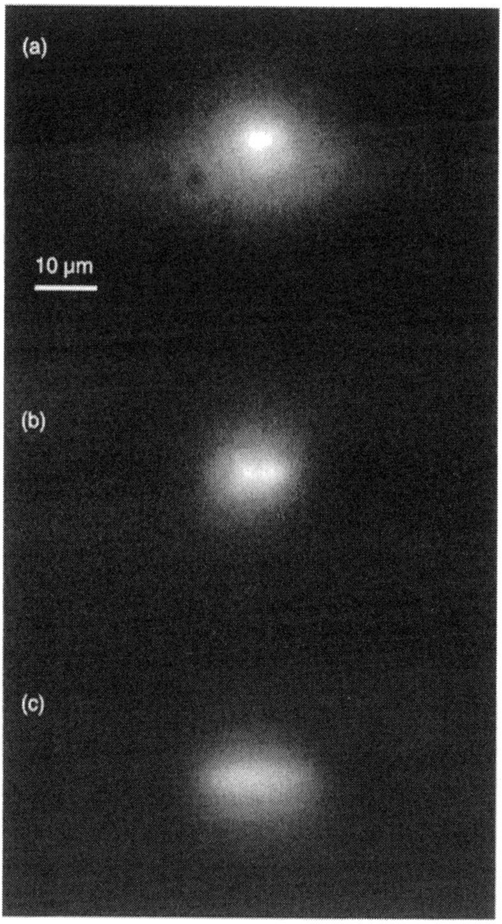

Fig. 8. The output intensity distribution at the output of the 15-nm-thick stripe waveguide for a 4- (a), 8- (b) and 12-µm-wide (c) stripe. (This figure is taken from Boltasseva et al., 2005b.)

MFD is expected to increase compared to the infinitely wide stripe of the same thickness (Berini, 2000), which is also seen from the experimentally obtained mode profiles (fig. 8). The LRSPP depth profile for an 8-µm-wide stripe together with the exponential fits is presented in fig. 9 showing quite good match except for around zero depth coordinate, where the intensity distribution was smoothened to Gaussian-like shape due to the limited resolution (1–1.5 µm) of the imaging system.

The lateral mode field profile was found to fit perfectly to a Gaussian distribution (see inset in fig. 5). The lateral MFD determined as a function of the 15-nm-thick stripe width (from 2 to 12 µm) is presented in fig. 5. It is seen that the lateral MFD decreases from ~12 µm for the stripe width

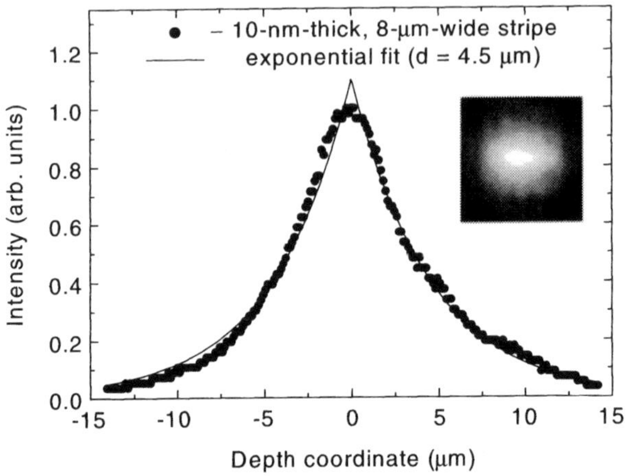

Fig. 9. The LRSPP depth profile for an 8-μm-wide stripe together with the exponential fits. The inset shows an example of the LRSPP stripe output intensity distribution. (This figure is taken from Boltasseva et al., 2005b.)

of 12 μm to ~10 μm for 6- to 8-μm-wide stripes, following the decrease in the waveguide width, and then starts to increase, reaching ~16 μm for a 2-μm-wide stripe waveguide, due to weaker light confinement for narrow stripes (Berini, 2000). This behavior is found to be in good agreement with the results of our simulations (fig. 5). The described features of the LRSPP mode profile in lateral and transverse directions provide thereby the possibility to significantly reduce the coupling loss between an LRSPP stripe waveguide and a standard single-mode fiber (down to $\leqslant 0.1$ dB) by choosing proper stripe dimensions and thus fitting the LRSPP mode profile to that of the fiber (Boltasseva et al., 2005b).

§ 4. Interferometric modulators and directional-coupler switches

In this section, design, fabrication and characterization of thermo-optic MZIMs and DCSs, whose operation utilizes the LRSPP waveguiding along thin gold stripes embedded in polymer and heated by electrical signal currents, are considered (Nikolajsen et al., 2004).

The LRSPP stripe waveguides were formed by 15-nm-thin and 8-μm-wide gold stripes (fabricated with UV lithography) sandwiched between 15-μm-thick layers of BCB supported by a silicon wafer as described in the previous section. Excitation (end-fire coupling with a single-mode fiber) and propagation of the fundamental LRSPP mode in these stripes has been characterized at telecom wavelengths (1.51–1.62 μm) using a

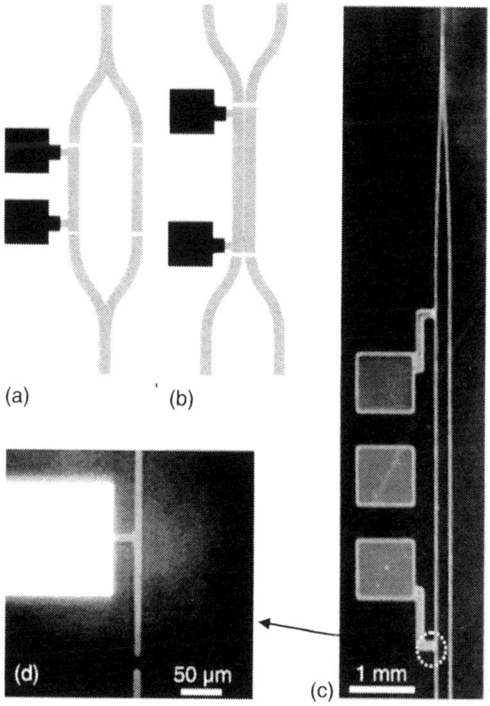

Fig. 10. Schematic layout of LRSPP-based (a) MZIM and (b) DCS. (c) Optical microscope image of the fabricated MZIM showing stripes with typical bends (curvature radius >20 mm to ensure low bend loss), 100-nm-thick contact pads and connecting electrodes. (d) Magnified image of the MZIM part containing an isolating 10-µm-long break in the waveguide stripe and a part of thick electrode connected with 20-µm-long (15-nm-thin) stripe to the waveguide. (This figure is taken from Nikolajsen et al., 2004.)

standard cutback technique (Nikolajsen et al., 2003), resulting in the propagation loss of ~6 dB/cm and coupling loss of ~0.5 dB per facet. It should be emphasized that, in the considered structures, the radiation is guided *along* the metal stripe with the field reaching its maximum right at the metal surface. Such a waveguiding principle thereby offers the *unique* possibility of using the *same* stripe as both a waveguide and a control electrode in the configuration that maximizes the influence of applied electrical signals. Here, this possibility is demonstrated with the dynamic components, whose schematic layout is shown in fig. 10, by making use of the (rather strong) thermo-optic effect in polymers (Ma et al., 2002).

Use of the waveguide stripe as an electrode poses the problem of electrical isolation of the active stripe region (i.e., used also for conducting electrical currents) from the rest of the stripe in order to selectively apply signal currents. Fortunately, as was demonstrated in the previous section, the fundamental LRSPP mode has a relatively large cross section and an effective index that is very close to the (surrounding) dielectric index. Experiments confirmed that micrometer-sized breaks (fig. 10) in the waveguide stripes did not introduce noticeable additional loss. Nevertheless, isolation breaks were introduced into both arms of the MZIM and DCS to preserve their symmetry (fig. 10). All fabricated components were tested with laser radiation at 1.55 μm being coupled using a PM fiber aligned with the dominant electric field component of the LRSPP. Reference measurements of the total insertion loss were performed employing straight waveguide stripes placed next to the tested component and the same coupling configuration.

4.1. Mach-Zehnder interferometric modulators

The generic operation principle of an MZIM is as follows. In the absence of a control signal, an input optical wave is split equally into two waves traveling along two identical arms (of a Mach-Zehnder interferometer), which are again joined together producing an output wave. Ideally, the two waves meeting in the output junction are identical in phase and amplitude. When a control signal is applied to one of the MZIM arms, the propagation of the corresponding wave is influenced (via one of the optical material effects), causing its phase to lag so that the phases of two recombining waves are different at the output junction. If the waves are exactly out of phase, they cancel each other and the result is zero MZIM output. Variation of the signal voltage results thereby in modulation of the MZIM output.

The operation of a thermo-optic MZIM is based on changing the LRSPP propagation constant in a heated arm resulting in the phase difference of two LRSPP modes that interfere in the output Y-junction. The fabricated MZIMs were 20 mm long in total with the arm separation of 250 μm achieved (with cosine bends) over the length of 5 mm and the active waveguide length $L = 5.7$ mm. Typically the total (fiber-to-fiber) insertion loss was the same (~13 dB) as that of the reference stripe. The MZIMs exhibited excellent dynamic characteristics: 8 mW of electrical power was sufficient to obtain an extinction ratio of >35 dB (fig. 11) with an exponential response time of ~0.7 ms (fig. 12). The achieved driving power is considerably lower than that of conventional thermo-optic MZIMs (Ma et al., 2002) because the control electrode is positioned *exactly* at the maximum of the LRSPP mode field, thereby inducing the maximum change in its effective index. Evaluating the dissipated power as $Q \sim 2\kappa \Delta T L w/d$ (where $\kappa \sim 0.2$ W/mK is the polymer thermal

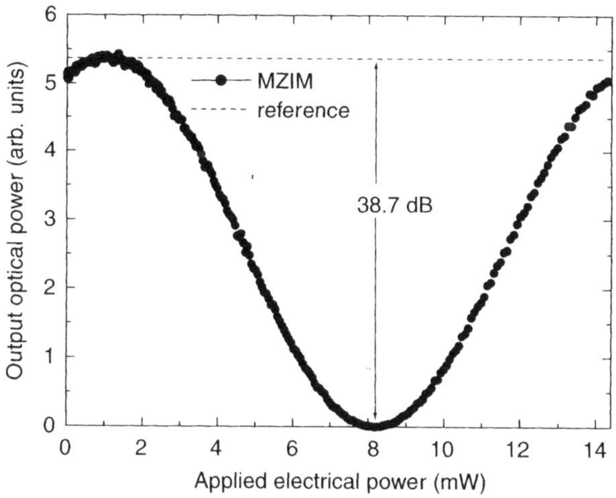

Fig. 11. The MZIM optical output as a function of the applied electrical power.

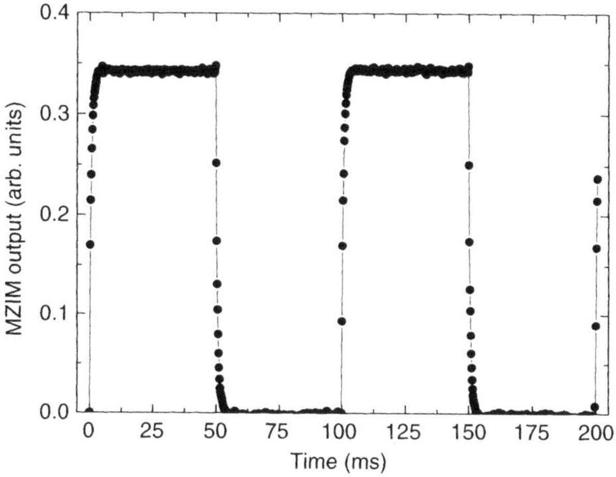

Fig. 12. The temporal response of the MZIM measured with an offset of 2 V and a peak-to-peak voltage of 3.8 V (the electrode resistance was ~1.6 kΩ). Fitting exponential dependencies to the rise and fall of the MZIM output power gives a response time constant of ~0.7 ms.

conductivity (Harper, 1970), ΔT is the temperature increase, $w = 8\,\mu$m is the stripe width, and $d = 15\,\mu$m is the cladding thickness) and the temperature increase needed for complete extinction at the MZIM output as $\Delta T = (\partial n/\partial T)^{-1}(\lambda/2L)$ (where $\partial n/\partial T \sim -2.5 \times 10^{-5}\,^{\circ}\mathrm{C}^{-1}$ is the thermo-optical coefficient of BCB and $\lambda = 1.55\,\mu$m is the light wavelength), one obtains the following estimate for the driving power: $P_{\pi} \sim (\partial n/\partial T)^{-1}\kappa w\lambda/d \approx 7$ mW (Nikolajsen et al., 2004). This estimate is close to the measured value and indicates that the driving power can be decreased even further by using polymers with larger thermo-optic coefficients (Ma et al., 2002). The corresponding time constant can be also evaluated in a simple manner by assuming the main dissipation to occur via the polymer cladding, resulting in $\tau \sim 0.5 c_{\rm p} \rho\, d^2/\kappa \approx 0.6\,{\rm ms}$, where $c_{\rm p} \sim 1\,{\rm J/gK}$ is the specific heat capacitance and $\rho \sim 1\,{\rm g/cm^3}$ is the specific mass density of BCB (Harper, 1970). The obtained value corresponds well to the measured response time of ~0.7 ms, indicating that one might easily gain more speed by using thinner cladding layers.

4.2. Directional coupler switches

Let us next turn to the operation principle of a generic DCS. In this device, two waveguides are in close proximity to each other over a portion of their length. As an input wave travels in one of the waveguides, it gradually tunnels into the other waveguide, which is identical in the absence of a control signal to the input side. The efficiency of this tunneling deteriorates if the two waveguides become different in the sense that the corresponding modes travel with different speeds. By controlling the propagation constant in one of the waveguides, one can completely stop the tunneling process. Hence a DCS can be used to efficiently switch radiation between the two waveguides at the output.

Proper operation of a DCS requires that the radiation injected into one arm at the DCS input is efficiently tunneling into another arm in the interaction region (where the arms are at a close distance) resulting in the complete power transfer (Hunsperger, 1995). Heating one of the arms induces phase mismatch between the LRSPP modes propagating in the coupled waveguides and thereby destroys the efficient tunneling. We found that, for the stripe separation of 4 μm, the power transfer is efficient (~20 dB) at the interaction length of 0.9 mm. The corresponding DCS was 15 mm long in total, and best performance was obtained when the waveguide carrying the coupled radiation was heated: ~66 mW of electrical power was needed to switch the optical power back to the excited waveguide achieving an extinction ratio of >20 dB (fig. 13). The total insertion loss of the device was measured to be slightly (~0.5 dB) larger than that (~11 dB) of the reference stripe and the temporal response was similar to that of the MZIM (fig. 12). The extinction ratio continued to increase for larger signal powers, reaching ~34 dB at 82 mW and stayed above 25 dB even at the first sidelobe (at 110 mW). This switching behavior implies that the considered DCS can be used as a digital-optical switch (DOS), which is a very attractive component for

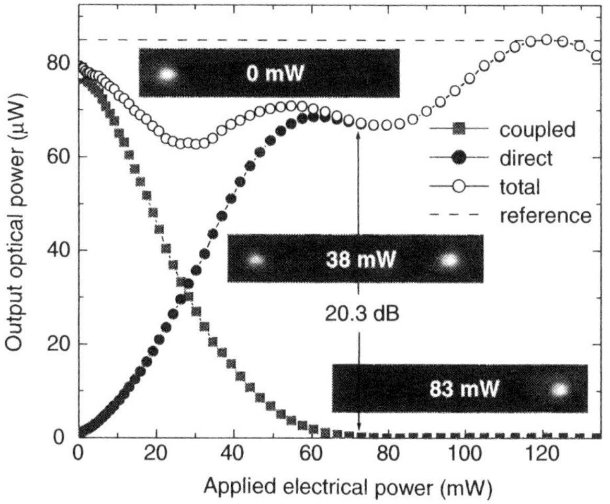

Fig. 13. Switching characteristics of the DCS. Without applied electrical current, the optical radiation is efficiently tunneled from the direct arm excited at the DCS input into the coupled arm. Inserts show microscope images of the intensity distributions at the output facet of the DCS (waveguide separation is 127 μm) for different values of the applied electrical power.

space-division switching in broadband photonic networks. Note that the driving power of the DCS was larger than that of the MZIM, because the DCS electrode was 6 mm long and significantly extended over the tunneling region, so as to decrease the total insertion loss when the electrode was heated. Finally, we would like to note that the DCS electrode length can be optimized reducing considerably the switching power.

§ 5. In-line extinction modulators

Optical modulators based on optical extinction, so-called cutoff (Hall et al., 1970) or mode-extinction (Ashley and Chang, 1984) modulators, are considered promising for usage in telecom networks as variable optical attenuators (VOAs) due to their simple and robust design, monotonic transfer characteristics with respect to electrical signals and weak wavelength dependence. In particular, VOAs based on thermo-optic extinction modulators (EMs) in polymers have recently attracted considerable attention because of their low cost, simple fabrication and easy integration with other polymer-based components (Ma et al., 2002). The general principle of EM operation relies on decreasing the refractive index in a waveguide core region (with externally applied electrical signals via, e.g., electro-, magneto- and thermo-optic effects) so that waveguide modes propagating in the core become progressively less confined and

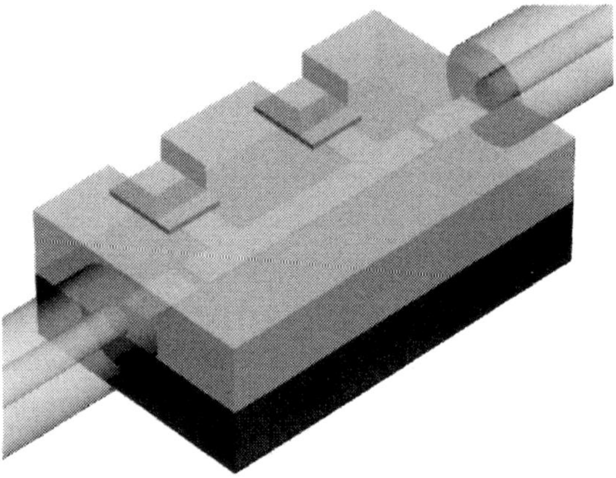

Fig. 14. Layout of the LRSPP-based ILEM along with the optical fibers used for the LRSPP excitation.

more leaky, i.e., coupled to radiation modes. In this section, an exceedingly simple ILEM consisting of a *single metal stripe embedded in dielectric*, with the same stripe being used to guide and control the LRSPP propagation, is considered (Nikolajsen et al., 2005).

The thermo-optic ILEMs (fig. 14) utilized 1-cm-long LRSPP stripe guides formed by 15-nm-thin and 8-μm-wide gold stripes sandwiched between 15-μm-thick BCB layers supported by a silicon wafer (see, for details, Section 3). The refractive index of BCB (and other polymers) is decreased when the polymer is heated, i.e., $\partial n/\partial T<0$ (Ma et al., 2002). This feature is advantageously exploited in the considered ILEM configuration, in which heating of the waveguide stripe decreases the refractive index of surrounding polymer, weakening the waveguiding effect of the metal stripe. Note that, similarly to the devices considered in the previous section, the effect of heating (affecting the LRSPP propagation) is most strong *exactly* where the LRSPP field reaches its maximum, enhancing thereby the influence of applied electrical signals. We used stripe pieces of different length (3–6 mm) as electrodes (resistance $\sim$0.48 kΩ/mm) by separating them from the rest of the stripe with 10-μm-wide breaks as in the above configurations (Section 4). All fabricated components were tested in the same manner as described in the previous section.

The investigated ILEMs exhibited the same insertion loss of $\sim$8 dB as that of the reference stripes in the absence of the applied electrical current. Typically, the ILEM optical output, when increasing the applied electrical power, was monotonously decreasing with the LRSPP mode intensity distribution gradually deteriorating into noisy background (fig. 15). The

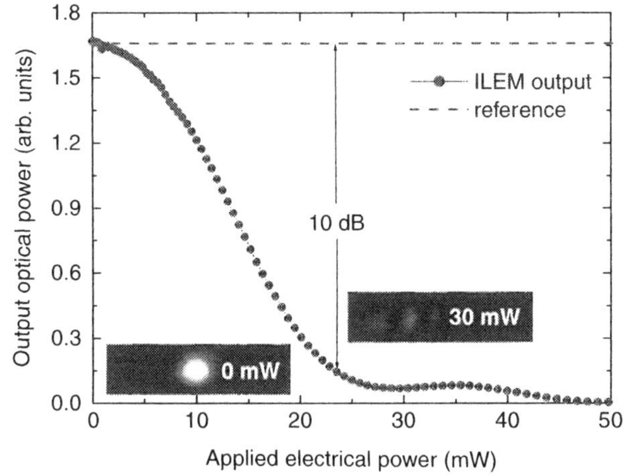

Fig. 15. The ILEM optical output as a function of the applied electrical power. Inserts show microscope images of the intensity distributions at the output facet of the ILEM for different values of the applied electrical power.

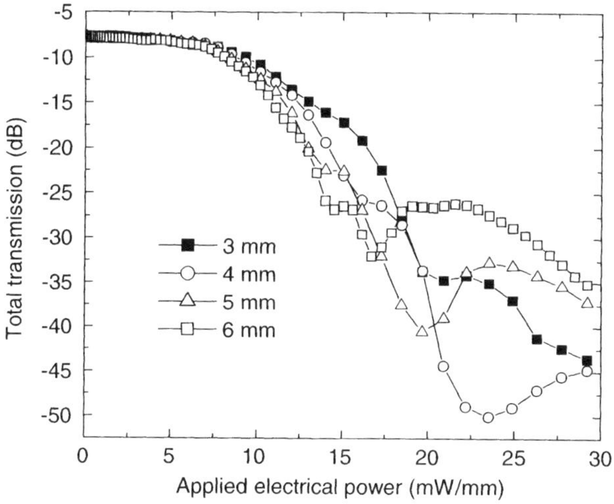

Fig. 16. Total fiber-to-fiber transmission of 1-cm-long ILEMs having the control electrodes of different lengths as a function of the applied electrical power per unit length. (This figure is taken from Nikolajsen et al., 2005.)

transmission characteristics measured for ILEMS with different electrode lengths indicate that the induced extinction is primarily determined by the power dissipated per unit length (not by the total power) though its effect is somewhat stronger for longer electrodes (fig. 16). It is also seen that the strongest variations of the insertion loss occur in the power interval from

10 to 20 mW/mm. The effective index of the LRSPP mode supported by a 15-nm-thick gold film embedded in BCB (refractive index $n \approx 1.535$ at 1.55 μm) was calculated as $N_{\mathrm{eff}} \approx 1.5366$ (Section 2). Assuming that the temperature increase needed to destroy the waveguiding can be evaluated as $\Delta T = (\partial n/\partial T)^{-1}(N_{\mathrm{eff}} - n)$, and using the same expression for the dissipated power per unit length as above (Section 4) one obtains the following power estimate: $Q/L \sim 2\kappa(\partial n/\partial T)^{-1}(N_{\mathrm{eff}} - n)w/d \approx 14$ mW/mm. The heat power Q is supplied via dissipation of the applied electrical power, and it is seen that the above estimate agrees well with the measured power levels inducing significant loss in the investigated ILEMs (fig. 16). Note that the power required decreases with the increase of the cladding thickness d due to the decrease in the temperature gradient, but this would occur at the expense of the increase in the response time (needed to heat the cladding to the same temperature). It should be noted that the above description is quite simplified and that, in principle, one should consider inhomogeneous temperature (and hence refractive index) distribution around the heated stripe and its effect on the extinction. However, such an analysis is rather complicated and has yet to be undertaken.

Another phenomenon contributing to the induced insertion loss is related to the circumstance that the heat dissipation in the ILEM is anisotropic because different media (air and silicon) are adjacent to the BCB layers with a gold stripe (fig. 14). The heat-induced difference between the top and lower BCB layers increases the propagation loss (Section 2) and changes the LRSPP field distribution causing increased light scattering at the junctions between the central and outer parts of the gold stripe. Simulations for a 15-nm-thick gold film embedded in BCB indicate that the critical index difference in this case is $\Delta n \sim 10^{-2}$, which causes the propagation loss increase from ~5.6 to 10.6 dB/cm and shifting the LRSPP mode field in the layer with a higher refractive index. Assuming that all power dissipates only to one side (the case of extreme anisotropy in the heat dissipation), one obtains the following power estimate $P/L \sim \kappa(\partial n/\partial T)^{-1}\Delta n\, w/d \approx 43$ mW/mm, which is ~3 times larger than that considered above implying that the anisotropy contribution is rather weak. However, at relatively large power levels (>20 mW/mm), this contribution should be taken into account and, for example, might be responsible for the complex behavior of transmission characteristics (fig. 16).

We have further characterized the temporal responses of the investigated ILEMs by applying an electrical square wave with different maximum levels of the applied power. In general, the responses were faster for larger applied electrical powers, and, for example, the ILEM with a 3-mm-long electrode exhibited the rise/fall times of <0.5 ms for the

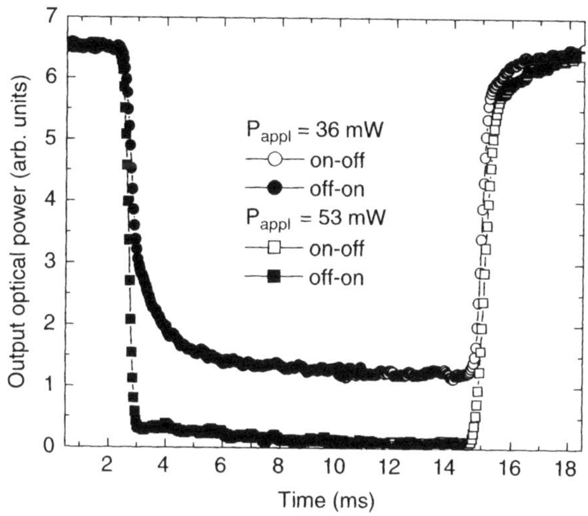

Fig. 17. The temporal responses of the ILEM having a 3-mm-long control electrode for different levels of the applied electrical power. (This figure is taken from Nikolajsen et al., 2005.)

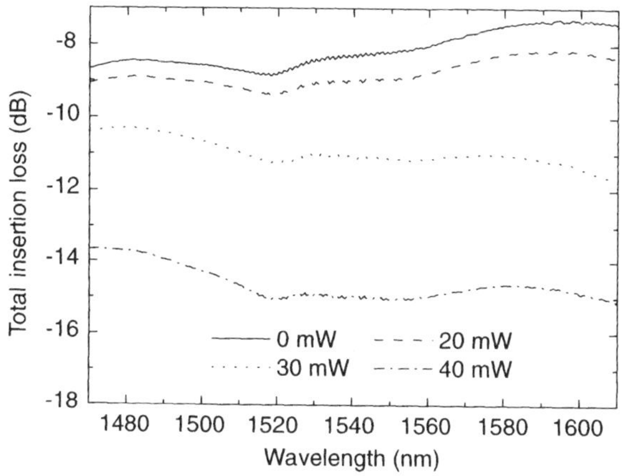

Fig. 18. Total fiber-to-fiber transmission of the 1-cm-long ILEM having a 3-mm-long control electrode for different levels of the applied electrical power. This figure is adapted from (Nikolajsen et al., 2005).

powers of $>50\,\mathrm{mW}$ (fig. 17). Finally, we have investigated the wavelength dependence of ILEM transmission within the main telecom interval of 1470–1610 nm covering S-, C- and L-bands. The total insertion loss was found to vary within 2 dB in this wavelength range (fig. 18). Note that the insertion loss of a gold stripe (no applied power) increases for shorter wavelengths due to increase in the absorption by gold and scattering (by inhomogeneities), whereas the loss induced by heating decreases, resulting in the overall decrease in the transmission for longer wavelengths at relatively large applied electrical powers (fig. 18). The latter feature confirms the given above explanation of the ILEM operation, since the waveguiding ability deteriorates with the wavelength increase, if the waveguide parameters are kept constant, making the ILEMs less power demanding for longer wavelengths. Note, that interference-type devices, such as MZIMs and DCSs, exhibit an opposite trend, since a modulation/switching condition relates an induced variation in the optical path length to the light wavelength.

§ 6. Integrated power monitors

In this section, it is shown that essentially the *same* metal stripes, which constitute the heart of LRSPP-based modulators and switches, can also be used to monitor the *transmitted* LRSPP power by means of measuring variations in the stripe resistance caused by heating (due to the LRSPP absorption). The design, fabrication and characterization of power monitors for LRSPPs excited at telecom wavelengths that can be used in LRSPP-based integrated photonic circuits (and also, due to relatively low insertion losses, as stand-alone infrared optical power monitors) is described in Bozhevolnyi et al., 2005b.

6.1. Design considerations

Let us consider a thin metal stripe (of thickness t and width w) embedded in dielectric (polymer) with the same thickness d of upper and lower cladding layers that transmits an LRSPP stripe mode having the power $P(x)$, with x being the coordinate along the stripe. In the steady-state regime, the optical power absorbed by the stripe is dissipated into the cladding. Evaluating the power dissipated per unit length as $Q{\sim}2\kappa\Delta Tw/d$, where κ is the dielectric thermal conductivity and ΔT is the temperature increase of the metal stripe due to absorption of the LRSPP power, one can estimate the latter as follows:

$$\Delta T(x) = \frac{d\alpha_{\mathrm{abs}}}{2\kappa w} P_{\mathrm{in}} \exp(-\alpha_{\mathrm{pr}} x). \qquad (6.1)$$

Here, α_{abs} is the coefficient of LRSPP absorption by the metal stripe, P_{in} is the power coupled in the LRSPP stripe mode and α_{pr} is the LRSPP attenuation coefficient that determines the LRSPP propagation loss. The temperature increase causes an increase in the metal resistivity and, consequently, in the stripe resistance, which can be expressed as

$$R(P_{\rm in}) \cong R(P_{\rm in}=0)\left\{1+\left[1-\exp(-\alpha_{\rm pr}L)\right]\frac{\alpha_{\rm th}d\alpha_{\rm abs}}{2\alpha_{\rm pr}\kappa wL}P_{\rm in}\right\}, \qquad (6.2)$$

where L is the resistance length and $\alpha_{\rm th}$ is the thermal resistance coefficient. It is seen from eq. (6.2) that the resistance increases linearly with the in-coupled and, thereby, transmitted LRSPP power, a feature that suggests monitoring of the stripe resistance as a method for evaluation of the transmitted LRSPP power. On the other hand, it is clear that the reference resistance $R(P_{\rm in}=0)$, i.e., the resistance in the absence of radiation, is influenced by the environment temperature and, therefore, difficult to control.

We suggested that, in order to reduce the influence of the environment, one should conduct measurements of the corresponding voltage drop in the Wheatstone bridge configuration with all conductors being stripes similar to that used to guide the LRSPP mode (fig. 19). For the bridge balanced in the absence of LRSPP radiation $R_1R_3=R_2R_4^0$ with $R_4^0=R_4(P_{\rm in}=0)$, the signal voltage is given by

$$V_{\rm s}=(P_{\rm in})=V_{\rm b}=\frac{R_2\left(R_4-R_4^0\right)}{\left(R_1+R_4^0\right)\left(R_2+R_3\right)}, \qquad (6.3)$$

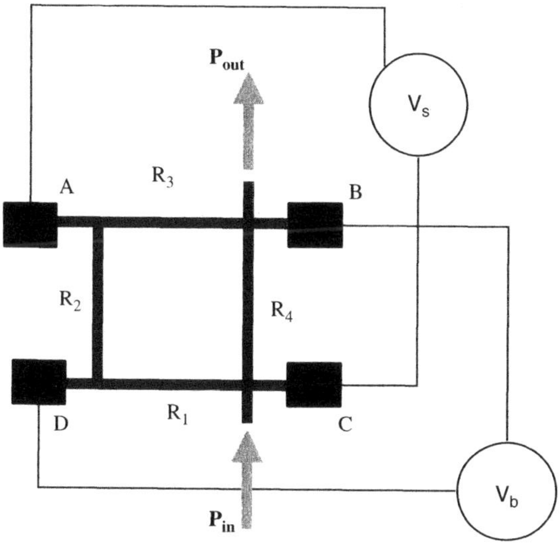

Fig. 19. Schematic layout (a) of a power monitor based on stripe resistance measurements using a Wheatstone bridge configuration, in which a bias voltage is applied to contact pads B and D whereas a signal voltage drop is measured between contact pads A and C. (This figure is taken from Bozhevolnyi et al., 2005b.)

where V_b is the bias voltage (fig. 19). It can easily be verified that the maximum response is expected when $R_2 = R_3$ and, consequently, $R_1 = R_4^0$, resulting in the final expression written for relatively low loss ($\alpha_{pr}L << 1$) as follows:

$$V_s = V_b \frac{\alpha_{th} d\alpha_{abs}}{8\kappa w} P_{in}. \tag{6.4}$$

The important feature of this arrangement is that the signal does not depend on the resistor length, implying that the power monitor can be made relatively short minimizing not only the device footprint but also the insertion loss.

6.2. *Fabrication and characterization*

The LRSPP stripe waveguides used in the corresponding power monitors (fig. 20) were formed by 15-nm-thin and 8-μm-wide gold stripes (fabricated with UV lithography) sandwiched between 15-μm-thick layers of BCB (Section 3). The LRSPP guides were excited and characterized in the same manner as those used in the LRSPP-based modulators and switches (Sections 5). We have fabricated and characterized several 1-mm-long LRSPP power monitors positioned at a distance of 200 μm from the sample input facet, so as to minimize the radiation power loss due to absorption. The results of our testing for three different but nominally identical monitors (fig. 20) are shown in fig. 21. It is seen that none of the

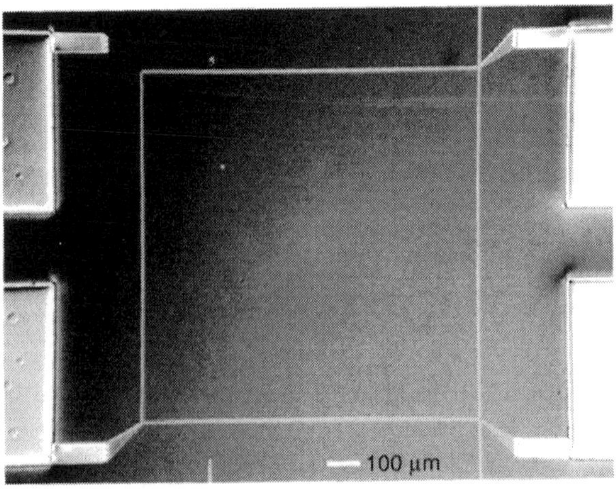

Fig. 20. Optical microscope image of a 1-mm-long power monitor showing 8-μm-wide stripes (including an LRSPP guiding stripe) and (partly) contact pads. (This figure is taken from Bozhevolnyi et al., 2005b.)

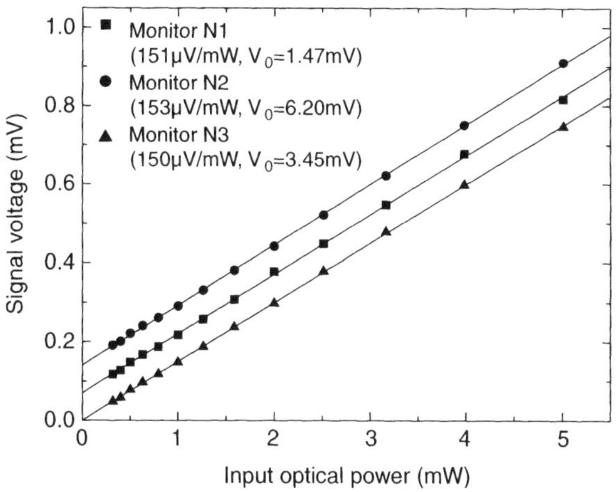

Fig. 21. Signal voltage measured with different 1-mm-long power monitors as a function of the optical input power for the bias voltage of 2 V. The experimental data are offset for clarity with zero-power offsets as indicated in the legend. Linear fits are calculated by the least-square method, and the resulting slopes are also indicated in the legend. (This figure is taken from Bozhevolnyi et al., 2005b.)

devices was perfectly balanced, most probably because of slightly different widths of (nominally identical) gold stripes causing their different resistances. Our measurements of the resistance of 12 similar 1-mm-long stripes resulted in the average value of 560 Ω with the standard deviation of 4 Ω, a dispersion that originates from the photolithographic process. Note that the responsivity of power monitors is better reproducible than their offsets, indicating that the power dissipation is not as directly governed by the stripe width (eq. (6.1)) as the resistance (probably because of the diffusion behavior of the heat dissipation).

The experimentally obtained (fig. 21) responsivity of 75 μV/(mW V) (i.e., per 1 V of the bias voltage) would be interesting to compare with the theoretical prediction. However, it should be borne in mind that the expression used in eq. (6.1) for the dissipation is rather simplified and valid only for the symmetrical environment with heat sinks positioned at distances (from the stripe) that are much smaller that the stripe width. For this reason, the relation obtained for the responsivity (eq. (6.4)) can be used only for approximate estimations. Furthermore, one has to find a suitable method for determination of the coefficient α_{abs} of LRSPP absorption by the stripe, which is one of the crucial parameters influencing the responsivity (eq. (6.4)). One can, for example, compare the *rate* of increase in the stripe resistance due to the absorption of transmitted optical power with that due to the direct heating by electrical current.

In our case, the first value was found from fig. 21 and eq. (6.3) as being ~0.165 Ω/mW, whereas the second one (~2.3 Ω/mW) was found by measuring changes in the stripe resistance as a function of the electrical power dissipated in the stripe. From these values we deduced that α_{abs} ~3 dB/cm, which is twice smaller than the measured LRSPP propagation loss and close to that expected from the calculations (Section 2). Using the thermal resistance coefficient $\alpha_{th} \sim 3 \times 10^{-3}\,K^{-1}$ and the polymer thermal conductivity $\kappa \sim 0.2$ W/mK (Harper, 1970) results in the responsivity of ~250 μV/(mW V). The difference between this value and that obtained from the measurements (fig. 21) can be due to approximate evaluation of heat dissipation (as mentioned above), as well as due to insufficient accuracy in the values of other parameters involved, viz., α_{th} and κ.

6.3. *Sensitivity*

The obtained responsivity of 75 μV/(mW V) is certainly large enough for monitoring of the LRSPP power in the mW range, which is typically the range of choice for practical applications. The sensitivity (the lowest detectable power) is fundamentally limited by thermal fluctuations (Johnson noise) that can be evaluated for a circuit with four equivalent resistors R (~560 Ω) as follows (Boyd, 1983): $\delta V \sim 4(kTR\Delta f)^{0.5} \sim 0.2\,\mu V$, where k is the Boltzmann's constant, $T \sim 300$ K is the resistor temperature and Δf is the measurement bandwidth that can be taken equal to 1 kHz, which is a typical bandwidth for thermal effects in these structures (Section 4). This estimation results in the sensitivity of ~3 μW per 1 V of the bias voltage, a value that seems quite sufficient for most practical applications. The upper power limit is related to the fact that the polymer cladding when heated (by a metal stripe) decreases its refractive index, increasing eventually the propagation loss when the LRSPP stripe mode approaches the cutoff (mode extinction). In our case, the electrical power needed to initiate the increase of propagation loss is ~3 mW/mm (Section 5), resulting in the upper optical power limit of ~40 mW (for the propagation loss due to absorption being ~3 dB/cm). Our measurements using a fiber amplifier to boost up the radiation power showed linear LRSPP transmission through a 5-mm-long gold stripe up to the power of ~50 mW. For higher power, significant deviations from the linear behavior were observed (fig. 22). For the same reason the bias voltage V_b should not be too large, since one should limit the stripe heating by the electrical current (driven by V_b). Requiring that the electrical power dissipated in the stripe, $V_b^2/(4R_4)$, does not exceed the electrical power, $P_{cr}L$, that initiates the mode extinction ($P_{cr} \sim 3$ mW/mm in our case), results in the following

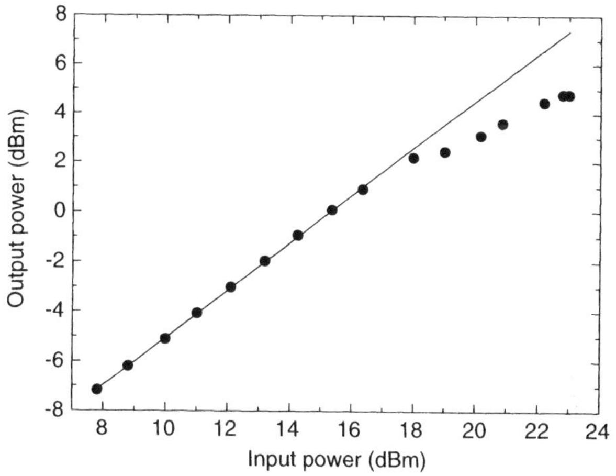

Fig. 22. Optical output power of a power monitor as a function of the input optical power.

limit: $V_b/L<2(R_0P_{cr})^{0.5}$ (with R_0 being the stripe resistance per unit length). In our case, the (measured) resistance is $R_0{\sim}560\,\Omega$/mm, leading to the bias voltage limit: $V_b/L<2.6$ V/mm. One can, therefore, apply higher bias voltages and obtain larger responses (eq. (6.4)) when using longer monitors. However, such an increase of the sensitivity would be at the expense of larger insertion losses increasing with the rate of ~0.6 dB/mm for the waveguides used here.

Finally, it should be noted that the responsivity depends also upon the wavelength used in accord with the dependence of the LRSPP propagation loss (Section 5) caused by metal absorption, which is wavelength dependent (Raether, 1988). Using a tuneable laser we have measured a signal decrease of ~100 μV for an optical input power of ~2.2 mW and a bias voltage of 3 V when the light wavelength was tuned from 1520 to 1580 nm, resulting in the average wavelength sensitivity of ~0.25 μV/(mW V nm). It should be mentioned that the coupling loss of ~0.5 dB is fairly constant in this wavelength range, so that all the signal variation can be attributed to the change in the optical power absorbed. Using the measured responsivity value of 75 μV/(mW V) results in the relative sensitivity of $\sim 3.4 \times 10^{-3}$ $(\text{nm})^{-1}$. Similar experiments with different structures resulted in different wavelength dependencies of the signal voltage. However, these dependencies when normalized with respect to the monitor length and bias voltage were similar (fig. 23), with small deviations that could be caused, e.g., by different in- and out-coupling adjustments. Considering the data obtained we suggest a conservative estimate of the relative wavelength sensitivity as $\sim 5 \times 10^{-3}$ $(\text{nm})^{-1}$ for LRSPP power

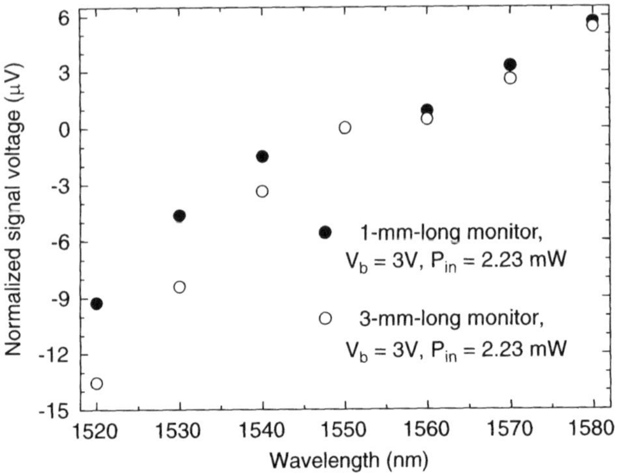

Fig. 23. Signal voltage measured with two different monitors as a function of the light wavelength and normalized with respect to the input optical power and the monitor length and bias voltage (all indicated in the legend). The normalized signal voltage is also offset for clarity so as to nullify the signal voltage at the wavelength of 1.55 μm. (This figure is taken from Bozhevolnyi et al., 2005b.)

monitors operating in the range of 1520–1580 nm. This means, for example, that such a monitor can be used (without changing its calibration) to evaluate the transmitted optical power with the accuracy of ~5% within the wavelength range of ~10 nm.

§ 7. Outlook

The dynamic components reviewed here represent first realizations of the unique feature inherent to LRSPPs that allows one to use the *same* metal stripe to guide and efficiently control the optical radiation. The very low power consumption exhibited by the considered devices is a direct consequence of this feature that, in turn, ensures that the effect of heating (affecting the LRSPP propagation) is most strong *exactly* where the LRSPP field reaches its maximum, warranting thereby the strongest influence of applied electrical signals. The required power can be decreased even further by using other polymers with stronger thermo-optic effects (Ma et al., 2002). For example, our preliminary experiments with BCB being replaced by inorganic polymer glass™ (IPG), which exhibits the thermo-optic coefficient ~10 times stronger that in BCB: $\partial n/\partial T \sim -3 \times 10^{-4}\,{}^{\circ}\mathrm{C}^{-1}$, indicated that the electrical power required to noticeably attenuate the output optical power in ILEMS decreased by the order of magnitude as expected. Taking also into account that the LRSPP propagation loss can be further reduced (to ~1 dB/cm) via design and processing optimization, we believe that the promising dynamic characteristics reported here can be significantly improved, making the demonstrated components attractive for use in telecommunications. For example, the realization of a 3.5-mm- long ILEM with the total insertion loss of <3 dB and the extinction ratio of >25 dB for the applied electrical power of <50 mW is feasible. Moreover, the same design principle can be easily applied to other configurations, e.g., to Y- and X-junction-based DOSs, and modified to employ other, e.g., electro-optic, effects. The possibility to seamlessly introduce (transmitted) power monitors, exhibiting linear response for up to 50 mW of input power (at telecom wavelengths) with the slope of ~0.2 mV/mW, in the considered dynamic devices increases further the application potential of the LRSPP-based photonic circuits. Finally, it should be emphasized that LRSPP components are based on true planar processing technology, which considerably simplifies development, large-scale integration and fabrication of photonic devices.

Acknowledgments

I am very grateful to my former colleagues at Micro Managed Photons A/S, T. Nikolajsen, K. Leosson and A. Boltasseva, who shared with me the excitement of finding out salient features of LRSPPs and trying out various configurations of LRSPP-based photonic components nearly all (!) of which did work. I acknowledge also support from the European Network of Excellence, PLASMO-NANO-DEVICES (FP6-2002-IST-1-507879).

References

Al-Bader, S.J., 2004, Optical transmission on metallic wires – fundamental modes, IEEE J. Quantum Electron. **40**, 325–329.

Ashley, P.R., Chang, W.S.C., 1984, Improved mode extinction modulator using a Ti in-diffused $LiNbO_3$ channel waveguide, Appl. Phys. Lett. **45**(8), 840–842.

Berini, P., 2000, Plasmon-polariton waves guided by thin lossy metal films of finite width: Bound modes of symmetric structures, Phys. Rev. B **61**, 10484–10503.

Boltasseva, A., Bozhevolnyi, S.I., Søndergaard, T., Nikolajsen, T., Leosson, K., 2005a, Compact Z-add-drop wavelength filters for long-range surface plasmon polaritons, Opt. Express **13**, 4237–4243.

Boltasseva, A., Nikolajsen, T., Leosson, K., Kjaer, K., Larsen, M.S., Bozhevolnyi, S.I., 2005b, Integrated optical components utilizing long-range surface plasmon polaritons, J. Lightwave Technol. **23**, 413–422.

Boyd, R.W., 1983, Radiometry and the Detection of Optical Radiation, Wiley, New York.

Bozhevolnyi, S.I., Boltasseva, A., Søndergaard, T., Nikolajsen, T., Leosson, K., 2005a, Photonic band gap structures for long-range surface plasmon polaritons, Opt. Commun. **250**, 328–333.

Bozhevolnyi, S.I., Nikolajsen, T., Leosson, K., 2005b, Integrated power monitor for long-range surface plasmon polaritons, Opt. Commun. **255**(1-3), 51–56.

Breukelaar, I., Charbonneau, R., Berini, P., 2006, Long-range surface plasmon-polariton mode cutoff and radiation, Appl. Phys. Lett. **88**(5), 051119.

Burke, J.J., Stegeman, G.I., Tamir, T., 1986, Surface-polariton-like waves guided by thin, lossy metal films, Phys. Rev. B **33**, 5186–5201.

Charbonneau, R., Berini, P., Berolo, E., Lisicka-Skrzek, E., 2000, Experimental observation of plasmon-polariton waves supported by a thin metal film of finite width, Opt. Lett. **25**, 844–846.

Charbonneau, R., Lahoud, N., Mattiussi, G., Berini, P., 2005, Demonstration of integrated optics elements based on long-ranging surface plasmon polaritons, Opt. Express **13**, 977–984.

Hall, D., Yariv, A., Garmire, E., 1970, Observation of propagation cutoff and its control in thin optical waveguides, Appl. Phys. Lett. **17**(3), 127–129.

Harper, C.A., 1970, Handbook of Materials and Processes for Electronics, McGraw-Hill, New York.

Hunsperger, R.G., 1995, Integrated Optics: Theory and Technology, Springer, Berlin.

Jetté-Charbonneau, S., Charbonneau, R., Lahoud, N., Mattiussi, G., Berini, P., 2005, Demonstration of Bragg gratings based on long-ranging surface plasmon polariton waveguides, Opt. Express **13**, 4672–4682.

Kogelnik, H., 1979, Topics in applied physics, in: Tamir, T. (Ed.), Integrated Optics, vol. 7, Springer, Berlin, pp. 64–66.

Ma, H., Jen, A.K.Y., Dalton, L.R., 2002, Polymer-based optical waveguides: materials, processing, and devices, Adv. Mater. **14**, 1339–1365.

Marcuse, D., 1974, Theory of Dielectric Optical Waveguides, Academic Press, New York.

Nikolajsen, T., Leosson, K., Bozhevolnyi, S.I., 2004, Surface plasmon polariton based modulators and switches operating at telecom wavelengths, Appl. Phys. Lett. **85**, 5833–5836.

Nikolajsen, T., Leosson, K., Bozhevolnyi, S.I., 2005, In-line extinction modulator based on long-range surface plasmon polaritons, Opt. Commun. **244**, 455–459.

Nikolajsen, T., Leosson, K., Salakhutdinov, I., Bozhevolnyi, S.I., 2003, Polymer-based surface-plasmon-polariton stripe waveguides at telecommunication wavelengths, Appl. Phys. Lett. **82**, 668–670.

Raether, H., 1988, Surface Plasmons, Springer, Berlin.

Sarid, D., 1981, Long-range surface-plasma waves on very thin metal films, Phys. Rev. Lett. **47**, 1927–1930.

Chapter 2

Metal strip and wire waveguides for surface plasmon polaritons

by

J.R. Krenn

Institute of Physics and Erwin Schrödinger Institute for Nanoscale Research, Karl-Franzens University, A-8010 Graz, Austria

J.-C. Weeber, A. Dereux

Laboratoire de Physique de l'Université de Bourgogne, Optique Submicronique, BP 47870, F-21078 Dijon, France

Nanophotonics with Surface Plasmons
Advances in Nano-Optics and Nano-Photonics
ISSN: 1871-0018

V.M. Shalaev & S. Kawata (Editors)

DOI: 10.1016/S1871-0018(06)02002-4

Contents

§ 1. Introduction

During the last decade the interest in surface plasmon polaritons (SPPs) has been considerably renewed. A great part of this revival is due to improved fabrication schemes that nowadays allow tailoring metal structures on the nanoscale. Doing so, the twofold potential of SPPs for advancing photonics has been brought within experimental control. On one hand, SPPs are surface waves and could enable to set up a two-dimensional 'flat' optical technology. The implications for miniaturization and integration are evident when thinking of microelectronics, which relies entirely on quasi-two-dimensional lithographic fabrication schemes. On the other hand, when sustained by properly designed metal structures, SPPs are not subject to the diffraction limit that restricts the miniaturization of conventional optical elements and devices to about half the effective light wavelength. SPPs hold thus the potential to overcome this limit and to enable the propagation and manipulation of light on the nanoscale. Before this background we discuss here waveguiding, one of the basic optical functionalities. Indeed, recent research has revealed that SPPs can be propagated along laterally confined metal structures, in analogy to wires and interconnects in electronics. We consider SPP waveguiding in laterally confined metal thin films, where the lateral dimensions range between some micrometers (strips) and subwavelength values (wires).

SPPs are propagating waves of resonant longitudinal electron oscillations at the interface between a metal and a dielectric, which are coupled to a light field (Raether, 1988). This field is maximum at the interface and decays exponentially in the perpendicular directions. As a resonant phenomenon, the SPP near field is considerably enhanced with respect to the exciting light field. More on the basics on SPPs can be found elsewhere in this volume. The attractive properties of SPPs are, however, accompanied by a major drawback. The high SPP damping due to ohmic losses in the metal restricts SPP propagation lengths in the visible and near-infrared spectral range to about 10–100 μm. These values are reached for silver or gold due to the low imaginary parts of the dielectric functions of these metals. In addition, the handling of these metals is convenient and the chemical stability (in the case of gold) is high. Most importantly, however, the achieved propagation lengths seem sufficiently high when thinking of highly miniaturized and integrated optical or electro-optical devices with correspondingly short signal paths.

A great number of publications have contributed to prepare the ground for plasmonic waveguides, starting with imaging SPP fields by near-field optical microscopy (Dawson et al., 1994), and investigating the interaction of SPPs with various nanoscale surface features (Smolyaninov et al., 1996, 1997; Bozhevolnyi and Pudonin, 1997; Krenn et al., 1997; Bozhevolnyi and Coello, 1998). This experimental work was complemented by a series of theoretical studies (Pincemin et al., 1994; Pincemin and Greffet, 1996; Sanchez-Gil, 1996, 1998; Schröter and Heitmann, 1998). The theoretical foundation of bound SPP modes in micrometer-wide metal strips was laid by Berini (1999, 2000, 2001). The experimental investigation of lithographically fabricated metal strips by scanning near-field optical microscopy was reported by Weeber et al. (2001, 2003). The same group succeeded in implementing passive elements as Bragg mirrors into SPP strips (Weeber et al., 2004, 2005). Again, experimental advances were accompanied by extended theoretical analysis (Sanchez-Gil and Maradudin, 2005; Zia et al., 2005a, b). On the other hand, SPP propagation along metal wires with sub-wavelength cross-section was proposed by Takahara (Takahara et al., 1997). Experimental work includes chemically (Dickson and Lyon, 2000; Ditlbacher et al., 2005) and lithographically (Krenn et al., 2002) fabricated structures.

It should be mentioned that waveguide schemes alternative to strips or wires have been proposed and investigated as well. SPP propagation along closely packed chains of metal nanoparticles was proposed in Quinten et al. (1998), but achievable propagation lengths turned out to be rather low, as experimentally shown in Maier et al. (2003). Another venue towards SPP waveguides exploits metal photonic band gap geometries (Barnes et al., 1996; Bozhevolnyi et al., 2001).

§ 2. Experimental aspects

2.1. Lithographic sample fabrication

SPP waveguides are fabricated by vacuum evaporation of thin gold or silver films with a thickness of typically 50–100 nm on a transparent substrate. Amongst the experimental methods for structuring these thin films laterally on the micro- and nanoscale, electron-beam lithography (EBL, McCord and Rooks, 1997) has proven to be the most versatile due to its flexibility and the availability of a variety of platforms, usually based on scanning electron microscopes (SEMs). The substrate for lithography has to be chosen with some care as EBL requires on one hand a conducting surface to prevent charging effects but, on the other hand optical experiments call for transparent substrates as glass which is electrically isolating. The standard solution to this problem is to cover glass substrates with a nanometer-thick transparent layer of indium–tin–oxide (ITO) providing weak ohmic conductivity sufficient for EBL. Onto the substrate a layer of an electron-sensitive resist (as, e.g., polymethyl metaacrylate, PMMA) is deposited by spin coating. The sample pattern is then transferred to the electron resist via exposure to a focused electron beam. Subsequent chemical development removes the resist from all regions exposed to the electron beam. Now metal is deposited on the sample by evaporation in high vacuum. Finally, the resist layer is removed at a chemical lift-off step, leaving the metal structures on the sample wherever they where deposited directly on the ITO–glass substrate through the openings in the resist mask.

2.2. Light/SPP coupling

For the efficient coupling of light to SPPs and vice versa, the SPP dispersion relation has to be considered. As the SPP wave vector k_{SPP} is larger than that of light, k, for any given frequency (Raether, 1988) appropriate means for matching both have to be taken. This can be accomplished using various methods as described briefly in the following. We are interested here in the propagation properties of SPPs in structures laterally confined to the microscale or below and focus thus on methods that allow *local* SPP excitation. In addition, these methods should be compatible with the imaging of SPP propagation with direct space methods as optical far-field or near-field microscopy.

Wave vector matching between light and SPPs can be achieved by the *Kretschmann* method (Raether, 1988). The SPP-sustaining metal is deposited as a thin film on a glass prism of refractive index n. The film thickness is chosen small enough (typically below 100 nm) to enable SPP excitation on the metal/air interface by light incident from the prism side (wave vector nk). Proper choice of the light incidence angle above the critical angle of total internal reflection allows matching k_{SPP} by the light wave vector component parallel to the interface. Using a focused laser beam ensures local SPP excitation, see fig. 1a.

Another spatially confined SPP excitation scheme relies on an opaque thin film (made from, e.g., aluminum) covering part of the substrate sustaining the SPP waveguide. Outside the aluminum film SPPs are coupled to the waveguide by the *Kretschmann* method using a collimated light beam. From there the metal waveguide is run over the aluminum

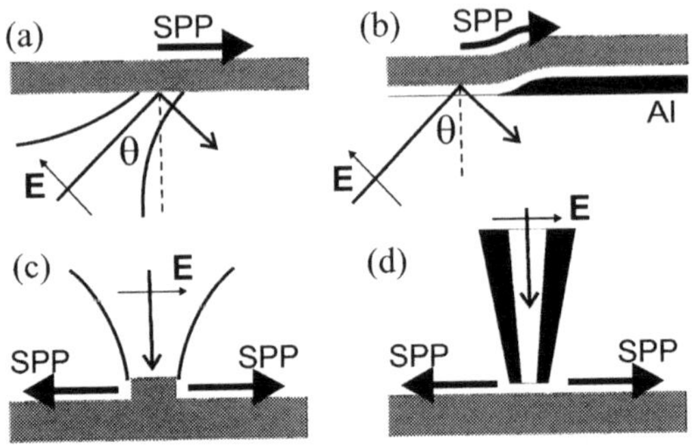

Fig. 1. Methods for local SPP excitation at a metal/air interface. (a) *Kretschmann* configuration. A laser beam polarized in the plane of incidence (**E** is the electric field vector) is focused onto the hypotenuse of the glass prism substrate under a mean angle θ. (b) A collimated laser beam is used for SPP excitation and the spatial confinement of the excitation region is achieved by an opaque aluminum (Al) shield that is separated from the SPP waveguides by a dielectric thin film spacer. (c) Local excitation by focusing a laser beam onto a nanoscale surface defect. (d) Local SPP excitation by the subwavelength aperture of a scanning near-field optical microscope tip.

film from which it is vertically seperated by a thin film dielectric spacer. Here, the aluminum screen blocks the excitation light coming from below and thus the SPPs propagate (and decay) freely, see fig. 1b. With properly chosen sample parameters, the curvature of the onset of the aluminum screen and its vicinity to the metal waveguide can be neglected to good approximation (Lamprecht et al., 2001).

The local coupling of light to SPPs can rely as well on light scattering at nanoscale surface structures. The scattered light exhibits a wave vector distribution according to the spatial Fourier spectrum determined by the structure geometry. The according evanescent components lead to SPP excitation, as sketched in fig. 1c. We note that this can be as well achieved by the evanescent components provided in the vicinity of the subwavelength aperture of the tip of a scanning near-field optical microscope (fig. 1d). Finally, as discussed elsewhere in this volume endfire coupling by optical fibers can be applied in the case of so-called long-range SPPs (Nikolajsen et al., 2003).

2.3. SPP imaging

2.3.1. Far-field microscopy

SPPs are bound to a metal/dielectric interface and thus specific techniques have to be applied for microscopic SPP imaging. Indeed, what is needed is a local probe within the SPP field that converts this field to propagating light fields to be measured by some kind of photodetector. This converter can be simply the inherent surface roughness of the metal film that inelastically scatters SPPs. As the intensity of the scattered light is proportional to the local SPP field intensity, imaging the scattered light distribution with a conventional optical microscope maps the SPP field intensity profile. This fast and experimentally straightforward technique allows, e.g., the quantitative measurement of SPP propagation lengths (Lamprecht et al., 2001). It suffers, however, from the drawbacks of rather low-signal levels and the spatially inhomogeneous distribution of surface roughness features.

Experimental work on SPPs relies almost exclusively on (structured) metal thin films deposited on a transparent substrate. The SPP modes sustained at the metal/air interface are thus leaky waves that radiate into the substrate in reversal of the *Kretschmann* excitation scheme (Raether, 1988). Each object point on the metal surface radiates in proportion to the local SPP intensity. Imaging this emitted radiation by means of an immersion objective optically coupled to the substrate (as leakage radiation is emitted at an angle larger than the critical angle of total internal reflection in the substrate) and thus imaging the SPP field intensity profile at the metal/air interface is the principle of leakage radiation microscopy (LRM) (Bouhelier et al., 2001; Hohenau et al., 2005).

Converting bound SPP fields to propagating light can as well rely on molecules as near-field probes via either fluorescence or (surface-enhanced) Raman scattering. The spatial pattern of the fluorescence or Raman intensity as observed with a conventional optical microscope yields information on the SPP field profile. In the case of fluorescence imaging, the molecules have to be seperated from the metal surface by a thin dielectric spacer layer or a polymer matrix to prevent quenching. Signal levels are usually high and straylight from the exciting light beam is strongly suppressed due to the wavelength converting measurement process (Ditlbacher et al., 2002a, b). However, molecular bleaching effects make quantitative measurements difficult. While these problems are not present in Raman imaging, the potential and limits of this method are yet to be explored (Laurent et al., 2005).

2.3.2. Near-field microscopy

While far-field microscopy is a fast and reliable technique it suffers from the drawback of diffraction-limited spatial resolution. Furthermore, there are experimental limitations to far-field techniques as discussed above.

Most of these limitations can be overcome by near-field optical microcopy in the so-called photon scanning tunneling microscope (PSTM) configuration (Courjon et al., 1989; Reddick et al., 1989). In a PSTM, the sample is illuminated by an evanescent light field generated by total internal reflection of a light beam inside a glass prism. We note that this illumination geometry complies with *Kretschmann* method related SPP excitation schemes (figs. 1a, b). The SPP fields are probed by a sharp glass fiber tip held in immediate vicinity of the sample, typically below 100 nm. If a force-detecting scheme (Karrai and Grober, 1995) is implemented, the tip–sample distance can be controlled in the range of a few nanometers and a topographic image of the sample is simultaneously acquired. The fiber tip scatters part of the intensity of the local optical near-field into the fiber that guides the signal to a photodetector. Raster scanning the tip over the sample and assigning the locally detected light intensity to each lateral scan position results in the near-field optical image. It has been shown that the signals acquired with purely dielectric fiber tips correspond closely to the local electric field intensity (Weeber et al., 1996; Krenn et al., 1999; Dereux et al., 2000). Such tips can therefore be considered noninvasive to good approximation, i.e., the near-field optical images are maps of the intensity distribution to be expected in absence of the tip. In the case of strongly scattering samples, propagating light fields are a severe problem as they might obscure the near-field signals. A chromium layer few nanometers in thickness was empirically found to suppress the scattered light level significantly while the proportionality of the detected signal to the local electric field intensity still holds true (Quidant et al., 2002).

§ 3. Metal strips

3.1. Field distribution of metal strip modes

By analogy to the modes of dielectric channel or planar waveguides, the properties of SPPs sustained by thin metal strips (MS) with widths of a few SPP wavelengths are expected to be different from those of SPPs supported by infinitely extended metal thin films. For example, it has been shown that the attenuation of SPP modes propagating along a metal strip (MS-SPP modes) depends upon the width of the waveguide (Lamprecht et al., 2001; Weeber et al., 2001) and, similarly, the field distribution of these modes is also expected to depend on this parameter (Berini, 2000, 2001). With the aim of characterizing the field distribution of MS-SPPs,

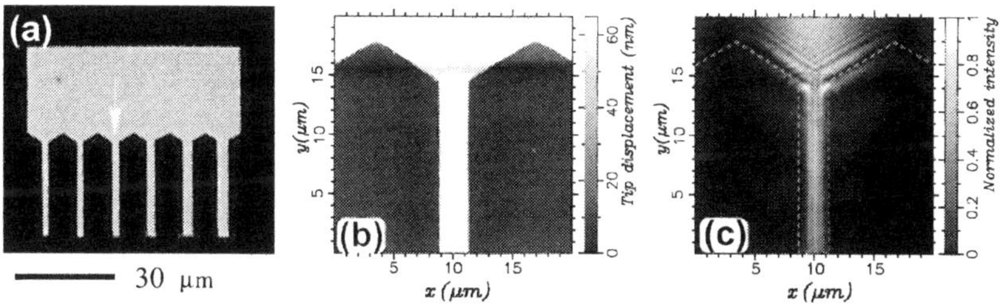

Fig. 2. SPP propagation along thin gold strips. (a) Scanning electron microscope image of an array of SPP strip waveguides connected to an extented thin film area for SPP launch. The arrow points at a 2.5-μm-wide strip shown in the atomic force microscope topographic image in (b). (c) Corresponding PSTM image. A SPP locally excited on the extended thin film area and propagating from top to bottom couples to the SPP sustained by the metal strip. The incident free-space wavelength is 800 nm.

the sample shown in fig. 2a has been fabricated by EBL on an ITO–glass substrate.

This sample comprises 60-nm-thick gold strips of different widths connected to a large thin film area. A titanium–sapphire laser injected into a lensed mono-mode fiber is used to obtain a focused incident spot with a radius of about 10 μm. The incident spot is adjusted on the thin film area to perform a local SPP excitation in the Kretschmann configuration (compare fig. 1a). This SPP propagates along the top interface of the extended thin film area and excites the MS-SPP modes through a tapered area provided that the plane of incidence is parallel to the longitudinal axis of the strip. A PSTM featuring a chromium coated multi-mode fiber tip has been operated to obtain the electric near-field intensity distribution of the MS-SPP modes. Figs. 2b and c show a topographic atomic force microscope and the corresponding PSTM image of a 2.5-μm-wide strip excited at a frequency corresponding to a wavelength of 800 nm in vacuum. Besides the very neat field confinement of the MS-SPP mode, a transverse three-peak structure is visible on the PSTM image. In order to deepen the insight into this MS-SPP mode structure, the influence of the strip width on the observed near-field distributions was investigated by recording PSTM images of strips with widths ranging from 4.5 to 1.5 μm. The transverse cross-cuts of these PSTM images are shown in fig. 3. Again, we observe that the MS achieves a very efficient lateral confinement of the SPP field. Indeed, for strips with a width equal or larger than about three times the incident free-space wavelength ($\lambda_0 = 800$ nm), the near-field intensity drops to zero within the width of the waveguides. In other words, the confinement factor is close to 100% for all the considered MS-SPP modes except for the strip 1.5 μm wide, see fig. 3. In

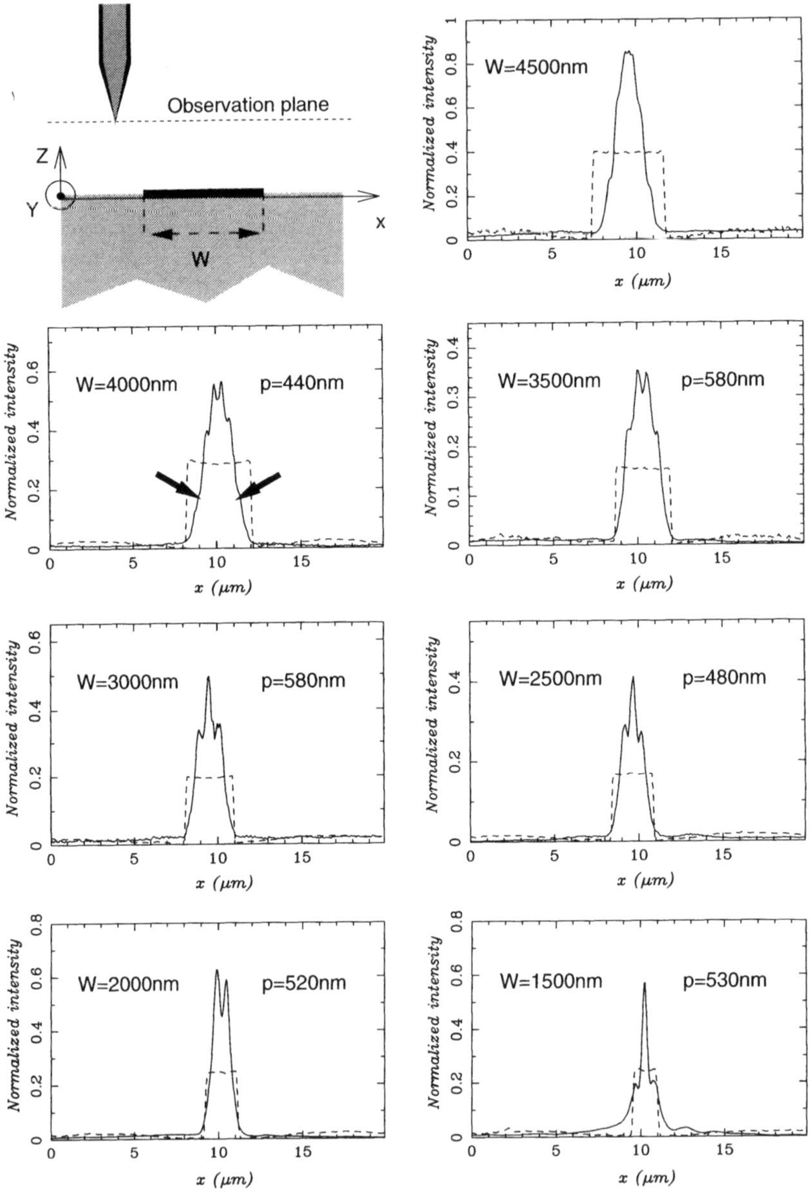

Fig. 3. Transverse cross-sections (solid lines) of the MS-SPP near-field in gold strips of different width *W*, as extracted from PSTM images. These results have been obtained for an incident free-space wavelength of 800 nm.The dashed lines indicate the topographic profiles of the strips. The *p* values correspond to the lateral distance between the peaks in the cross-sections. Reprinted with permission from Weeber et al., 2005 © The American Physical Society.

addition, it can be seen that for decreasing strip width the number of peaks in the optical profiles and/or the lateral distance between these peaks decreases. It has been shown recently that a MS strip lying on a glass substrate can support SPP leaky modes with phase constants which are close to those of a SPP traveling along an extended thin film (Zia et al., 2005a, b). At a given frequency and for sufficiently wide MSs (which sustain several SPP eigenmodes), our MS illumination configuration could then achieve a multi-mode SPP excitation provided that the MS-SPP eigenmode field distribution lead to a non-zero overlap with that of the incident SPP. By analogy to a multi-mode channel waveguide, the multi-mode excitation of a wide MS is thus expected to create complex field distributions similar to those we observe experimentally. Unfortunately, although very appealing, the hypothesis of a multi-mode excitation fails to explain the MS-SPP near-field optical profiles observed experimentally as a single-mode SPP waveguide (such as, e.g., a strip 2.5 μm wide at $\lambda_0 = 800$ nm) leads as well to a multi-peak optical profile. For the explanation of these observations the important contribution of the strip edges to the MS-SPP field distributions has been unambiguously demonstrated by imaging a SPP traveling parallel to the abrupt step of a semi-infinitely extended thin film (Weeber et al., 2003). These observations lead to the conclusion that the MS-SPP modes could be hybrid modes resulting from the coupling of a finite-interface SPP mode with a Gaussian profile (corresponding to the fundamental leaky MS-SPP mode) and oscillating boundaries or edges modes sustained by either the vertical walls or the edges of the strips. Although not fully understood yet, the near-field structure of MS-SPP modes is nevertheless of interest for future applications. In particular, the very high lateral field confinement of the MS-SPP modes inhibiting crosstalk between closely packed waveguides should enable to integrate these guiding devices at high density in coplanar geometries.

3.2. Microstructured metal strips

The metal strips considered in the previous paragraph can be viewed as passive SPP waveguides in the sense that they can confine the SPP field. However, to be useful for practical applications an SPP waveguide should be also able to control the SPP propagation direction. To achieve this goal, the most basic solution consists in designing bent metal strips. Unfortunately, bent strips exhibit rather large bend losses that prohibit their use for an efficient SPP guiding. On the other hand, it has been shown that a SPP sustained by an extended thin film is efficiently Bragg reflected (Ditlbacher et al., 2002b) by a grating provided that the SPP wavevector obeys

$$k_{\text{SPP}} = \frac{k_g}{2\cos\delta}, \tag{3.1}$$

where $k_g = 2\pi/P$ denotes the grating Bragg vector magnitude (P is the grating constant) and δ the SPP angle of incidence on the grating. Thus an alternative solution to bent waveguides for SPP propagation control could be the integration of tilted Bragg mirrors into metal strips. However, prior to the characterization of tilted mirror equipped MS, it is necessary to check that similarly to extended thin film SPPs, an MS-SPP mode can be Bragg-reflected by micrograting. Indeed, due to the lateral field confinement of the MS-SPP modes, their plane-wave spectrum is wider than that of a thin film SPP. As a result, for a given grating, the reflection efficiency could be significantly smaller for an MS-SPP mode as compared to that of an extended thin film mode since not all plane-wave components of the MS-SPP mode spectrum match the Bragg condition given above. In order to gain insight into the interaction of MS-SPP modes with micrgratings, the sample shown in figs. 4a and b has been fabricated. The micrograting consists of indentations engraved into a 2.5-μm-wide strip by focused ion beam milling.

The PSTM image of the microstructured strip recorded for an incident free-space wavelength of 800 nm is shown in fig. 4c. A neat standing wave pattern is visible all along the strip. However, some care has to be taken in interpreation as this standing wave pattern cannot be attributed a priori to a surface-wave interference. In fact, it could as well be generated by the incident MS-SPP mode interacting with a back-scattered light wave propagating at a near-grazing angle. Therefore, the physical origin of this pattern has been investigated by taking PSTM images of the same strip for different tip-sample distances. From these images, it has been shown that the depth of modulation of the interference fringes defined as $C = (I_{\max} - I_{\min})/(I_{\max} + I_{\min})$ (where $I_{\max}$ and $I_{\min}$ are, respectively, the intensity at a given standing wave maximum and the following minimum, compare fig. 4d) does not depend on the tip-sample distance. Consequently, the standing wave pattern is established by the interaction of two contra-propagating evanescent waves (Weeber et al., 2004). With the aim of showing that the micrograting of slits acts on the MS-SPP modes as a Bragg mirror, we now analyze the influence of structural parameters on the reflection efficiency. Figures 5a–c show, respectively, the evolution of the SPP fringe contrast as a function of the number of slits in the mirror, the SPP wavelength for a given grating period, and the ratio of grating period to SPP wavelength. From fig. 5a, it can be seen that

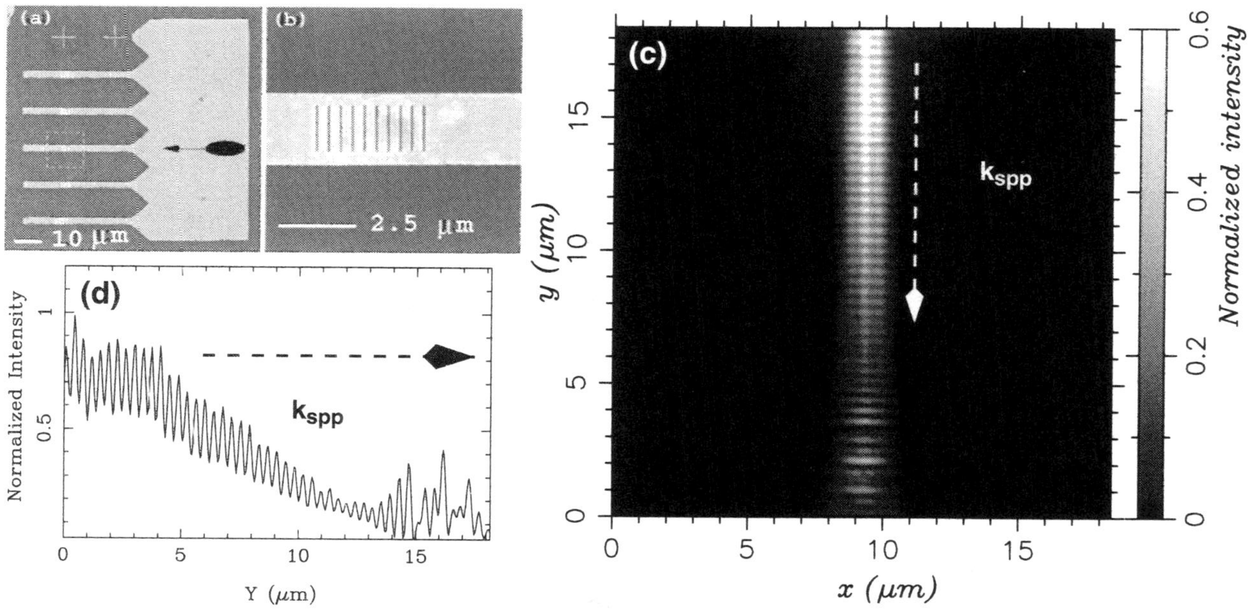

Fig. 4. Bragg mirror equipped MS. (a), (b) SEM images of the Bragg mirror consisting of a slit grating with a grating constant of 400 nm and a slit width of 150 nm. (c) PSTM image recorded for an incident free-space wavelength of 800 nm.(d) Cross-cut of the PSTM image shown in (c), taken along the strip axis. k_{SPP} defines the direction of MS-SPP propagation.

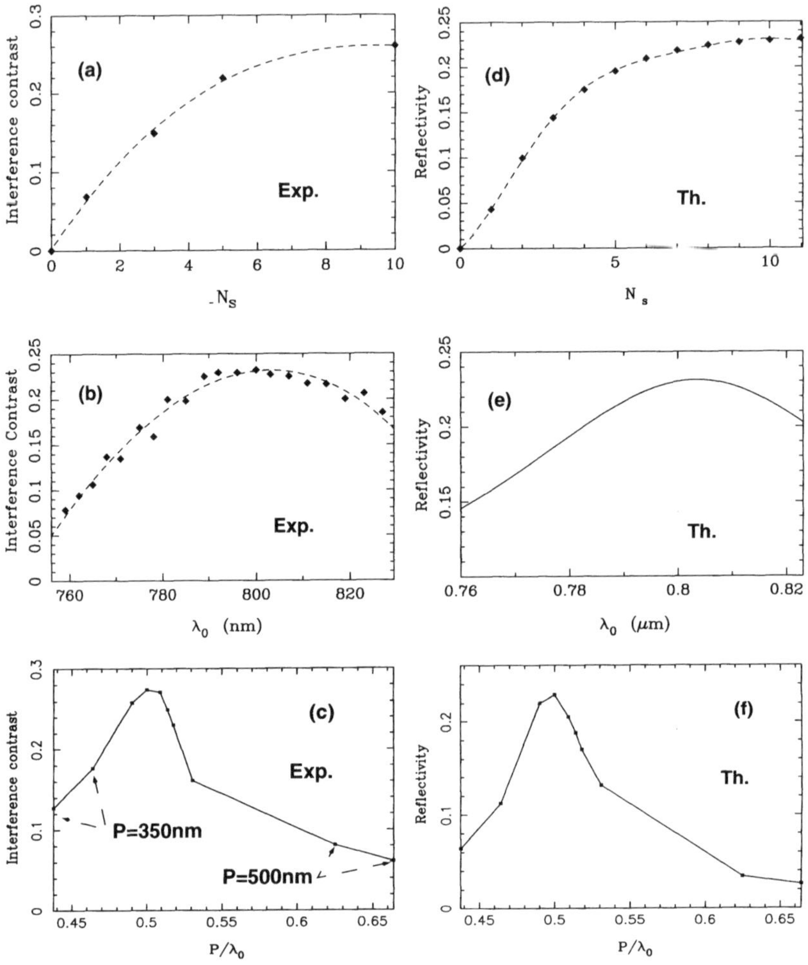

Fig. 5. Experimental and simulated data on parametric studies of MS Bragg mirrors. Measured depth of modulation of the interference fringes as a function of (a) the number of slits N_s in the Bragg mirror (period = 400 nm, $\lambda_0 = 800$ nm), (b) the incident wavelength (period = 400 nm, $N_s = 10$), and (c) the ratio period/λ_0. The lower row of panels shows simulations qualitatively corresponding to (a)–(c), computing the reflectivity of an equivalent Bragg mirror consisting of a stack of two materials (A) and (B) as described in the text as a function of (d) the number of Bragg mirror layers N_s (period = 400 nm, $\lambda_0 = 800$ nm), (e) the incident wavelength (period = 400 nm, $N_s = 10$), and (f) the ratio period/λ_0.

the fringe contrast saturates for a rather low number of slits ($N_s = 10$), showing that only a tiny fraction of the incident SPP reaches the last slits of the mirror. Figure 5b shows the fringe contrast for a grating period of 400 nm and a set of incident free-space wavelengths within the tunability range of the used laser source (760–820 nm). The optimum reflectivity is obtained for a free-space wavelength equal to two times the grating period. This result is further confirmed by the curve displayed in fig. 5c showing that the optimum mirror efficiency corresponds to a period to incident wavelength ratio of 0.5. From these last two results we conclude that for frequencies in the near—infrared region, the effective index of the MS-SPP mode traveling along a 2.5-μm-wide strip is very close to 1.0. For the purpose of mirror design, the MS-SPP wave-vector modulus can therefore be approximated by $2\pi/\lambda_0$, where λ_0 denotes the incident free-space wavelength. As shown in figs. 5d–f, the properties of the micrograting of slits can be qualitatively simulated by a stack of two materials (A) and (B) with indexes of refraction $n_{(A)} = 1.0 + i0.3$ and $n_{(B)} = 1.01$ for modeling, respectively, the air slits and the metal in between two slits (Weeber et al., 2004). From both the experimental and the equivalent simulated Bragg model results, we conclude that a micrograting integrated into an MS acts on the MS-SPP modes as a lossy Bragg mirror. Due to these losses (mainly scattering losses), the number of surface defects (playing the role of an interface between two media in a standard Bragg mirror) involved in the reflection of the MS-SPP modes is necessarily quite low, leading to a rather poor spectral selectivity of this kind of SPP mirrors.

3.3. *Routing SPPs with integrated Bragg mirrors*

From the results of the previous study, we conclude that MS-SPP modes can be Bragg-reflected by microgratings integrated into the waveguides and featuring grating constants given by eq. (3.1). For waveguides with a width equal to a few times the SPP wavelength, the most efficient mirrors are obtained assuming that the SPP mode wave-vector modulus is practically equal to that of the incident light wave. In principle, it is possible to quantitatively measure the reflection efficiency of the integrated mirrors by analyzing the depth of modulation of the standing wave pattern observed in the near-field images. However, from an experimental point of view such a measurement is difficult as the finite size of the optical tip used to detect the near-field can lead to an underestimate of the fringe contrast and thus lead to erroneous values of the reflection efficiency. In any case, however, for the purpose of MS-SPP mode guiding, it is highly desirable to assess these efficiencies. Therefore, the sample shown in fig. 6 has been fabricated for the measurement of an MS-SPP mode traveling along a sharp 90° bend in a 2.7-μm-wide gold strip. In the bend the strip is equipped with a Bragg mirror constituted by gold ridges. As discussed below, the measured bend loss allows deducing a minimum reflection

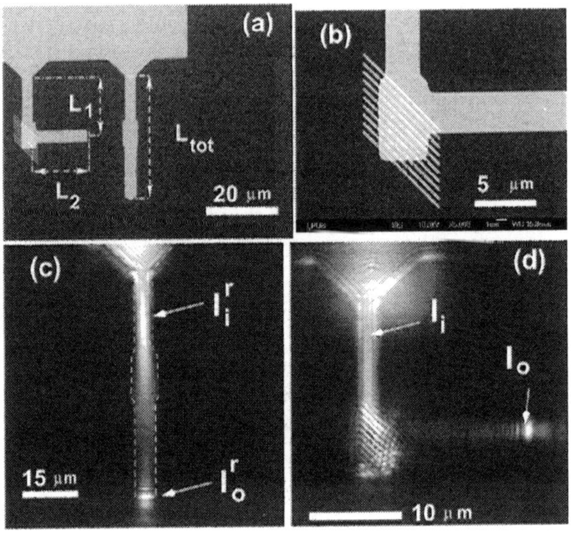

Fig. 6. 90° bent strip equipped with tilted Bragg mirrors. (a) SEM image of the bent and a reference straight strip connected to an extented gold thin film for SPP launch. (b) Magnified SEM image of the tilted Bragg mirror. (c) PSTM image of the reference straight strip and (d) PSTM image of the right angle bent strip, $\lambda_0 = 800$ nm.

efficiency of the mirror. This sample was fabricated in a two-step EBL process. The mirror consisting of 10 gold ridges (height 60 nm, period 550 nm) and the strips have been, respectively, fabricated during the first and the second lithography step.

The sample comprises a bent and a straight strip having the same width and the same total length ($L_{\text{tot}} = L_1 + L_2$, see fig. 6a). The bent strip is equipped with a Bragg mirror designed to deflect the incident SPP traveling along the input strip (length L_1), see fig. 6b. Figure 6c (resp. 6d) shows the SPP propagation along the reference straight (resp. bent) strip. In order to assess the bend losses, we compare the intensity of the scattering spot at the end of the bent strip (I_{o}) with that at the end of the reference strip (I_{o}^r). Normalizing these output intensities by their corresponding input levels, the bend losses are computed according to the equation:

$$\mathscr{L}(\text{dB}) = -10 \times \log\left[\frac{I_{\text{o}}}{I_{\text{o}}^r} \times \frac{I_i^r}{I_i}\right]. \tag{3.2}$$

In this way, we obtain bend losses characterizing the reflection efficiency of the tilted Bragg mirror and the insertion loss of the reflected SPP mode into the output strip (length L_2). By taking small scan PSTM images of both reference and bent strips (not shown here), the relevant

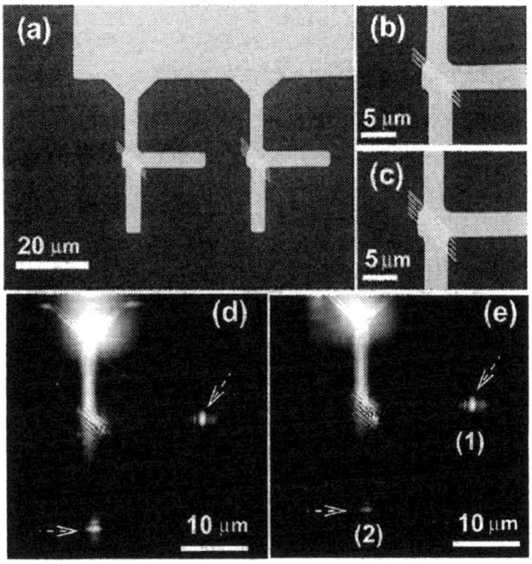

Fig. 7. Gold strip based SPP splitters. (a) SEM image of two SPP splitters connected to an extented gold thin film for SPP launch. (b) and (c) SEM images of Bragg mirrors constituted by 3 and 5 ridges, respectively. (d) and (e) show the respective PSTM images, $\lambda_0 = 800\,\mathrm{nm}$.

intensities have been accurately measured and bend losses as low as $1.9 \pm 0.6\,\mathrm{dB}$ have been obtained (Weeber et al., 2005). Such losses seem acceptable for a variety of applications and therefore sharply bent MS could be conveniently used for the design of many useful SPP-based optical devices such as splitters of interferometers.

The samples shown in figs. 7a–c have been fabricated to demonstrate that the splitting ratio of MS-SPPs can be tuned by using appropriately tuned 45° tilted Bragg mirrors. By comparing scattering spot intensities at the end of each output branch, a 50/50 MS-SPP splitter was found to be obtained by a Bragg mirror comprising only three ridges (see figs. 7b, d). If the number of ridges in the mirror in increased to five, the SPP is now mainly guided along the output branch perpendicular to the input strip leading to 70/30 splitter (figs. 7c, e).

§ 4. Metal nanowires

A reduction of the lateral strip width leads to a waveguide with sub-wavelength lateral extension, a nanowire. Metal nanowires are of particular interest as their cross-section can in principle be scaled down below the length scale of the light wavelength, rendering the guiding of quasi-one-dimensional SPP beams possible (Takahara et al., 1997). First results on SPP propagation following local excitation along gold

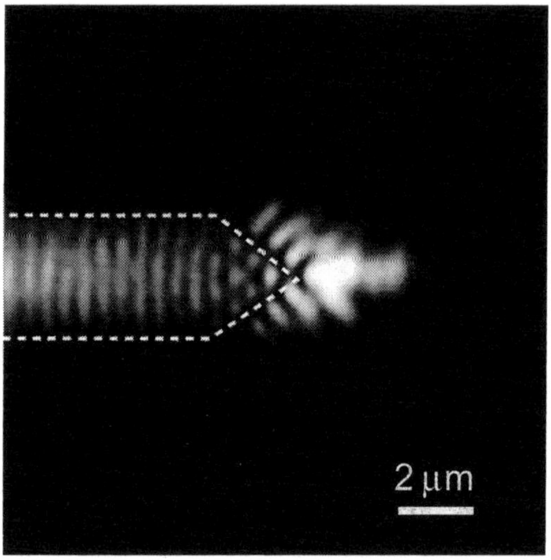

Fig. 8. PSTM image of the triangular termination of a 3.5-μm-wide and 60-nm-thick silver strip. The SPP propagates from left to right; the wavelength of the exciting light is 633 nm.The dashed line outlines the strip geometry.

and silver nanowires have been reported in Dickson and Lyon (2000). Here, we discuss two types of nanowires. First, a lithographically fabricated gold nanowire with a cross-section of $200 \times 50\,\mathrm{nm}^2$. Second, chemically prepared silver nanowires with a cross-section diameter of about 100 nm.

As has been shown above, SPPs propagate efficiently in micrometer-wide metal strips. An evident approach to achieve metal waveguides with submicrometer width is thus to laterally taper a strip waveguide. Figure 8 shows the PSTM image of a 3.5-μm-wide and 60-nm-thick silver strip ending in a triangular termination. The observed fringe pattern indicates that the SPP incident on the strip termination is partly reflected. In addition, a bright spot is indicative of SPP forward scattering. This spot is confined to about 1 μm (full width at half maximum, FWHM) in the horizontal direction, which is direct proof of efficient SPP focusing by the triangular strip termination (Weeber et al., 2001).

4.1. Lithographically fabricated nanowires

A taper as in fig. 8 can be used to feed a strip SPP into a metal nanowire, i.e., a SPP waveguide with subwavelength cross-section. Figure 9a shows a 3 μm wide and 50 nm thick gold strip, which is laterally tapered to a

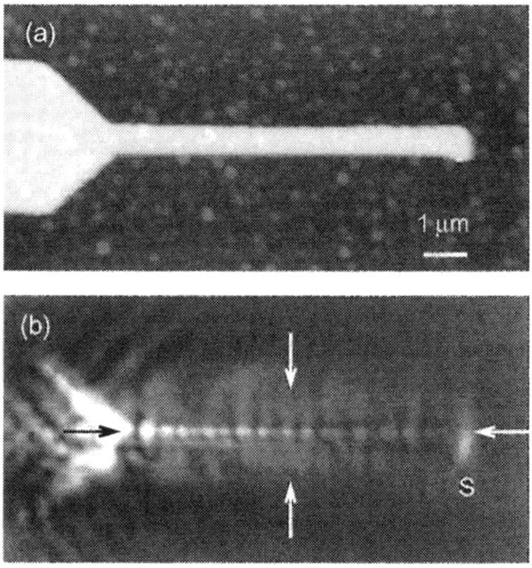

Fig. 9. SPP propagation in a lithographically fabricated silver nanowire. (a) Topography, (b) simultaneously acquired PSTM image. *S* marks SPP scattering at the nanowire termination. The arrows indicate the lines where the cross-cuts in fig. 10 were taken.

width of 200 nm. SPP excitation in the 3-µm-wide gold strip is achieved by the opaque shield method (compare fig. 1b). The PSTM image ($\lambda_0 = 633$ nm) in fig. 9b shows that SPP propagation along the nanowire is indeed taking place (Krenn et al., 2002). Besides the bright spot at the nanowire end (marked by *S*), which is again indicative of SPP scattering, an intensity modulation along the nanowire is observed. For closer analysis we extract intensity cross-sections from fig. 9b. Figure 10a displays the cross-section along the nanowire axis as marked by the arrows in fig. 9b. Apart from some distortion along the first µm of the nanowire due to the presence of the taper, we find an overall decrease in SPP intensity with distance (solid line) which can be fitted as an exponential intensity decay with a corresponding value of the SPP propagation length $L_{SPP} = 2.5$ µm (dashed line). Furthermore, the intensity along the nanowire is found to be modulated with a periodicity of half the SPP wavelength expected for an extcnted gold/air interface (Raether, 1988). This result thus demonstrates that within experimental error the SPP wavelength of the SPP mode in the gold nanowire 200 nm wide equals the SPP wavelength on micrometer-wide strips. As in the case of the silver strip in fig. 8 this modulation thus corresponds to interference between the SPP propagating along the nanowire from left to right and a counterpropagating SPP due to reflection at the nanowire termination. This interpretation is further supported by the SPP intensity profile along a 20-µm-long gold

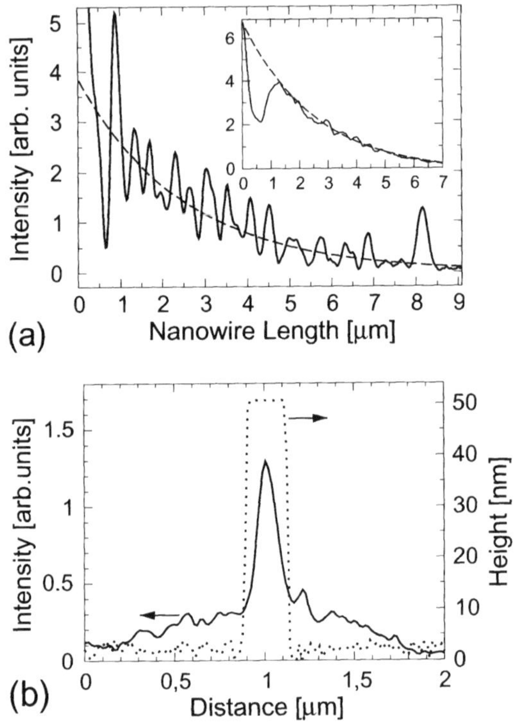

Fig. 10. Cross-cuts from fig. 9b. (a) Cross-cut (solid line) along the nanowire as indicated by the horizontal arrows in fig. 9b. The dashed line is an exponential fit with $L_{SPP} = 2.5\,\mu m$. The inset shows a corresponding cross-cut (solid line) from a 20-μm-long nanowire. Again, the dashed line is an exponential fit with $L_{SPP} = 2.5\,\mu m$. (b) Solid line: cross-cut as indicated by the vertical arrows in fig. 9b. Dotted line: topography profile of the nanowire taken from SEM data, combined with height information from fig. 9a.

nanowire. As shown in the inset of fig. 10a, L_{SPP} is again found to be 2.5 μm but in this case no interference pattern is observed. This finding is readily explained by the length of the nanowire leading to SPP intensity at the nanowire termination that has already strongly decayed so that no detectable SPP reflection occurs.

An intensity cross-cut in the direction perpendicular to the wire axis is shown in fig. 10b. Apart from a central peak we observe a rather broad and shallow intensity profile which depends on the geometry of the taper (other taper geometries not shown here) and which is consequently assigned to SPP scattering at the taper edges. The sharp central peak with a FWHM of only 115 nm, corresponding to $\approx\lambda_0/7$, represents thus the lateral width of the nanowire SPP mode. The intensity drops sharply near the nanowire edges so that crosstalk between closely packed nanowires can be expected to be accordingly low (Krenn et al., 2002).

4.2. Chemically fabricated nanowires

While the results discussed in the preceding paragraph clearly demonstrate SPP waveguiding in a metal nanowire we are left with the question if propagation lengths are limited to the observed value or higher values could be achieved. In fact, EBL fabrication relies on metal vacuum evaporation leading to polycristalline samples with a certain surface roughness, factors that could contribute to SPP scattering and thus loss. For a better understanding of the ultimate limits of SPP propagation in metal nanowires, we thus turn to nanowires with a well-defined crystalline structure. Such nanowires can be made of silver with cross-section diameters around 100 nm and lengths up to 70 μm by a chemical reduction method of silver ions in an aqueous electrolyte solution. High-resolution transmission electron microscopy reveals the nanowires to consist of a lattice aligned bundle of five monocrystalline rods of triangular cross-section forming an almost regular pentagonal cross-section (Graff et al., 2005). Casting the purified electrolyte on a glass slide and letting it dry under ambient conditions yields well-separated individual wires on the slide. We note that essential differences in the SPP propagation characteristics due to the type of metal chosen, silver or gold, can be largely excluded when working in the red spectral range.

SPP propagation along a nanowire can be straightforwardly demonstrated by local optical excitation (Dickson and Lyon, 2000; Krenn and Weeber, 2004), see fig. 11a. The according experimental microscopic image is shown in fig. 11b where a focused laser beam ($\lambda_0 = 785$ nm), as visible on the left-hand side, is positioned over one end of a 18.6-μm-long nanowire with a diameter of 120 nm. The polarization of the exciting laser beam is oriented along the nanowire axis. Part of the incident laser intensity is scattered into a SPP mode, which propagates towards the right-hand end of the wire. There, part of the plasmon intensity is scattered to light, which is collected with a microscope objective and detected by a charge-coupled-device (CCD) camera (fig. 11b). As expected for SPP modes for a light field polarization normal to the wire axis the distal end remains dark.

Light emission from the nanowire in fig. 11b is constricted to the end face due to the strongly bound character of the SPP field, which couples to far-field light only at wire discontinuities. Direct imaging of the SPP field along the nanowire can thus only be accomplished by a near-field technique such as PSTM. A PSTM image over the sample area defined by the box in fig. 11b is shown in fig. 11c. This image reveals the modulation of the SPP near-field along the nanowire due to plasmon reflection at the

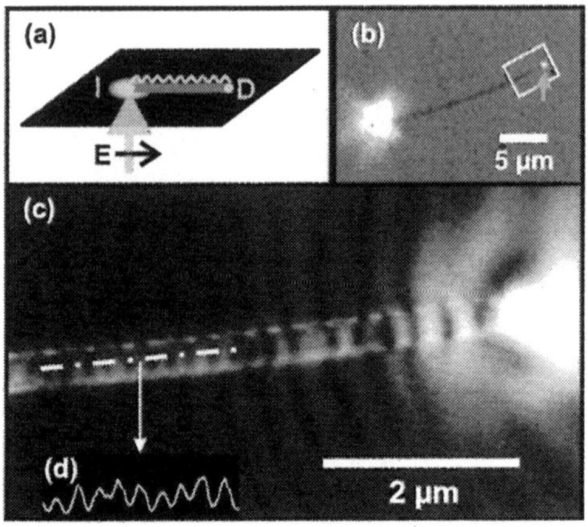

Fig. 11. SPP propagation along a silver nanowire 18.6 μm long. (a) Sketch of optical excitation; I (input) and D (distal) mark the two end faces of the wire. (b) Microscopic image, the bright spot to the left is the focused exciting light. The arrow indicates light scattered from the right hand side (distal) wire end. (c) PSTM image, the image area corresponds to the white box in (b). (d) Cross-cut along the chain dotted line in (c). Reprinted with permission from Ditlbacher et al., 2005 © The American Physical Society.

wire end face, as further illustrated by the cross-cut in fig. 11d. Similar patterns have been observed before on metal strips with widths ranging from a few μm down to 200 nm, compare fig. 10a (Weeber et al., 2001; Krenn et al., 2002). In all these cases, SPP wavelengths (two times the observed modulation pitch) closely matching those observed for a flat extended surface were deduced. Here, however, for a silver nanowire with a diameter of 120 nm the SPP wavelength is 414 nm, which is considerably shorter than the exciting light wavelength of 785 nm. The ratio of these two wavelengths shows that the SPP mode cannot directly couple to far-field light neither in air nor in the glass substrate (refractive index 1.5). This finding implies that SPP propagation along this wire is not radiation damped.

Deepened insight into the nanowire SPP can be gained by optical spectroscopy when analyzing white light scattered from the nanowire end faces. The nanowire is uniformly illuminated with a collimated white light beam (diameter 1 mm) under total internal reflection from the substrate side (fig. 12a). The illumination scheme is thus a dark field setup that offers the advantage of supressed background (excitation) light. The scattered light is selectively picked from one of the two nanowire end faces with a microscope objective. While now both nanowire ends are illuminated, the asymmetry in the illumination scheme still defines an input (I) and a distal (D) wire end face, see (fig. 12a). Indeed, the exciting

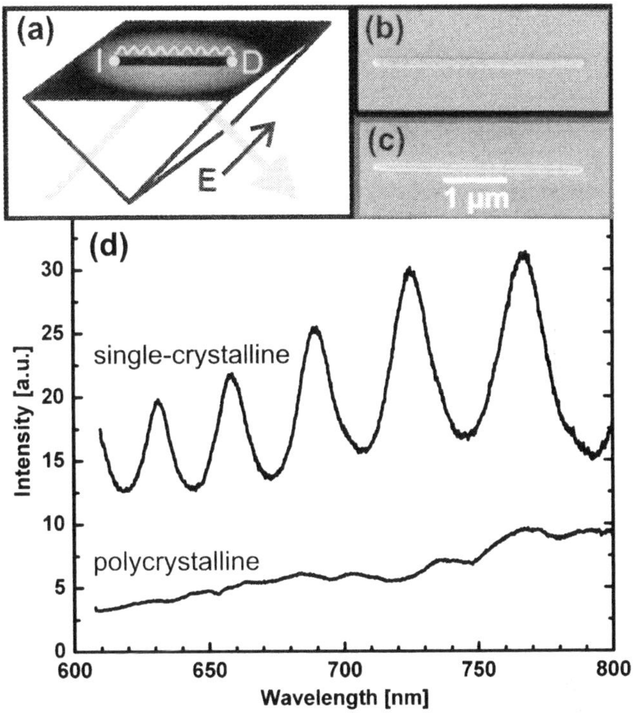

Fig. 12. Scattered light spectra of 3.3-µm-long silver nanowires. (a) Sketch of optical excitation. The exciting light propagation direction projected onto the substrate plane is parallel to the nanowire axis defining an input (I) and a distal (D) end face. The polarization is fixed in the plane of incidence. (b) and (c) SEM images of a chemically and an EBL fabricated silver nanowire, respectively. (d) Scattered light spectra from the distal nanowire end face of the chemically fabricated wire (single-crystalline, upper curve) and the lithographically fabricated wire (polycrystalline, lower curve). Reprinted with permission from Ditlbacher et al., 2005 © The American Physical Society.

light field couples to the SPP mode (with a noteworthy efficiency) only at the input end face of the wire (Ditlbacher et al., 2005).

The upper curve in fig. 12d shows the spectrum taken from the distal end of a 3.3-µm-long nanowire with a diameter of 90 nm, see fig. 12b. The scattered light intensity is modulated as a function of wavelength, with the distinct line shape of Fabry–Perot resonator modes. This is indicative of multiple reflections of the SPP and thus a propagation length considerably larger than the wire length. Indeed, provided that the wire end faces reflect an incident SPP, a nanowire can be turned into a SPP resonator. Then resonator modes, i.e., standing SPP waves along the nanowire axis exist whenever an integer of half the SPP wavelength equals the wire length. The maximum achievable resonator length is limited by the metallic damping of the SPP mode. For comparison, we applied EBL to fabricate a nanowire geometrically closely matching the wire in fig. 12b. This sample (fig. 12c) is polycrystalline and shows some surface roughness. The corresponding scattered light spectrum (lower curve in fig. 12d) displays no regular signal modulation. This underlines the importance of a highly ordered crystalline structure of the nanowire to achieve large SPP propagation length.

Analyzing spectra taken from nanowires of different lengths (not shown here) reveals the following aspects. First, with increasing wire length the linewidth of the Fabry–Perot resonator modes increases. This is due to an increase of SPP losses with increasing wire length. Second, signal minima from the input end correspond to maxima from the distal wire end, and vice versa. This finding underlines the interpretation of optically excited nanowires as Fabry–Perot resonators, i.e., the scattered light intensities from the input and distal nanowire ends correspond to the transmission and reflection intensities, respectively, of a resonator constituted by two mirrors facing each other. We note that this result confirms our former assumption that SPPs are not directly excited by the incident light at the distal wire end with a noteworthy efficiency. Third, the modulation depth of the spectra varies with both wavelength (reflecting the wavelength-dependent dielectric function of silver) and nanowire length. Analyzing the modulation depth from nanowires of different lengths in terms of the Fabry–Perot model retrieves the SPP propagation length L_{SPP} and the nanowire end face reflectivity R. For the investigated silver nanowires with diameters around 100 nm this analysis gives $L_{SPP} = 10.1 \pm 0.4$ and $R = 0.25 \pm 0.01$ (Ditlbacher et al., 2005). The propagation length is well above the propagation lengths on metal nanowires reported before. This result can be mainly assigned to the well-developed crystal and surface structure of the chemically fabricated samples. In addition, however, we deal with a plasmon mode that does not lose energy via radiation damping, as illustrated by the nanowire plasmon dispersion relation in fig. 13, deduced from the spectral data (Ditlbacher et al., 2005). The dispersion relation lying to the right-hand side of both light lines in air and in the glass substrate clearly illustrates that no direct coupling of SPP and light can occur.

§ 5. Summary and future directions

Gold or silver strips and wires are waveguide geometries capable of propagating SPPs over distances of typically some micrometers to some tens of micrometers. While the modal behavior of SPP strips is still

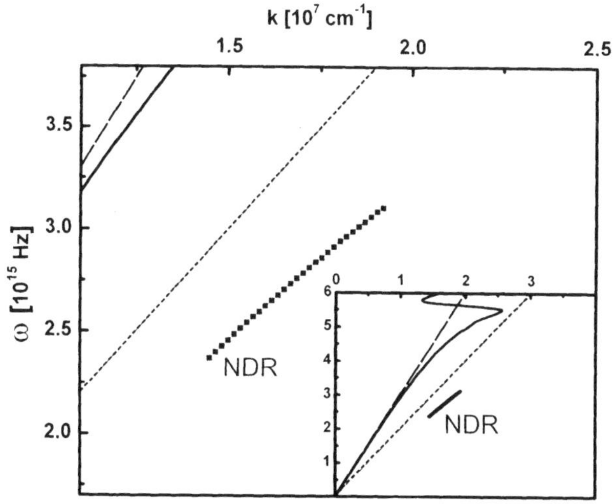

Fig. 13. SPP dispersion relation of a silver nanowire (nanowire dispersion relation, NDR), as derived from spectral data as shown in fig. 12c. The solid black line is the SPP dispersion relation at a silver/air interface. The long and the short dashed curves define the light lines in air and the glass substrate, respectively. and inset show the same curves over different axes ranges. Reprinted with permission from Ditlbacher et al., 2005 © The American Physical Society.

subject to some debate, the two-dimensional character and high confinement factor of such waveguides have been clearly demonstrated. This could enable a 'flat' optics approach relying on densly packed integration in coplanar geometry, paving the way for photonics to the whole set of quasi-two-dimensional lithography-based fabrication schemes as readily used in microelectronics. While a wealth of applications can be envisioned in this context, SPP waveguides might as well turn out as simple and cheap means to incorporate optical interconnects into electronic and optoelectronic systems. Finally, the waveguiding and routing capabilities of SPP strip waveguides might complement existing SPP sensor concepts that mainly rely on functionalized gold surfaces. On the other hand, metal nanowires can sustain short-wavelength SPP modes and thus hold the potential for the optical addressing of subwavelength volumes or individual nano-objects as molecules or quantum dots.

Acknowledgments

The authors thank F.R. Aussenegg, H. Ditlbacher, A. Drezet, A. Hohenau, B. Lamprecht, A. Leitner, B. Steinberger, and A.L. Stepanov for fruitful discussions and advice.

References

Barnes, W.L., Preist, T.W., Kitson, S.C., Sambles, J.R., 1996, Physical origin of photonic energy gaps in the propagation of surface plasmons on gratings, Phys. Rev. B **54**, 6227–6244.

Berini, P., 1999, Plasmon-polariton modes guided by a metal film of finite width, Opt. Lett. **24**, 1011–1013.

Berini, P., 2000, Plasmon polariton waves guided by thin lossy metal films of finite width: Bound modes of symmetric structures, Phys. Rev. B **61**, 10484–10503.

Berini, P., 2001, Plasmon polariton waves guided by thin lossy metal films of finite width: Bound modes of asymmetric structures, Phys. Rev. B **63**, 125417.

Bouhelier, A., Huser, T., Tamaru, H., Guntherodt, H.J., Pohl, D.W., Baida, F.I., Van Labeke, D., 2001, Plasmon optics of structured silver films, Phys. Rev. B **63**, 155404.

Bozhevolnyi, S.I., Coello, V., 1998, Elastic scattering of surface plasmon polaritons: Modeling and experiment, Phys. Rev. B **58**, 10899–10910.

Bozhevolnyi, S.I., Erland, J.E., Leosson, K., Skovgaard, P.M.W., Hvam, J.M., 2001, Waveguiding in surface plasmon polariton band gap structures, Phys. Rev. Lett. **86**, 3008–3011.

Bozhevolnyi, S.I., Pudonin, F.A., 1997, Two-dimensional micro-optics of surface plasmons, Phys. Rev. Lett. **78**, 2823–2826.

Courjon, D., Sarayedine, K., Spajer, M., 1989, Scanning tunneling optical microscopy, Opt. Comm. **71**, 23–28.

Dawson, P., de Fornel, F., Goudonnet, J.P., 1994, Imaging of surface plasmon propagation and edge interaction using a photon scanning tunneling microscope, Phys. Rev. Lett. **72**, 2927–2930.

Dereux, A., Girard, C., Weeber, J.C., 2000, Theoretical principles of near-field optical microscopies and spectroscopies, J. Chem. Phys. **112**, 7775–7789.

Dickson, R.M., Lyon, L.A., 2000, Unidirectional plasmon propagation in metallic nanowires, J. Phys. Chem. B **104**, 6095–6098.

Ditlbacher, H., Hohenau, A., Wagner, D., Kreibig, U., Rogers, M., Hofer, F., Aussenegg, F.R., Krenn, J.R., 2005, Silver nanowires as surface plasmon resonators, Phys. Rev. Lett. **95**, 257403.

Ditlbacher, H., Krenn, J.R., Felidj, N., Lamprecht, B., Schider, G., Salerno, M., Leitner, A., Aussenegg, F.R., 2002a, Fluorescence imaging of surface plasmon fields, Appl. Phys. Lett. **80**, 404–406.

Ditlbacher, H., Krenn, J.R., Schider, G., Leitner, A., Aussenegg, F.R., 2002b, Two-dimensional optics with surface plasmon polaritons, Appl. Phys. Lett. **81**, 1762–1764.

Graff, A., Wagner, D., Ditlbacher, H., Kreibig, U., 2005, Silver nanowires, Eur. Phys. J. D **34**, 263–269.

Hohenau, A., Krenn, J.R., Stepanov, A., Drezet, A., Ditlbacher, H., Steinberger, B., Leitner, A., Aussenegg, F.R., 2005, Dielectric optical elements for surface plasmons, Opt. Lett. **30**, 893–895.

Karrai, K., Grober, R.D., 1995, Piezoelectric tip–sample distance control for near field optical microscopes, Appl. Phys. Lett. **66**, 1842–1844.

Krenn, J.R., Weeber, J.C., 2004, Surface plasmon polaritons in metal stripes and wires, Phil. Trans. R. Soc. Lond. A **326**, 739–756.

Krenn, J.R., Dereux, A., Weeber, J.R., Bourillot, E., Lacroute, Y., Goudonnet, J.P., Schider, G., Gotschy, W., Leitner, A., Aussenegg, F.R., Girard, C., 1999, Squeezing the optical near–field zone by plasmon coupling of metallic nanoparticles, Phys. Rev. Lett. **82**, 2590–2593.

Krenn, J.R., Lamprecht, B., Ditlbacher, H., Schider, G., Salerno, M., Leitner, A., Aussenegg, F.R., 2002, Non diffraction limited light transport by gold nanowires, Europhys. Lett. **60**, 663–669.

Krenn, J.R., Wolf, R., Leitner, A., Aussenegg, F.R., 1997, Near-field optical imaging the surface plasmon fields around lithographically designed nanostructures, Opt. Commun. **137**, 46–50.

Lamprecht, B., Krenn, J.R., Schider, G., Ditlbacher, H., Salerno, M., Felidj, N., Leitner, A., Aussenegg, F.R., Weeber, J.C., 2001, Surface plasmon propagation in microscale metal stripes, Appl. Phys. Lett. **79**, 51–53.

Laurent, G., Felidj, N., Lau Truong, S., Aubard, J., Levi, G., Krenn, J.R., Hohenau, A., Leitner, A., Aussenegg, F.R., 2005, Imaging surface plasmons on gold nanoparticle-arrays by far-field Raman scattering, Nano Lett. **5**, 253–258.

Maier, S.A., Kik, P.G., Atwater, H.A., Meltzer, S., Harel, E., Koel, B.E., Requicha, A.A.G., 2003, Local detection of electromagnetic energy transport below the diffraction limit in metal nanoparticle plasmon waveguides, Nature Mater. **2**, 229–232.

McCord, M.A., Rooks, M.J., 1997, Handbook of Microlithography, Micromachining and Microfabrication, in: Rai-Choudhury, P. (Ed.), SPIE and The Institution of Electrical Engineers, vol. 1, Bellingham, Washington, pp. 139–249, ch. 2.

Nikolajsen, T., Leosson, K., Salakhutdinov, I., Bozhevolnyi, S.I., 2003, Polymer-based surface-plasmon-polariton stripe waveguides at telecommunication wavelengths, Appl. Phys. Lett. **82**, 668–670.

Pincemin, F., Greffet, J.J., 1996, Propagation and localization of a surface plasmon polariton on a finite grating, J. Opt. Soc. Am. B **13**, 1499–1509.

Pincemin, F., Maradudin, A.A., Boardman, A.D., Greffet, J.J., 1994, Scattering of a surface plasmon polariton by a surface defect, Phys. Rev. B **50**, 12261–12275.

Quidant, R., Weeber, J.C., Dereux, A., Peyrade, D., Chen, Y., Girard, C., 2002, Near–field observation of evanescent light wave coupling in subwavelength optical waveguides, Europhys. Lett. **57**, 191–197.

Quinten, M., Leitner, A., Krenn, J.R., Aussenegg, F.R., 1998, Electromagnetic energy transport via linear chains of silver nanoparticles, Opt. Lett. **23**, 1331–1333.

Raether, H., 1988, Surface Plasmons, Springer, Berlin.

Reddick, R.C., Warmack, R.J., Ferrell, T.L., 1989, New form of scanning optical microscopy, Phys. Rev. B **39**, 767–770.

Sanchez-Gil, J.A., 1996, Coupling, resonance transmission, and tunneling of surface – plasmon polaritons through metallic gratings of finite length, Phys. Rev. B **53**, 10317–10327.

Sanchez-Gil, J.A., 1998, Surface defect scattering of surface plasmon polaritons: Mirrors and light emitters, Appl. Phys. Lett. **73**, 3509–3511.

Sanchez-Gil, J.A., Maradudin, A.A., 2005, Surface-plasmon polariton scattering from a finite array of nanogrooves/ridges: Efficient mirrors, Appl. Phys. Lett. **86**, 251106.

Schröter, U., Heitmann, D., 1998, Surface-plasmon-enhanced transmission through metallic gratings, Phys. Rev. B **58**, 15419–15421.

Smolyaninov, I.I., Mazzoni, D.L., Davis, C.C., 1996, Imaging of surface plasmon scattering by lithographically created individual surface defects, Phys. Rev. Lett. **77**, 3877–3880.

Smolyaninov, I.I., Mazzoni, D.L., Mait, J., Davis, C.C., 1997, Experimental study of surface-plasmon scattering by individual surface defects, Phys. Rev. B **56**, 1601–1611.

Takahara, J., Yamagishi, S., Taki, H., Morimoto, A., Kobayashi, T., 1997, Guiding of a one–dimensional optical beam with nanometer diameter, Opt. Lett. **22**, 475–477.

Weeber, J.C., Bourillot, E., Dereux, A., Goudonnet, J.P., Chen, Y., Girard, C., 1996, Observation of light confinement effects with a near-field optical microscope, Phys. Rev. Lett. **77**, 5332–5335.

Weeber, J.C., Gonzalez, M.U., Baudrion, A.-L., Dereux, A., 2005, Surface plasmon routing along right angle bent metal stripes, Appl. Phys. Lett. **87**, 221101.

Weeber, J.C., Krenn, J.R., Dereux, A., Lamprecht, B., Lacroute, Y., Goudonnet, J.P., 2001, Near-field observation of surface plasmon polariton propagation on thin metal stripes, Phys. Rev. B **64**, 045411.

Weeber, J.C., Lacroute, Y., Dereux, A., 2003, Optical near-field distributions of surface plasmon waveguide modes, Phys. Rev. B. **68**, 115401.

Weeber, J.C., Lacroute, Y., Dereux, A., Devaux, E., Ebbesen, T., Girard, C., Gonzalez, M.U., Baudrion, A.-L., 2004, Near-field characterization of Bragg mirrors engraved in surface plasmon waveguides, Phys. Rev. B. **70**, 235406.

Zia, R., Chandran, A., Brongersma, M.L., 2005a, Dielectric waveguide model for guided surface polaritons, Opt. Lett. **30**, 1473–1475.

Zia, R., Selker, M.D., Brongersma, M.L., 2005b, Leaky and bound modes of surface plasmon waveguides, Phys. Rev. B **71**, 165431.

Chapter 3

Super-resolution microscopy using surface plasmon polaritons

by

Igor I. Smolyaninov

Department of Electrical and Computer Engineering, University of Maryland, College Park, MD 20742

Anatoly V. Zayats

Nano-Optics and Near-Field Spectroscopy Group, Centre for Nanostructured Media, IRCEP, The Queen's University of Belfast, Belfast BT7 1NN, UK

Nanophotonics with Surface Plasmons
Advances in Nano-Optics and Nano-Photonics
ISSN: 1871-0018

V.M. Shalaev & S. Kawata (Editors)
© 2007 Published by Elsevier B.V.
DOI: 10.1016/S1871-0018(06)02003-6

Contents

§ 1. Introduction

Optical microscopy is one of the oldest research tools. It dates back to 1609 when Galileo Galilei developed an *occhiolino* or compound microscope with convex and concave lenses. Although various electron and scanning probe microscopes had long surpassed it in resolution power, optical microscopy remains invaluable in many fields of science. With the development of nanosciences, our ability to achieve optical imaging on the nanoscale is imperative for further progress in material science, chemistry and biology. Investigations of optical properties and processes on the micro- and nanoscales provide information on electronic structure and chemical and biological specificity. They also lead to new photolithographic techniques. However, spatial resolution of regular optical imaging is limited by the loss of information contained in large wavevector, non-propagating electromagnetic field components k which are sensitive to small features a of a probed object ($ka \sim 1$).

The reason for limited resolution of an optical microscope is diffraction of light waves and, ultimately, uncertainty principle: a wave cannot be localised much tighter than half of its wavelength $\lambda_0/2$ in vacuum. Immersion microscopes (see, for example, Kingslake, 1983) have slightly improved resolution (down to $\lambda_0/2n$) due to shorter wavelength λ_0/n of light in the medium with refractive index n. Nevertheless, immersion microscope resolution is limited by the narrow range of refractive indices n of available transparent materials, which can be used as immersion media. One of the ways to achieve nanometre-scale spatial resolution is to detect optical field in sub-wavelength proximity to a studied surface where large-wave-vector evanescent field components of the diffracted light are present. This can be done by using a scanning near-field optical microscope (Pohl and Courjon, 1993; Richards and Zayats, 2004). Although many fascinating results have been obtained in near-field optics, near-field microscopes are not as versatile and convenient to use as conventional far-field optical microscopes. For example, an image in a near-field optical microscopy is obtained by point-by-point raster scanning, which is a rather slow process and requires cumbersome distance regulation techniques to keep the probe tip at a few nanometres distance from the surface.

Very recently, some alternative approaches to achieve sub-diffraction limited resolution have been proposed. One of them is a superresolution imaging technique based on a "perfect lens" (Pendry, 2000) made from an artificial material with simultaneously negative dielectric permittivity and magnetic permeability. In theory, when losses are neglected such a "lens" is capable of "enhancing" evanescent field components (only negative permittivity is needed in the case of a thin slab). The "perfect lens" schemes have not yet achieved optical magnification in the experiment. However, a theoretical scheme which exhibits magnification in the electrostatic near-field regime has been described by Ramakrishna and Pendry (2004). Successful 60 nm resolution photolithography based on the silver film superlenses has been reported by Fang et al. (2005) and Melville and Blaikie (2005). The main drawback of the "perfect lens" in microscopy applications is that the losses in optical materials severely limit performance of such a lens. Until metamaterials with very low losses and negative permeability in the optical spectral range are developed, an image formed by a "perfect lens" can be observable only in the near-field of the lens using an auxiliary near-field scanning optical microscope.

Another recently developed far-field microscopy approach is based on the saturated depletion of stimulated emission (Westphal and Hell, 2005). This method is, however, applicable only to fluorescent objects. It relies on raster scanning similar to near-field optical microscopy. Image acquisition in such a microscope is sequential, hence it is slow. Direct high-resolution far-field optical imaging still remains an attractive dream for numerous potential applications in material science, biology, photolithography and many other fields of science.

In order to improve the spatial resolution of optical imaging, the large wavevector components of the field diffracted by the object have to be detected. In conventional three-dimensional (3D) optics, the largest wavevector of propagating light wave supported by a medium with the dielectric constant $\varepsilon(\omega)$ is $k_0 = \varepsilon^{1/2}\omega/c$, where ω is the light frequency and c is the speed of light in vacuum. The light waves with larger wave-vectors are evanescent. They rapidly decay with the distance from the object resulting in diffraction-limited resolution of far-field optical instruments, as described by the Rayleigh–Abbe resolution criterion.

To overcome the limitations of 3D optics, a situation can be considered when light is confined in a two-dimensional (2D) geometry. In this case, which can be a surface wave or a guided wave in a 2D waveguide, a large-wavevector propagating waves may exist in 2D [(x,y)-plane] with the dispersion relation

$$k_{xy} = \left(\frac{\omega^2}{c^2}\varepsilon(\omega) - \kappa^2\right)^{1/2} \tag{1.1}$$

where κ is the coefficient describing the field behaviour in z-direction perpendicular to the surface. If the wave is confined to the surface, κ is imaginary (only the evanescent field exists on both sides of the surface), leading to the increased wavevector for the in-plane wave propagation. Thus, using such large-wave-vector (short-wavelength) surface waves for image formation, it is possible to achieve very high spatial resolution in accordance with the Rayleigh–Abbe criterion.

Such a situation can be realised using surface plasmon polaritons (SPPs). SPPs are surface electromagnetic waves on a metal–dielectric interface that can have short wavelengths in the spectral range near the surface plasmon resonance frequency. Recent overviews of the basic properties of SPPs can be found in the recent reviews by Zayats and Smolyaninov (2003) and Zayats et al. (2005b) as well as in many contributions to this book. SPPs may be used to form a sufficiently magnified image of an object in 2D plane of the metal surface, which may be large enough to be visualised and resolved with a conventional optical microscope. This task may be performed with the help of various recently developed 2D plasmon-optical components and devices that mimic conventional 3D optics, including planar, focusing and magnifying mirrors, lenses, resonators etc.

A simplified consideration of SPP optics in the short-wavelength regime may be described as follows (Smolyaninov et al., 2005c, d). The SPP wavevector corresponding to the frequency ω of the surface electromagnetic wave propagating along an interface between semi-infinite metal and dielectric is given by

$$k_{\mathrm{SP}} = \frac{\omega}{c}\left(\frac{\varepsilon_{\mathrm{d}}\varepsilon_{\mathrm{m}}}{\varepsilon_{\mathrm{d}} + \varepsilon_{\mathrm{m}}}\right)^{1/2}, \tag{1.2}$$

where ε_{m} and ε_{d} are the frequency-dependent dielectric constants of the metal and dielectric respectively. Under the condition $\varepsilon_{\mathrm{m}}(\omega) \to -\varepsilon_{\mathrm{d}}(\omega)$, the wavelength $\lambda_{\mathrm{SP}} = 2\pi/k_{\mathrm{SP}}$ of the SPP modes becomes very small if the frequency is near the frequency of surface plasmon resonance ω_0 defined by

$$\varepsilon_{\mathrm{m}}(\omega_0) = -\varepsilon_{\mathrm{d}}(\omega_0), \tag{1.3}$$

and we neglect the imaginary part of ε_{m}. In other words, in this frequency range the effective refractive index $n_{\text{eff}} = \lambda_0/\lambda_{\text{SP}}$ for SPPs becomes very large. While this dispersion relation is modified in the general case of a finite-thickness lossy metal film (Burke et al., 1986), the case of low losses in a symmetric configuration (in which the permittivities of the dielectric on both sides of the thin metal film coincide) is also described by eq. (1.2) in the short-wavelength ($\omega \to \omega_0$) limit. Since n_{eff} may reach extremely large values up to 10^2, the theoretical diffraction limit of resolution with such SPP waves, determined by $\lambda_{\text{SP}}/2$, may reach a scale of a few nanometres. However, in reality the Ohmic losses in a metal limit the wavelength and the propagation length of the short-wavelength SPPs, and thus severely limit the attainable resolution. We will concentrate on possible strategies to overcome this limitation later in this chapter.

The large wavevector SPPs cannot be excited in the Kretschmann configuration using total internal reflection of light in a prism (in this case the largest wavevector is determined by the prism dielectric constant and is similar to the one achievable in immersion microscopy). Instead, the diffraction grating should be used to efficiently couple illuminating light to surface polaritons with large wavevectors (Zayats et al., 2005b). In such a case the periodic lattice provides the wavevector conservation for light coupling to SPPs.

Due to a high effective refractive index of SPPs near the frequency ω_0, an appropriately shaped 2D dielectric layer deposited onto the metal film surface behaves as a very strong lens for surface polaritons propagating through it. On the other hand, the edge of the dielectric layer becomes an efficient mirror for surface polaritons excited inside it at almost any angle of incidence due to the total internal reflection (fig. 1). (This leads to the "black hole" analogy for a dielectric droplet placed on a metal surface described by Smolyaninov, 2003.)

If SPPs are excited by an object located inside a dielectric (or if this object is illuminated by SPPs excited by other means), a 2D image of the object can be produced by SPPs in the appropriate location on the metal interface due to SPP reflection from the dielectric boundary which can be shaped as a parabolic, elliptical or spherical mirror. Because of the metal surface roughness and the Raleigh scattering in the dielectric, SPPs are partially scattered into photons and can be observed with a conventional optical microscope. The image brightness should far exceed the background of scattered SPPs in other areas of the surface due to higher energy concentration around the image. A fluorescent scheme of the SPP field visualisation may also be used (Ditlbacher et al., 2002).

Thus, the goal of a 2D SPP mirror design is to provide sufficiently high 2D image magnification, so that all the 2D image details would be larger

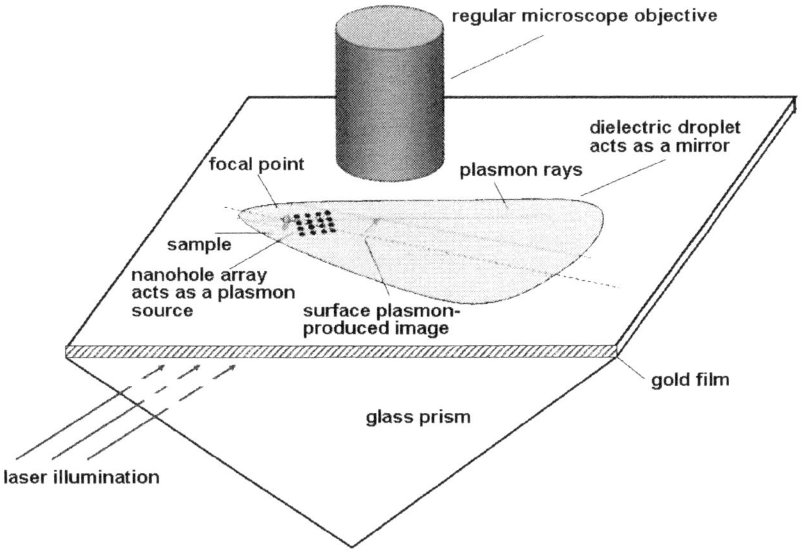

Fig. 1. Schematics of SPP-assisted optical microscopy: SPPs are excited by external illumination and propagate inside a parabolically shaped 2D mirror. Placing an object near the focus of a parabola produces a magnified image formed by SPPs on the surface, which is viewed from the top with a conventional optical microscope.

than the $\lambda_0/2$ resolution limit of the conventional far-field optical microscope. In this way, the combined resolution of such a two-stage microscope arrangement (fig. 1) will be defined by the resolution provided by short-wavelength SPPs on the metal–dielectric interface. As a result, a far-field, non-scanning optical microscopy with the sub-wavelength resolution can be realised. The resolution down to $\sim\lambda_0/10$ has been achieved in such a scheme, which is comparable to the resolution achieved in the lithographical experiments based on the "perfect lens" concept.

The simplified description of the SPP-based imaging discussed above requires some revision if losses in metal are taken into account. The magnification and hence the resolution of the SPP-based microscope depend critically on the propagation length of the short-wavelength SPPs, which is determined by the imaginary part of ε_m. Near ω_0 a typical value of the SPP propagation length would not exceed a few wavelengths on the surface of semi-infinite metal. This limitation can be overcome using thin metal films in symmetric environment where the so-called long-range SPPs can be excited. SPP waves can also be "regenerated" due to optical power influx from a neighbouring phase-matched dielectric waveguide (Lee and Gray, 2005). Such phase matching may be achieved by periodic modulation of either the metal film or the waveguide. The combination of these effects is responsible for the performance of the SPP-microscope discussed below where coupling of SPP modes to waveguiding modes in the dielectric can occur. The rigourous model of the SPP-based microscopy should be based on the properties of SPP photonic crystals. Such a model will be described below.

In the remaining sections of this chapter, after the description of main parameters of short-wavelength SPPs and dielectric SPP mirrors, we will discuss the operation and the basic principles of sub-diffraction imaging using 2D surface polaritonic crystals. First, we will overview the details of the experimental approach and discuss the ways to validate the resolution of the SPP-based microscope. Next, we will consider imaging properties of 2D photonic and SPP crystals that are used in the image formation. The sub-wavelength optical resolution can be achieved due to coupling of evanescent components of the diffraction field generated by the object to the propagating Bloch modes of the photonic crystal. Next, we will describe imaging with surface polaritonic crystal mirrors in both positive and negative refractive index regimes. Finally, optical visualisation of a T4 phage virus will be described as an example of application of this technique to biological samples. The digital image processing further improves the resolution down to 30 nm level, which makes the SPP-based microscope a useful tool for biological studies.

§ 2. Principles of SPP-assisted microscopy

In this section we will describe experimental realisation of curved dielectric SPP mirrors needed to achieve in-plane image magnification by surface polaritons. We will also discuss image formation is such mirrors and properties of short-wavelength SPPs.

2.1. Experimental realization of dielectric SPP mirrors

Magnifying dielectric mirrors for microscopy applications of short-wavelength SPPs can be realised using glycerine microdroplets. The dielectric constant of glycerine $\varepsilon_g = 2.161$ is ideally suited for imaging experiments performed on gold surfaces and thin films using a set of wavelengths of an argon-ion laser (fig. 2b). At the $\lambda_0 = 502$ nm wavelength the real part of the gold dielectric constant is $\varepsilon_m = -2.256$ (Weast, 1987) and the light around this wavelength is located very close to the surface plasmon

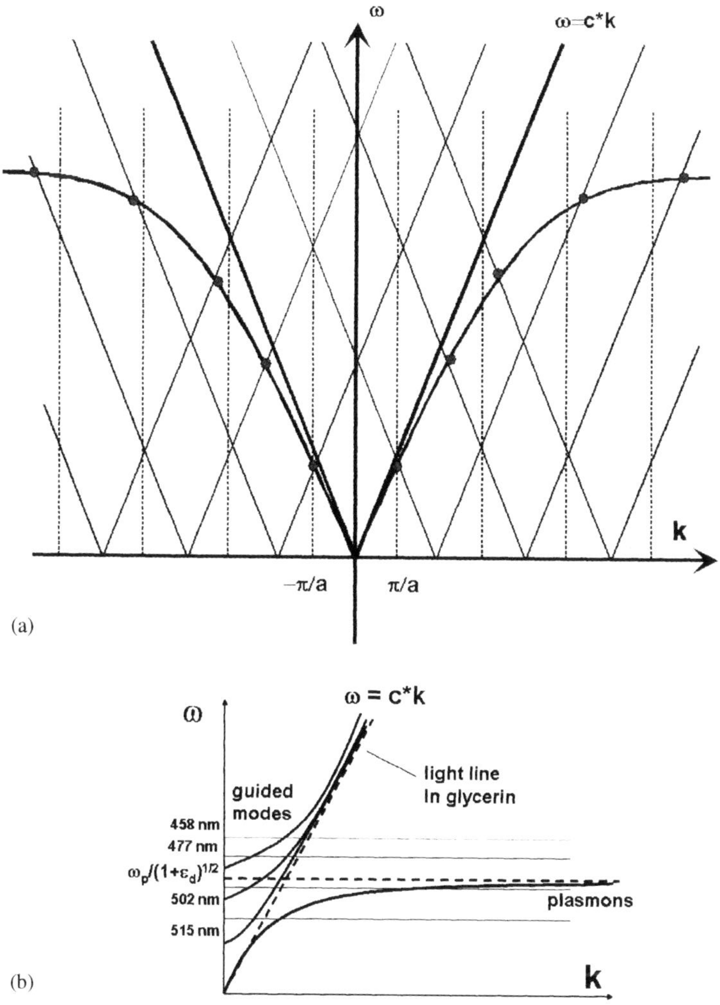

Fig. 2. (a) The dispersion of SPPs and photons propagating inside the dielectric at small angles along the metal–dielectric interface. The intersections between the modes are shown by the dots. They correspond to the efficient excitation of short-wavelength SPPs. (b) Sketch of the Ar-ion laser line positions with respect to the dispersion curve of SPPs on the gold–glycerine interface. Also shown are the approximate locations of the guided optical modes inside the thin layer of glycerine.

resonance. According to eq. (1.2), the corresponding SPP wavelength inside the dielectric is $\lambda_{SP} \sim 70\,\text{nm}$, and the effective refractive index for SPP waves is $n_{eff} = \lambda_0/\lambda_{SP} \sim 7$ if losses are neglected. On the other hand, the dielectric constant of glycerine is close to the dielectric constant of the glass substrate, providing symmetrical environment for a thin gold film. This leads to longer propagation length of SPPs when the Ohmic losses in gold are taken into account.

Magnifying SPP mirrors were formed in desired locations (on top of the objects to be imaged so that they are totally or partially immersed inside glycerine) by bringing a small probe (fig. 3(a)) wetted in glycerine close to a metal surface. The probe was prepared from a tapered optical fibre, which has a seed epoxy microdroplet near its apex. Bringing the probe to a surface region covered with glycerine led to a glycerine microdroplet formation under the probe (fig. 3b). The size of the glycerine droplet was determined by the size of the seed droplet of epoxy. The glycerine droplet under the probe can be moved to a desired location under the visual control, using a conventional microscope. Such droplet deposition procedure allowed us to form SPP mirror shapes, which were reasonably close to parabolic. In addition, the liquid droplet boundary may be expected to be rather smooth because of the surface tension, which is essential for the proper performance of the droplet boundary as a SPP mirror. Since the SPP wavelength is much smaller than the droplets, the image formation in such a mirror can be analysed by simple geometrical optics in 2D. It should be noted that both SPPs and guided modes of the thin dielectric layer (fig. 2b) participate in the image formation. Since the 2D shape of the droplet, as perceived by both kinds of modes, is basically the same, the SPP and the guided mode images are formed in the same location according to the laws of geometrical optics.

2.2. Properties of short-wavelength SPPs

Let us discuss two main parameters, wavelength and propagation length, of SPP waves with respect to imaging applications (Zayats et al., 2005a). In the case of SPPs on a surface of semi-infinite metal, the SPP wavevector is defined by eq. (1.2). Its value is limited by the Ohmic losses resulting, under the conditions when illumination light frequency $\omega_0 = \omega_{\mathrm{SP}}$ is chosen so that $\mathrm{Re}\varepsilon_{\mathrm{m}}\ (\omega_{\mathrm{SP}}) = -\varepsilon_{\mathrm{d}}$, in the largest possible SPP wavevector

$$k_{\mathrm{SP}}^{(0)} = \frac{\omega_{\mathrm{SP}}}{c}\varepsilon_{\mathrm{d}}^{1/2}\left(1 + i\frac{\varepsilon_{\mathrm{d}}}{\mathrm{Im}\varepsilon_{\mathrm{m}}(\omega_{\mathrm{SP}})}\right). \tag{2.1}$$

This wavevector lies in the vicinity of the light line and limits the resolution of the optical elements based on such surface waves. The situation

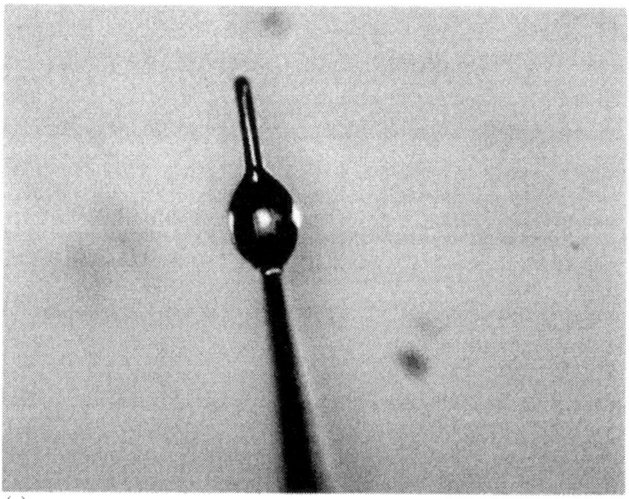

(a)

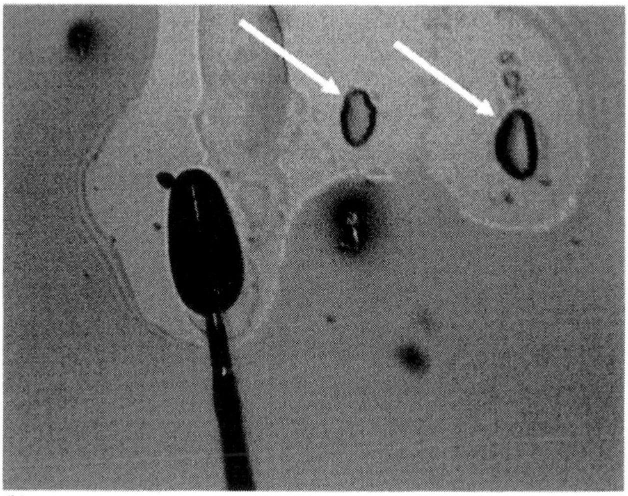

(b)

Fig. 3. (a) The image of the fibre probe with a seed epoxy droplet. (b) The image of the fibre probe wetted in glycerine and glycerine droplets (indicated by arrows) formed in the desired locations by bringing the wetted probe in contact to a surface.

is different in the case of thin metal films ($|\kappa h| \lesssim 1$, where h is the film thickness) in a symmetric environment (Raether, 1988). Owing to the interaction between the SPP modes on the opposite interfaces of the film, the SPP modes are split into high- and low-frequency modes with different symmetry of the field distribution in the film. We can use non-retarded calculations to estimate the largest possible wavevector in this case as

$$\mathrm{Re}k_{\mathrm{SP}}^{(0)} \approx \frac{1}{2h}\ln\left(1+\frac{4\varepsilon_{\mathrm{d}}^{2}}{(\mathrm{Im}\varepsilon_{\mathrm{m}}\ (\omega_{\mathrm{SP}}))^{2}}\right). \tag{2.2}$$

(Please note that eq. (2.2) is not the dispersion relation, but the k_{SP} dependence on the film thickness for given resonant frequency.) Thus, thin enough films can support very short-wavelength SPPs. Moreover, the losses associated with the film SPP mode with anti-symmetric electric field distribution also dramatically decrease with the film thickness as $\mathrm{Im}k_{\mathrm{SP}}{\sim}h^2$ (Raether, 1988). These considerations show that using 2D optical elements based on the film SPP modes, imaging with short-wavelength SPPs can be achieved in principle: both wavelength and propagation losses can be optimised by appropriate choice of film thickness, refractive indeces of substrate and dielectric forming a SPP mirror and illumination frequency. However, practical limits on the 2D microscope magnification and resolution are set by the material constants of metal and dielectric, and may be quite restrictive.

The SPP behaviour discussed above can be illustrated with numerical modelling of metal–dielectric system using finite-element method (fig. 4). On the interface of semi-infinite metal (the situation described by eq. 2.1) one can propagate short-wavelength SPPs only if the losses are small. Consideration of the Ohmic losses of a real metal immediately results in the limitation on the SPP wavelength and strong absorption leading to fast decay of the SPP wave during propagation along the interface (fig. 4a). However, the situation changes significantly for a thin metal film with a strong coupling between SPPs on both interfaces of the film (fig. 4b). In this case, the short-wavelength SPPs exist and propagate reasonably well along the film, and thus can be used for imaging applications.

If the top medium is a thin film of dielectric (as in a real experimental arrangement), one can observe formation of additional waveguiding modes in the dielectric film, and the mode coupling between these modes and the SPPs during SPP propagation along the metal interface (fig. 4c). These modes originate from the finite thickness of the dielectric film forming a SPP mirror. They depend on the film thickness (fig. 2), and typically have very small propagation losses. In the experiment, SPP interaction with the edges of the dielectric, defects on the interface, height and width variation of the dielectric layer etc. might lead to coupling of SPPs to these waveguiding modes.

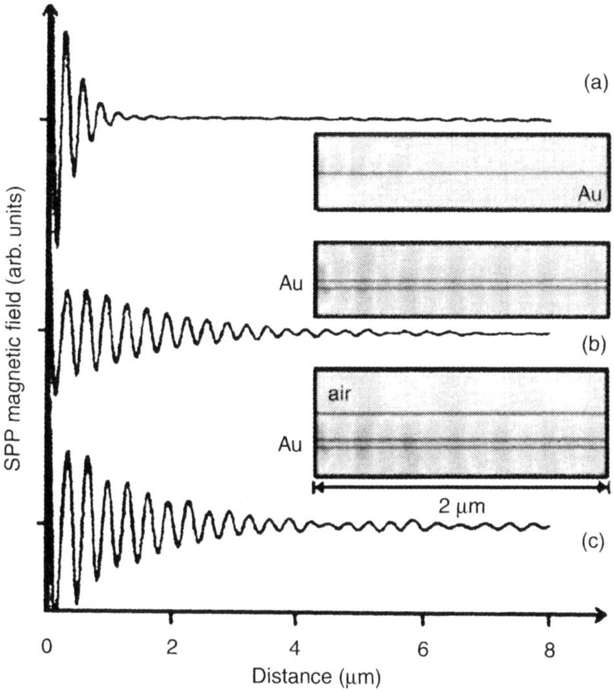

Fig. 4. Distance dependencies of the transverse magnetic field of SPP propagating along the gold–dielectric interface: (a) the interface of semi-infinite media, (b) the thin Au film ($h = 40$ nm) in semi-infinite symmetric environment, (c) the thin Au film ($h = 40$ nm) in symmetric environment with finite thickness of the dielectric ($H = 150$ nm). Illuminating light at wavelength 515 nm; the substrate and superstrate have the same refractive index $n = 1.47$. SPPs propagate from left to right on all images. Inserts show the respective magnetic field distribution in the vicinity of the interface.

The SPP propagation length over the gold–glycerine interface at 502 nm has been measured using two complementary techniques: the near-field imaging technique described by Smolyaninov et al. (1996) and the fluorescent SPP imaging similar to the one described by Ditlbacher et al. (2002). Both techniques provide similar results. In the experiments the artificial pinholes in the gold film were produced inside a thin glycerine droplet (which was stained with the Bodipy dye) by touching the gold film with a sharp scanning tunneling microscope (STM) tip. Such pinholes are known to emit SPP beams (Smolyaninov et al., 1996). The characteristic exponentially decaying SPP beam profile (excited from the right side of the image) observed in the experiments is shown in fig. 5(a), which has been obtained using fluorescent imaging. The cross section of this beam shown in fig. 5(b) indicates the SPP propagation length of the order of 3 μm at the 502 nm wavelength. In some cases it is possible to image the

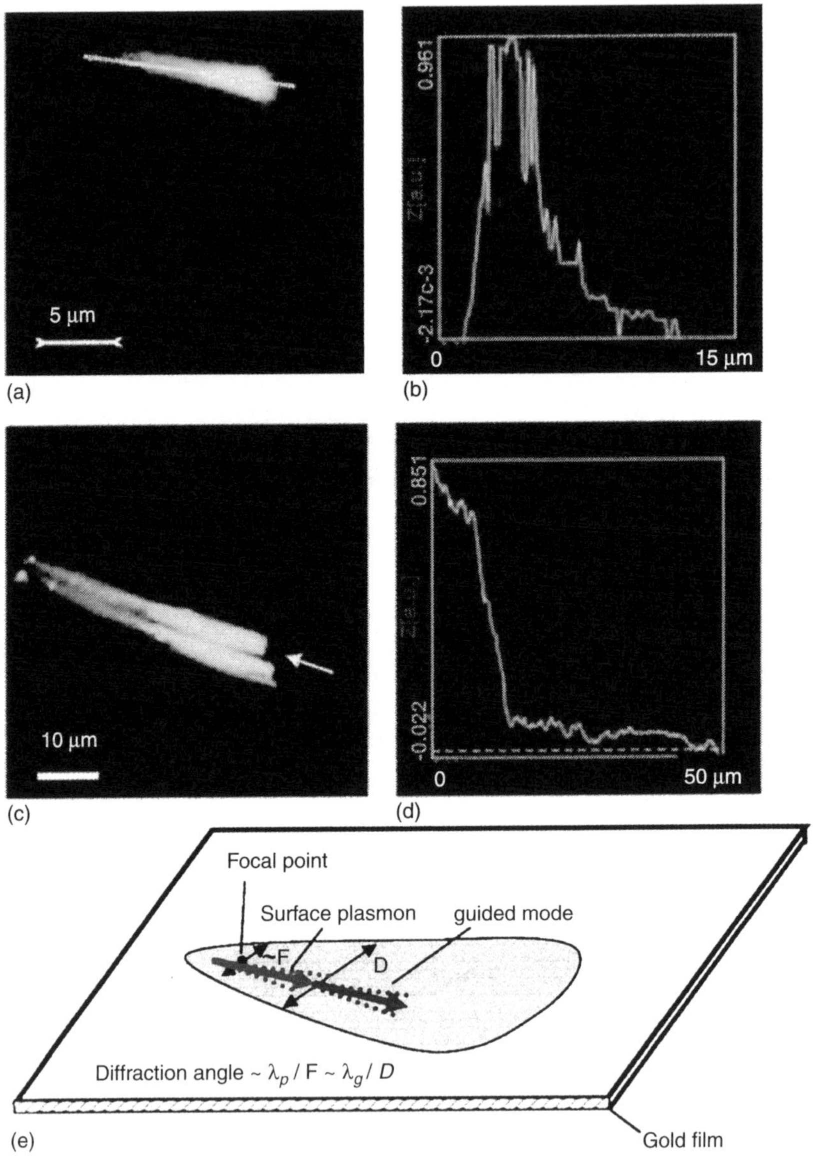

Fig. 5. (a) Exponentially decaying SPP beam emitted from an artificial pinhole in the 50 nm thick gold film immersed in a thin glycerine droplet stained with the Bodipy dye. The cross section of the beam is shown in (b). Image in (c) and its cross section (d) show the effect of the mode coupling due to the slowly varying shape of the glycerine droplet: fast decaying SPP beams emitted by two pinholes give rise to the weaker guided mode beams, which have much larger propagation length. The direction of the cross section is indicated by an arrow in (c). Sketch in (e) illustrates how the mode coupling effect conserves the angular resolution.

coupling of SPP to the guided modes of the thin glycerine layer, as shown in fig. 5(c, d). Image (fig. 5(c)) and its cross section (fig. 5(d)) show the effect of mode coupling due to the slowly varying shape of the glycerine droplet: quickly decaying SPP beams emitted by two pinholes give rise to the weaker guided modes that exhibit much slower decay and longer propagation length as predicted by modelling (fig. 4). From fig. 5(d) the conversion efficiency of this process may be estimated as ~8%. This conversion efficiency should be reasonably close to the efficiency of SPP conversion into photons which are collected by the microscope objective, (if we assume that scattering of SPPs by sub-wavelength-scale roughness of the metal surface is isotropic).

2.3. *Image formation in focusing SPP mirrors*

The image formation in curved SPP-mirrors has been studied using both the geometric and the diffractive optics approaches (Smolyaninov et al., 2005a). The elliptical mirror with the focal distance $P = A^2/B$ can be described as $y = x^2/(2P) + y^2/2C$, where at $C = \infty$ a parabolic mirror with the same focal distance is obtained. If the shape of the 2D mirror (the droplet edge) is given by the exact parabolic dependence, the point (x_1, y_1) of the object is reflected into the point (x_2, y_2) of the image according to the following expressions

$$x_2 = -\frac{P}{x_1}\left[\left\{(y_1 - P/2)^2 + x_1^2\right\}^{1/2} - (y_1 - P/2)\right], \tag{2.3}$$

$$y_2 = \left(\frac{P^2}{2x_1^2} - \frac{1}{2}\right)\left[\left\{(y_1 - P/2)^2 + x_1^2\right\}^{1/2} - (y_1 - P/2)\right] + P/2. \tag{2.4}$$

These expressions allow us to calculate the images of the object in the geometric optics approximation via ray tracing in the SPP-mirrors.

For diffractive optics modelling, the simplified scalar approach can be used which produces excellent agreement with the experimentally measured images of the SPP field distribution. It is assumed that a given source produces a circular wave of the form $e^{ik_{SP}r}e^{-r/L_{SP}}/r^{1/2}$, where k_{SP} is the SPP wavevector and L_{SP} is the SPP propagation length. When the primary circular wave reaches the mirror boundary (the droplet edge), each point of the boundary produces secondary circular waves of the same form, similar to Huygens–Fresnel–Kirchoff principle (Born and Wolf,

1999). The field at each point inside the dielectric droplet is calculated as a superposition of all primary and secondary waves from all sources (fig. 6).

These two approaches were used to trace the transition from the diffractive to the geometrical optics limit of image formation as a function of the mirror size to SPP wavelength ratio (fig. 7). For the sake of convenience, the size of the parabolic droplet was kept constant in this simulation, while the effective refractive index of the droplet was varied. If the lossless ($L_{SP} = \infty$) approximation is used, these calculations indicate that for short SPP wavelengths (large effective refractive indices of the droplet) an image is formed by the droplet boundary in locations, which are consistent with the simple rules of geometrical optics. When losses are introduced, some remaining interference pattern in the background is removed, and resemblance to geometrical optics improves even further (fig. 8).

If additional primary sources are added in order to model complex objects emitting SPP waves, interference effects introduce some deviations from the geometrical optics picture. This can be seen in the images of an array of four-point sources arranged in a square (fig. 9). Nevertheless, general agreement between the geometrical ray tracing and diffractive optics image calculations remains fair. For comparison, in fig. 9(b) the image obtained using geometrical ray tracing is superimposed onto the negative of the image obtained with diffraction modelling in order to show good agreement between these approaches. This figure also demonstrates the image shape deformation due to imaging using a curved mirror. The degree of deformation depends on the position of the object with respect to the mirror focus, the same as in conventional 3D optics.

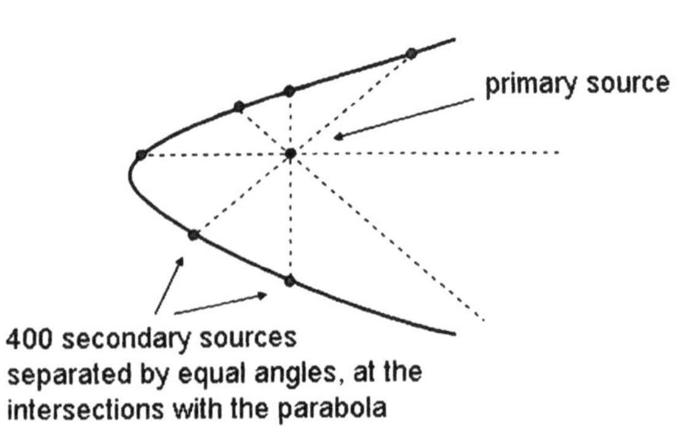

Fig. 6. Diffractive optics calculation geometry: the field is calculated as a superposition of circular waves from the primary source and the secondary sources located at the parabolic mirror boundary.

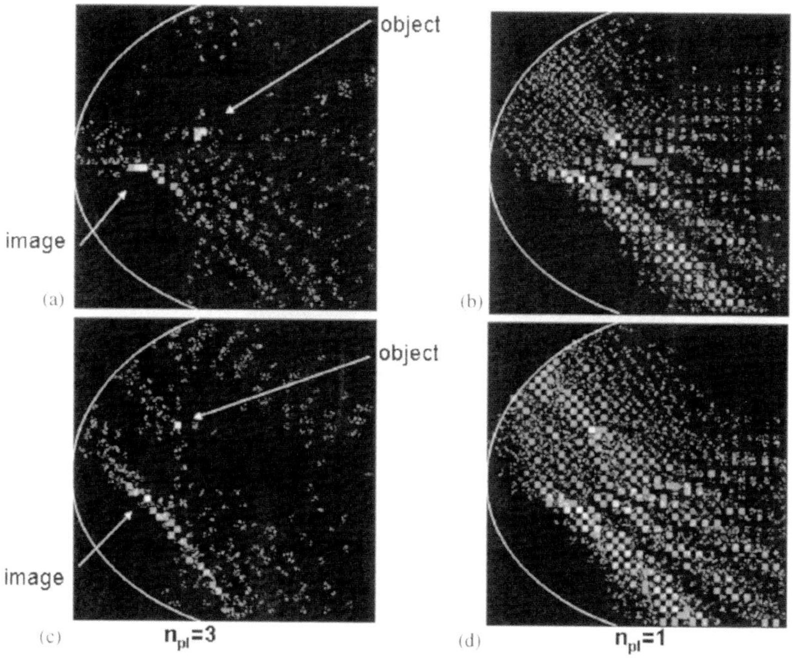

Fig. 7. Calculated images of a point object produced by SPPs in a parabolic mirror: (a, c) $n_{SP} = 3$, (b, d) $n_{SP} = 1$. The pair of images (a,b) and (c,d) are obtained for different position of the object with respect to the focal point of the mirror. Lossless approximation ($\mathrm{Im}\,\varepsilon_m = 0$, $L_{SP} = \infty$) has been used. At larger SPP refractive index, the geometrical optics approximation is recovered. The image sizes are $10 \times 10\,\mu m^2$.

Finally, we have studied the effect of mirror shape on the SPP image formation. In order to perform a fair comparison, the images of the same square source pattern as in fig. 9 were calculated using parabolic and elliptical mirrors, which have the same focal distance in paraxial geometrical optics approximation as in fig. 9. It appears that the imaging properties of the mirrors vary drastically. The elliptical mirror in fig. 10(b) appears to have lost any useful imaging properties. On the other hand, the mirror in fig. 10(c) produces very clear image, which is quite similar to the shape of the original object.

These theoretical data indicate that 2D SPP-based imaging is possible. On the other hand, any practical SPP imaging device would require a very good control of the dielectric mirror shape and position of the object with respect to the mirror. The geometrical optics approximation appears to work reasonably well in reconstructing the imaging properties of SPP mirrors in the short-wavelength limit, where the diffraction effects are not

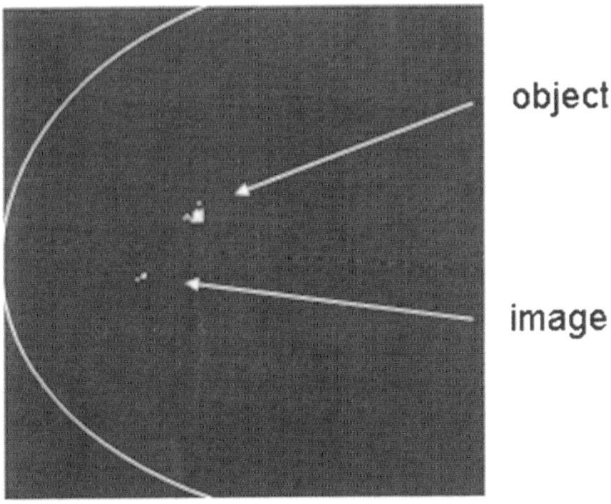

Fig. 8. Calculated image of a point object produced by SPPs in a parabolic mirror as in fig. 7(a) with the Ohmic losses included: $n_{SP} = 3$ and $L_{SP} = 5\,\mu m$. The effect of the finite SPP propagation length in the presence of Ohmic losses improves image quality due to suppressed background. The image size is $10 \times 10\,\mu m^2$.

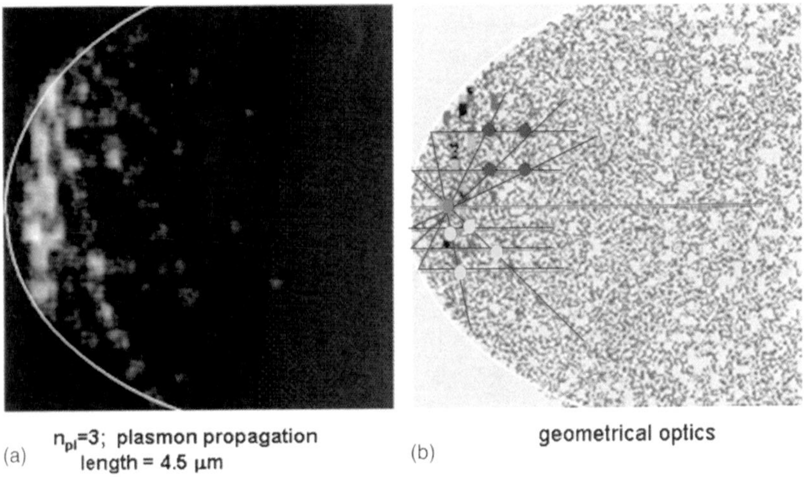

Fig. 9. Calculated images of a square pattern object obtained using (a) diffractive ($n_{SP} = 3$, $L_{SP} = 5\,\mu m$) and (b) geometrical SPP optics. In (b) geometrical construction is superimposed on negative of (a). The focal point of the mirror is shown with a blue point, the object is a square array formed by four red points, the resulting geometrical optics image is shown with yellow points. The image sizes are $10 \times 10\,\mu m^2$.

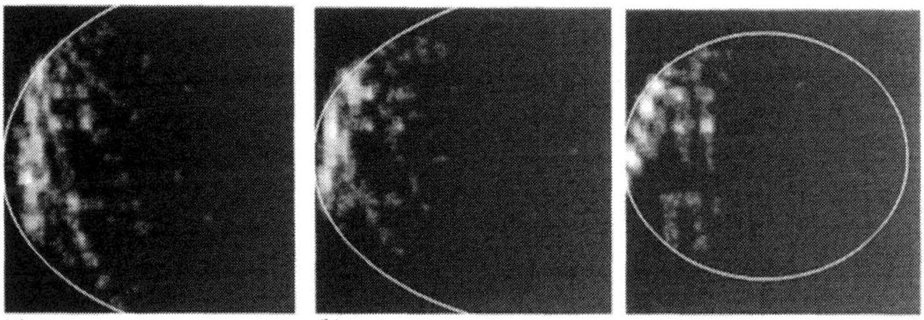
(a) (b) (c)

Fig. 10. Comparison of the imaging properties of parabolic (a) and elliptical (b, c) mirrors using the diffraction optics approximation and the object and SPP parameters as in fig. 9: (a) $P = 3\,\mu m$ and $C = \infty$, (b) $P = 3\,\mu m$ and $C = 10\,\mu m$, (c) $P = 3\,\mu m$ and $C = 5\,\mu m$. The image sizes are $10 \times 10\,\mu m^2$.

too important. However, the effects of scattering, diffraction and mirror imperfections limit the spatial resolution of practical devices based on SPP.

§ 3. Imaging through photonic crystal space

The exact theory of SPP-assisted microscopy can be developed by considering surface plasmon polaritonic crystals. Let us consider the transmission of electromagnetic waves with various spatial frequencies generated by a luminous object immersed in an infinite "photonic crystal space". In the following simple numerical example, we use a test object which consists of two luminous dots separated by a gap (fig. 11(a)). The Fourier spectrum of this object is shown in fig. 11(b). If the angular spectrum of spatial frequencies available for probing in the far-field of the object with an optical apparatus (e.g., microscope objective lens) is limited by some maximum wavevector k_{max} (represented by a circle of radius k_{max} shown in fig. 11(b)), the free (empty) space between the object and the lens serves as a spatial frequency filter, which removes the spatial frequencies corresponding to evanescent waves. Thus, whatever optical design is implemented to collect the electromagnetic waves propagating from the object, the best image in the far-field region would result from the inverse Fourier transformation of the portion of the spectrum falling inside the circle in fig. 11(b). As a result, a smeared image in fig. 11(c) would be obtained. The two-dot structure of the original object is lost in this image.

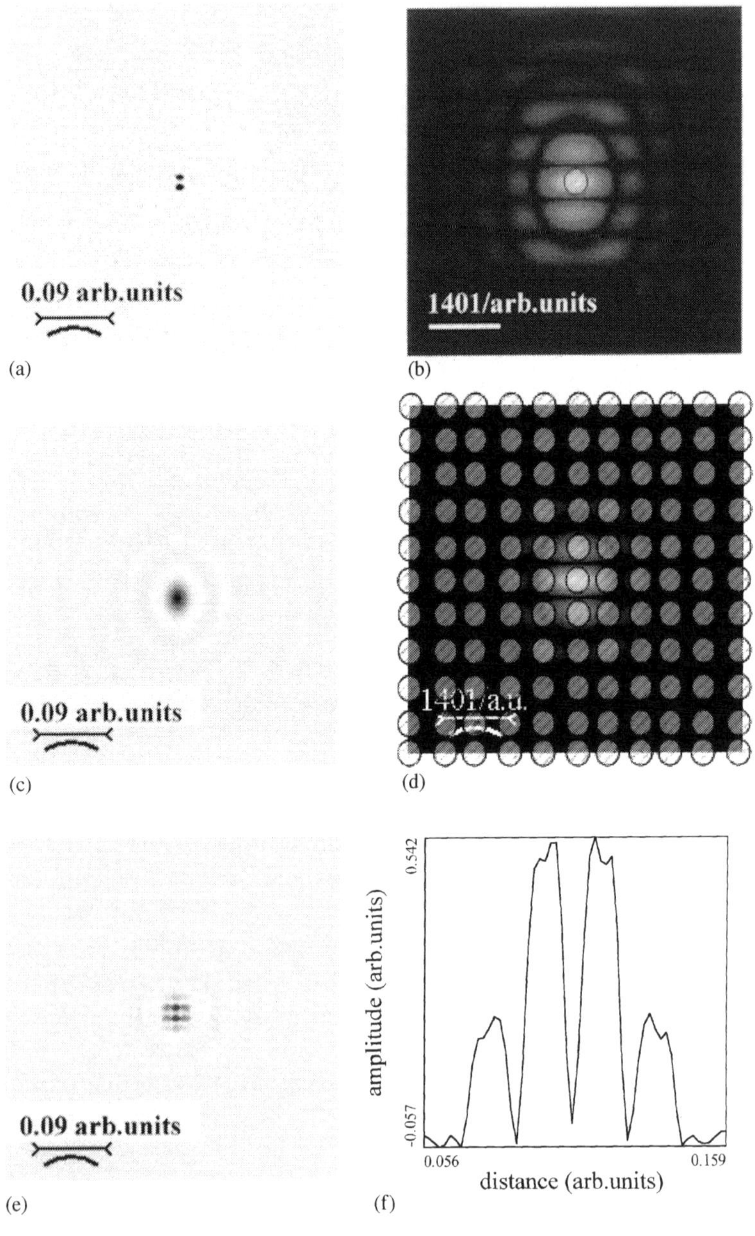

Fig. 11. (a) A test object consists of two luminous dots separated by a gap. (b) The Fourier transform of the object shown in (a). An area of the spatial frequency spectrum of electromagnetic waves available to probe is shown by a circle. (c) The inverse Fourier transform of the portion of the spectrum inside the circle shown in (b). (d) The pass band of the "photonic crystal space" with square photonic crystal lattice. (e) The inverse Fourier transform of the spatial frequency spectrum inside the photonic crystal pass band. (f) Cross section of the image in (e) indicates recovery of the two-dot structure of the object.

If the same object is placed inside a "photonic crystal space", while the same part of the spatial frequency spectrum (limited by the k_{max} value) is available for far-field probing outside the photonic crystal, the points in the Fourier space separated by integer multiples of the inverse lattice vectors become equivalent to each other because of the photonic crystal periodicity. If we assume that the photonic crystal has a square lattice (fig. 11(d)) and acts as a spatial filter, an image shown in fig. 11(e) can be recovered, which is obtained by the inverse Fourier transformation of the portion of the original spectrum inside all the circles in fig. 11(d). The original information about the two-dot structure of the object is recovered in this case (fig. 11(e)) as seen from the cross section (fig. 11(f)).

This simple numerical example demonstrates that a far-field optical microscope with resolution beyond the $\lambda_0/2$ diffraction limit of conventional far-field optics can be built using photonic crystal materials. However, in order to achieve sub-diffraction-limited resolution, an object should be placed inside or very near the photonic crystal. Such a microscope can be dubbed "an immersion microscope based on photonic crystal material".

The imaging properties of a 2D periodic structure, such as a photonic crystal slab or surface polaritonic crystal, can be described by considering properties of electromagnetic Bloch waves in a periodic structure:

$$\psi_{\vec{k}} = \sum_{\vec{K}} C_{\vec{k}-\vec{K}} e^{i(\vec{k}-\vec{K})\vec{r}}, \tag{3.1}$$

where $\vec{k}$ is the wavevector defined within the first Brillouin zone, and $\vec{K}$ represents all the inverse lattice vectors. The Bloch wave is capable of carrying spatial frequencies of an object, which would be evanescent in free space. It does not matter if the dispersion of some particular Bloch wave is negative or positive. What is important for microscopy is that the Bloch waves should have sufficiently large $C_{\vec{k}-\vec{K}}$ coefficients at large $\vec{K}$. The Fourier spectrum $F_{\vec{\kappa}}$ of the test object $f(\vec{r})$ described in the example above can be written in the usual way as

$$f_{(\vec{r})} = \sum_{\vec{\kappa}} F_{\vec{\kappa}} e^{-i\vec{\kappa}\vec{r}}, \tag{3.2}$$

where $\vec{\kappa}$ is the wavevector in free space. The same spectrum of the object in terms of the Bloch waves (eq. (2.3)) is given by

$$F_{\vec{k}} = \sum_{\vec{K}} F_{-\vec{k}+\vec{K}} C_{\vec{k}-\vec{K}}. \tag{3.3}$$

Thus, high spatial frequencies of the object shape $F_{-\vec{k}+\vec{K}}$ are carried into the far-field zone of the object by $C_{\vec{k}-\vec{K}}$ components with large $\vec{K}$. The limit $C_{\vec{k}-\vec{K}} \approx const$ is the most beneficial for super-resolution imaging. It corresponds to the photonic Bloch waves obtained in the tightly bound approximation, in which the photonic bands are flat.

In the case of imaging with surface plasmon polaritonic crystals based on SPP optics, this Bloch wave description agrees well with the model of short-wavelength SPPs which are excited by the periodic nanohole array: near the surface plasmon resonance the SPP dispersion is almost flat.

In order to be useful in far-field microscopy, a given photonic crystal geometry must exhibit image magnification to the extent that the image size should surpass the $\lambda_0/2$ diffraction limit of usual far-field optics. Such a magnified image can be transferred to a free space region and viewed by a regular microscope. This means that some curved photonic crystal boundary should be used. Since refraction of photonic crystals depends very strongly on frequency, propagation direction and other parameters (a superprism effect is well known in photonic crystal geometries, see for example Zayats et al. (2003b) and Chung and Hong (2002)), a reliable photonic crystal lens geometry would be very difficult to predict theoretically and realise in the experiment. On the other hand, a reflective optics geometry seems to be a good practical solution. The law of reflection is observed for almost all wavevectors $\vec{k}$ within the first Brillouin zone for Bloch wave interacting with a planar photonic crystal boundary (fig. 12). The umklapp processes, which occur in the corners of the Brillouin zone, do not spoil the geometrical optics reflection picture because in a periodic lattice the $\vec{k}_r$ and $\vec{k}_r^*$ directions are physically equivalent, and they correspond to the same Bloch wave. If the reflecting boundary is slightly curved (so that the radius of curvature is much larger than the period of the photonic crystal lattice) geometrical optics picture of reflection remains valid. Thus, a magnifying mirror can be designed using photonic crystal materials. This idea has been realised in the experiments with SPPs in which the role of the mirror is played by the boundary of the dielectric droplet, which is placed on the surface of the periodic nanohole array in a gold film (see § 4).

Once the image is magnified, it should be projected into free space outside the photonic crystal so that it can be viewed. At this stage the refractive properties of the photonic crystal play an important role in image formation. As we shall see in the following sections of this chapter, the sign of the effective refractive index of the photonic crystal defines the character of image magnification of the 2D optical system based on photonic crystal mirror.

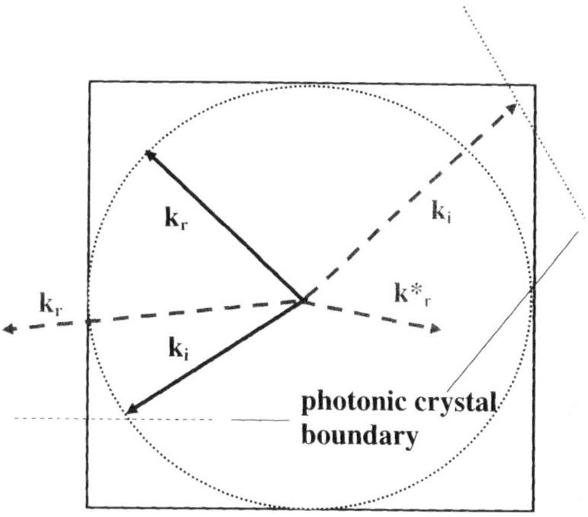

Fig. 12. A Bloch wave reflection from a photonic crystal boundary. (Solid line) For incident wavevectors $\vec{k}_i$ inside the dashed circle the wavevector of the reflected Bloch wave $\vec{k}_r$ obeys the law of reflection. (Dashed line) In the case of incident wavevectors located in the corners of the first Brillouin zone (outside the dashed circle) the $\vec{k}_r$ vector obtained according to the law of reflection must be shifted inside the first Brillouin zone by an addition of an inverse lattice vector. However, the obtained $\vec{k}_r^*$ and $\vec{k}_r$ directions of the Bloch wave propagation are physically equivalent in a periodic lattice.

We should also point out that while 2D configuration of the SPP-assisted microscopy based on 2D photonic or SPP crystal mirrors offers some important advantages, such as relative ease of the structure fabrication, strong interaction between biological samples and SPP Bloch waves etc., a 3D configuration of a microscope based on a photonic crystal mirror is also possible. In this case, the evanescent components of the diffraction field generated by the object are coupled to the 3D photonic crystal Bloch modes but in contrast to 2D case, the image is formed out of the surface plane. One of potentially interesting configurations is shown in fig. 13. In this configuration, a photonic crystal mirror would consist of two parts: a substrate and a cover part, which would work together as a photonic crystal mirror. The plane separating these two photonic crystal parts should be close to the focal plane of the mirror and filled with a very thin layer of index matching gel. If an object is positioned on the substrate and covered with the top mirror part, a magnified image of the object would be formed in free space, which may be viewed by a regular optical microscope. Operation of such microscope would be very similar to a regular microscope. However, practical realisation of

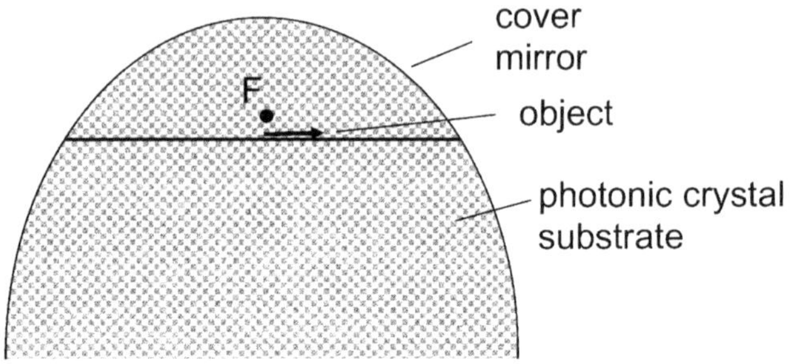

Fig. 13. Schematic of a 3D magnifying photonic crystal mirror, which may be used in a 3D configuration of an immersion microscope based on photonic crystal materials.

this 3D microscope idea in the optical frequency range would require fabrication of high quality 3D photonic crystal materials.

§ 4. Imaging and resolution tests

Periodic nanohole arrays in metal films appear to be ideal test samples for SPP-based microscopy. Illuminated by laser light, such arrays efficiently produce propagating SPPs, which explain the anomalous transmission of such arrays at optical frequencies. Figure 14 shows various degrees of 2D image magnification obtained with a $30 \times 30\,\mu m^2$ rectangular array of nanoholes made in the gold film using focused ion beam milling. This test sample consists of 300 nm diameter holes with 600 nm spacing. In general, smaller glycerine droplets produced higher magnification of the images. The reconstruction of the images using 2D geometrical optics (via ray tracing) is shown next to each experimental image. However, the edges of the droplets in the experiments may only approximately be represented by parabolas, and the damping of the SPP field over varying propagation lengths has not been included in these simulations. This limits the precision of the image reconstruction based on geometric optics. Nevertheless, an impressive qualitative agreement between the experimental and modelled images has been achieved. In all the calculated images described below, the individual nanoholes of the test samples are shown as individual dots in the calculated images. Comparison of fig. 14(c) and fig. 14(d) indicates that the rows of nanoholes separated by 500 nm may have been resolved in the image (c) obtained using only a $10\times$ objective of the conventional microscope, while comparison of fig. 14(e) and fig. 14(f) obtained using a $50\times$ objective indicates that individual 150 nm diameter

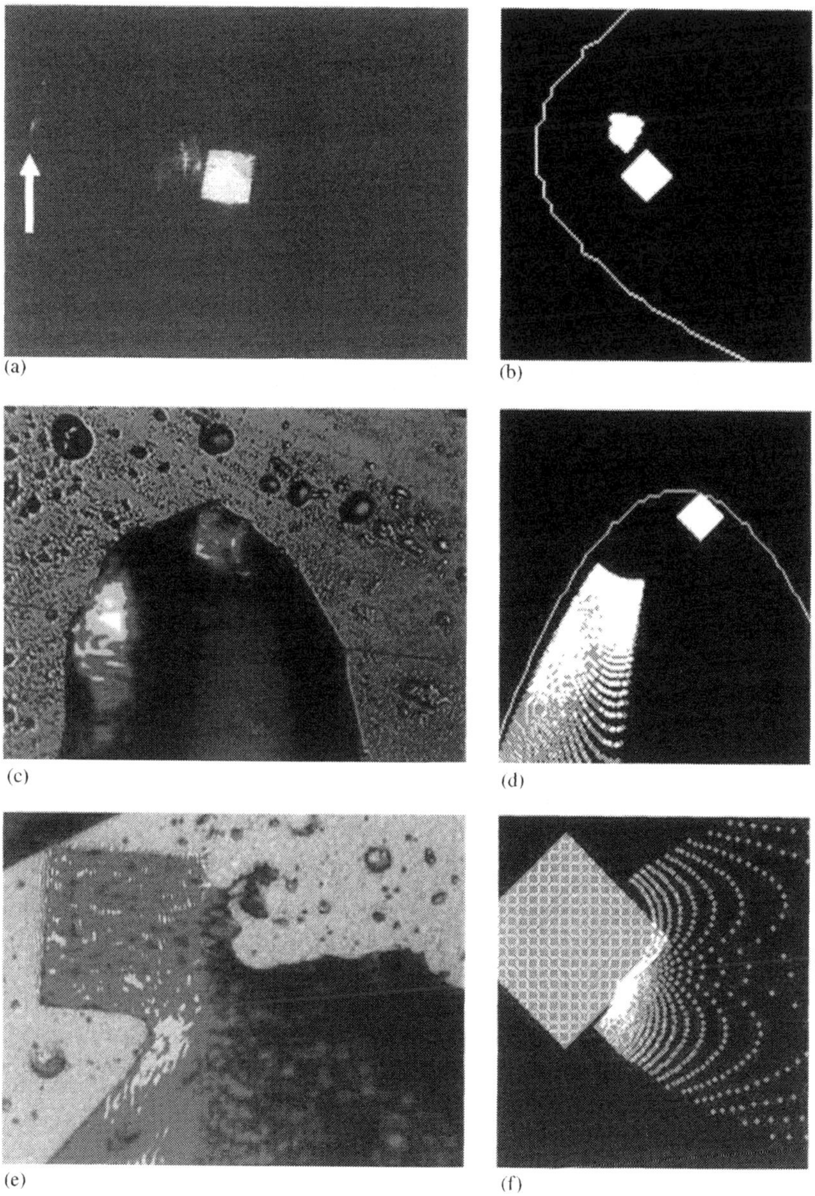

Fig. 14. The images of a $30 \times 30\,\mu m^2$ rectangular nanohole array with 500 nm hole spacing, which are formed in a 2D mirrors of various shapes and sizes. The arrow in (a) indicates the glycerine edge acting as a mirror visible due to SPP scattering. Reconstructions of the images via ray tracing are shown to the right of each experimental image. Individual nanoholes of the arrays are shown as individual dots in the calculated images. The $10\times$ microscope objective was used to obtain images (a) and (c), while the $50\times$ objective was used in (e). The illuminating light wavelength is 502 nm.

nanoholes separated by 500 nm gaps are resolved in the image (e) obtained at the 502 nm wavelength. These individual nanoholes are located in close proximity to the focus of the mirror, and hence experience the highest image magnification. This also leads to significant image distortion, which is usually the case when a curved mirror is used for imaging. Even though the exact role of mode coupling in formation of each image in fig. 14 is not clear, it seems certain that the 2D images in figs. 14(a, c) are formed with considerable participation of the guided modes, since the distance travelled by the electromagnetic modes is of the order of 100 μm in this case. While the image in fig. 14(a) does not contain any evidence of high spatial resolution, the image in fig. 14(c) seems to demonstrate that the SPP to guided mode coupling preserves high angular and spatial resolution of the image. It should also be pointed out that the high spatial resolution (of at least 150 nm at the 502 nm wavelength) obtained in the 2D imaging experiments in fig. 14 (which are dominated by regular guided modes) has been confirmed by Challener et al. (2005). This paper reported the use of parabola-shaped dielectric waveguides in 2D focusing of optical energy down to 90 nm spots at 413 nm laser wavelength. This tight focusing is achieved due to very high numerical aperture of the 2D parabola-shaped waveguide used as a focusing mirror.

A resolution test of the microscope has been performed using a $30 \times 30\,\mu m^2$ array of triplet nanoholes (100 nm hole diameter with 40 nm distance between the hole edges) shown in fig. 15(c). This array was imaged using a glycerine droplet shown in fig. 15(a). The image of the triplet array obtained at 515 nm using a $100\times$ microscope objective is shown in fig. 15(b) (compare it with an image in fig. 15(d) calculated using the 2D geometrical optics). Even though some discrepancy between the experimental and model images can be seen (the image pattern observed in fig. 15(b) looks convex from the left, compared to the concave pattern observed in the calculation (d)), the overall match between these images is impressive. The most probable reason for the observed convex/concave discrepancy is the fact that the droplet shape is not exactly parabolic, which produces some image aberrations. Although the expected resolution of the microscope at the 515 nm illumination wavelength is somewhat lower than at 502 nm, the 515 nm laser line is brighter, which allowed us to obtain higher contrast in the SPP-formed image. The least-distorted part of fig. 15(b) (far from the droplet edge, yet close enough to the nanohole array, so that SPP decay does not affect resolution) is shown in fig. 15(e, f). These images clearly visualise the triplet nanohole structure of the test sample.

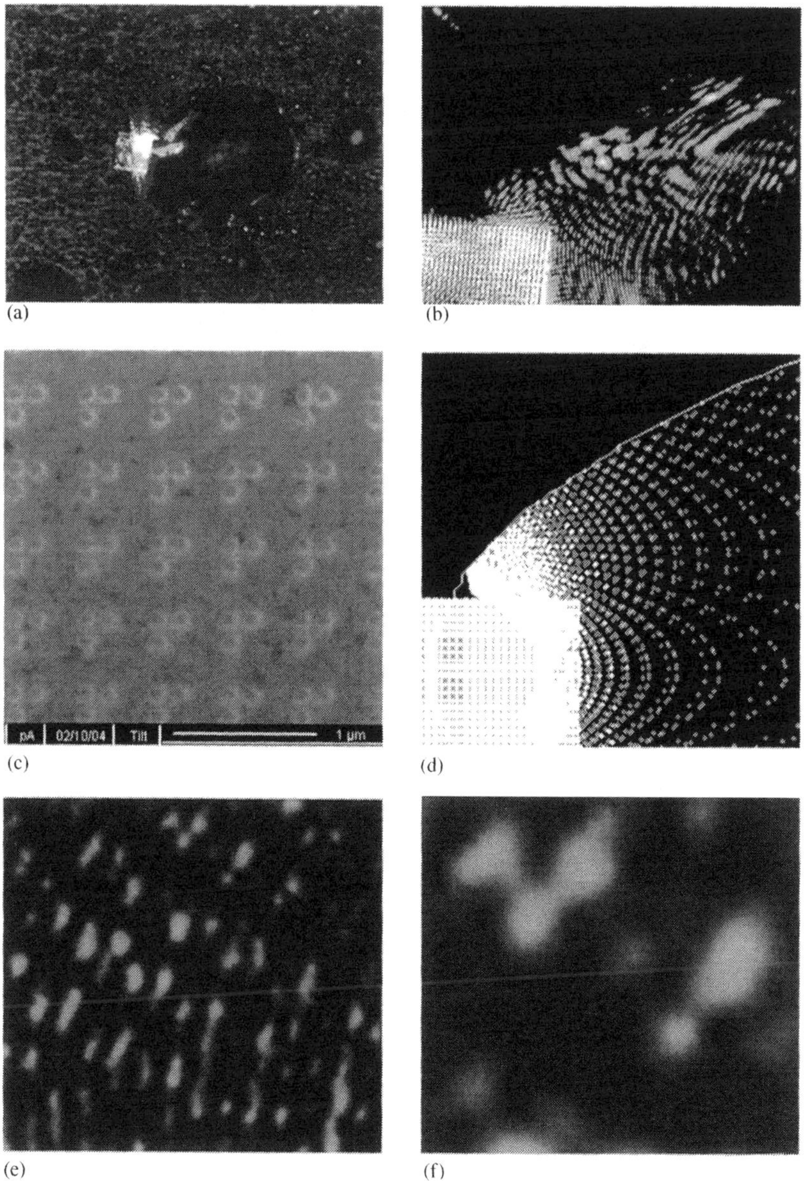

Fig. 15. The resolution test of the SPP-based microscope. The array od triplet nanoholes (the SEM image is shown in (c)) is imaged using a SPP mirror with (a) 10x and (b) 100x microscope objectives. (d) The geometrical optics reconstruction of the image in (b). (e,f) Two successive zooms in the least-distorted part of the image (b). The illuminating light wavelength is 515 nm.

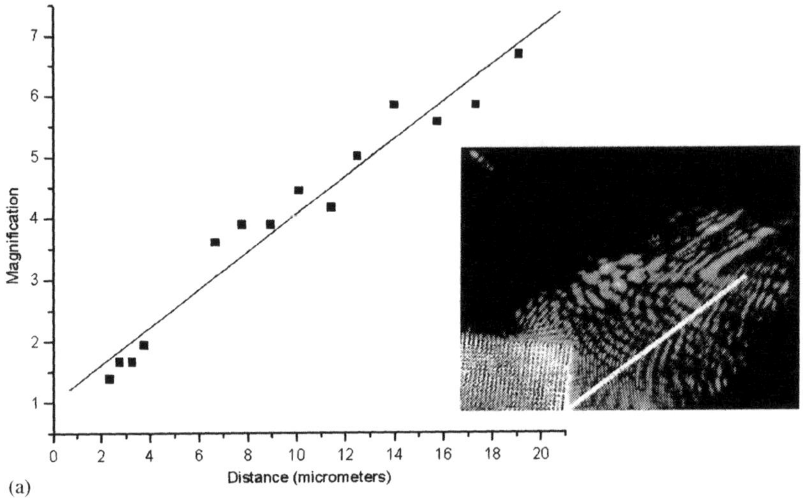

(a)

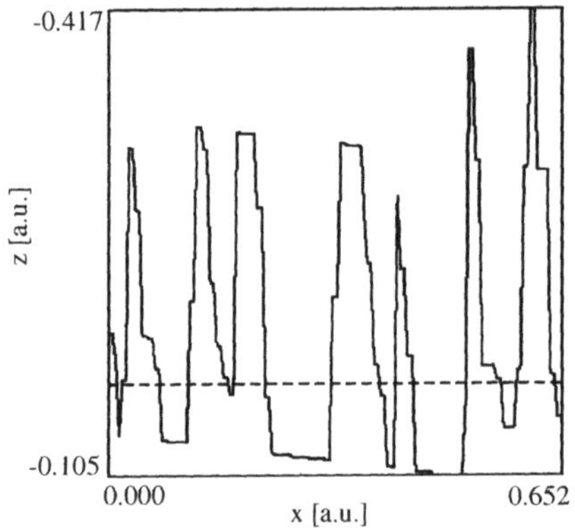

(b)

Fig. 16. (a) The image magnification measured in the SPP-formed image of the triplet nanohole test sample as in fig. 15 along the line shown in the inset, which is parallel to the optical axis of the mirror. The dots in the graph show the distance between the neighbouring triplets in the image as a function of the triplet position measured along the optical axis. (b) The cross section through the line of double holes in the image of the triplet array. The illuminating light wavelength is 515 nm.

According to the geometrical optics, the image magnification M is supposed to grow linearly with distance along the optical axis of the SPP mirror

$$M = \frac{2y}{P} - 1. \tag{4.1}$$

The measurements of the image magnification indeed exhibit such linear dependence (fig. 16a). The dots in the graph show the distance between the neighbouring triplets in the image as a function of triplet position measured along the optical axis of the mirror. At small distances, individual nanoholes are not resolved within the triplet. At larger distances (where the triplets are resolved, see the cross section in fig. 16(b) measured through the line of double holes in the image of the triplet array) the data points represent the positions of the triplets centres. The gap in the data corresponds to the intermediate area of the image in which the feature identification in the image is difficult. The slope of the measured linear dependence in fig. 16(a) corresponds to $P = 7\,\mu m$, which agrees reasonably well with the value of the focal distance of the order of $P \sim 10\,\mu m$, which can be determined from the mirror dimensions in fig. 15(a).

In order to prove that the SPP microscope is capable of aperiodic samples visualisation, the images of small gaps in the periodic nanohole arrays were studied (fig. 17). The electron microscope image of one of the gaps in the periodic array of nanoholes is shown in fig. 17(a). Two wider mutually orthogonal gaps were made in the array along both axis of the structure as shown in the theoretical reconstruction in fig. 17(c) (see also Zayats et al., 2003b). The SPP-formed image in fig. 17(b) and its cross section in fig. 17(d) obtained at the 502 nm wavelength show both the periodic nanohole structure and the gap in the structure indicated by the arrows in the images. The width of the gap in the image grows linearly with the distance from the sample in agreement with the SPP-mirror magnification (fig. 16).

In order to evaluate the SPP-assisted microscopy resolution at the optimised 502 nm wavelength, the cross sections of the images of the triplet structure similar to the one described earlier in fig. 15 were analysed. The most magnified triplets, which are still discernible in the experimental image in fig. 18(a), are shown by the arrow. These triplets are shown at a higher zoom in fig. 18(c). The cross section through two individual nanoholes in the triplet clearly shows the 40 nm gap between the nanoholes. The distance between the centres of the nanoholes is 140 nm by design. Comparison of the cross sections of the SPP-formed images in fig. 18(c) and fig. 16(b) with the designed parameters of the triplet nanohole array indicates the resolution of about 50 nm at the 502 nm wavelength of illuminating light, at least 3 times better than at the 515 nm illuminating light. At the same time, the resolution is lost at the 488 nm wavelength illumination at which no SPPs can be excited at the structure.

The Fourier spectra of the SPP-formed optical image and the SEM image of the triplet test pattern are compared in fig. 19. The highest Fourier orders still resolved in the optical image corresponds to periodicity ~90 nm. Thus, the measured spatial resolution of the SPP-based microscopy is in the 50–90 nm range, which is consistent with the 70 nm

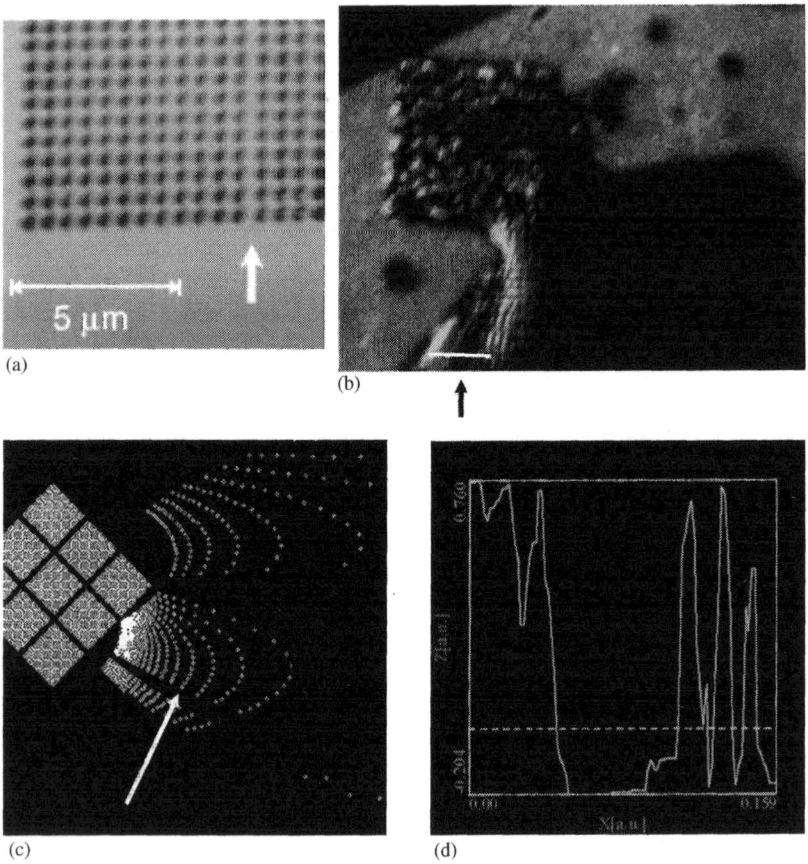

Fig. 17. (a) SPP-formed images of the gaps in the $30 \times 30\,\mu m^2$ nanohole array and (b) geometrical optics reconstruction. (c) One of the gaps is indicated by an arrow in the electron microscope image of the structure. (d) The cross section of the SPP-formed image obtained along the line is shown in (b).

resolution reported in lithographic experiments performed using SPPs in thin silver films (Fang et al., 2005).

§ 5. The role of effective refractive index of the SPP crystal mirror in image magnification

A typical dispersion of the SPP modes on a periodically modulated surface of a metal film in the vicinity of the surface plasmon resonance is shown schematically in fig. 20. The examples of surface polariton dispersion on various 2D SPP crystals may be found in Darmanyan and

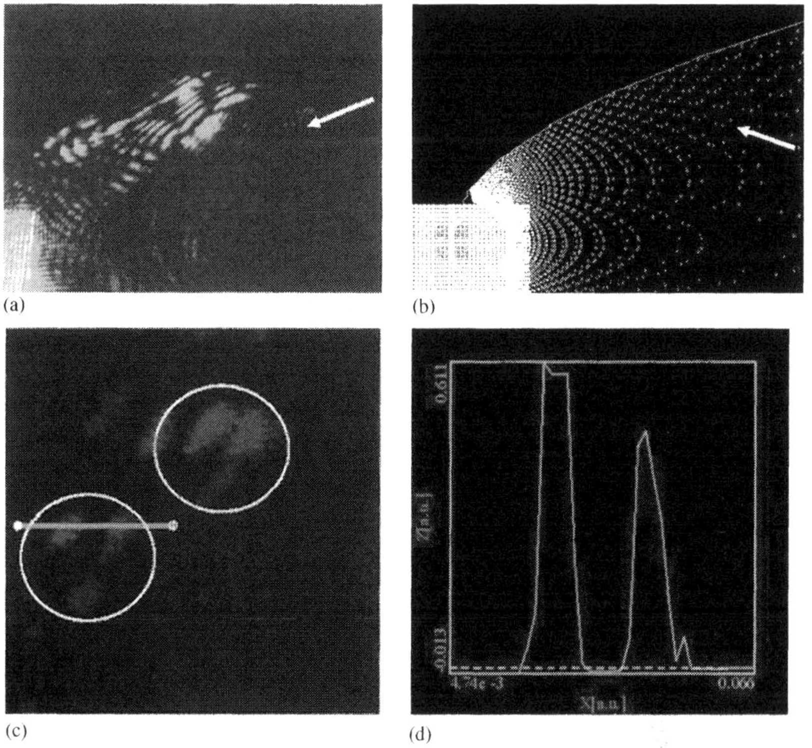

Fig. 18. Evaluation of the SPP-based microscope resolution at the 502 nm wavelength. The triplet holes visible in the images are indicated by the arrows in the measured (a) and calculated (b) images. The same triplets are shown at a higher magnification in the experimental image (c). The cross section through two individual nanoholes in the triplet along the line shown in (c) is presented in (d).

Zayats (2003) and Kretschmann et al. (2003). It appears that the sign of the SPP group velocity may be either positive or negative depending on the Brillouin zone structure and SPP Bloch wave frequency (in fact, the SPP dispersion can be almost flat, especially in higher Brillouin zones which lie near the surface plasmon resonance frequency). As shown above, the latter circumstance $C_{\vec{k}-\vec{K}} \approx const$ is advantageous for high-resolution imaging applications. On the other hand, the sign of the SPP group velocity defines the effective refractive index of the photonic crystal material in this frequency range (Agranovich et al., 2004). According to calculations by Kretschmann et al. (2003), the sign of the SPP group velocity for a particular SPP Bloch mode branch is rather insensitive to the propagation angle, which means that the model of geometrical optics

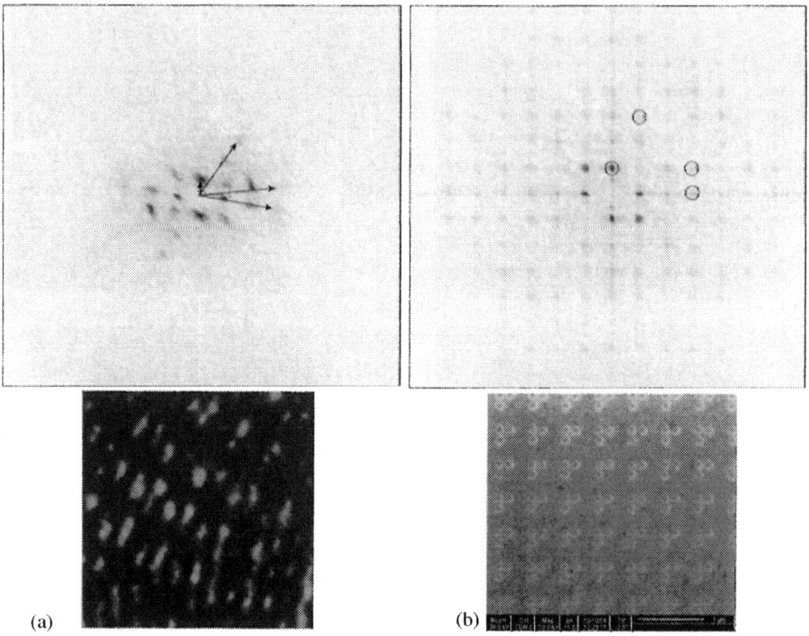

Fig. 19. Comparison of the Fourier spectra of the SPP-formed image (a) and the electron microscope image (b) of the same triplet test pattern. The highest Fourier orders still resolved in the optical image are shown by arrows in (a) and by circles in (b).

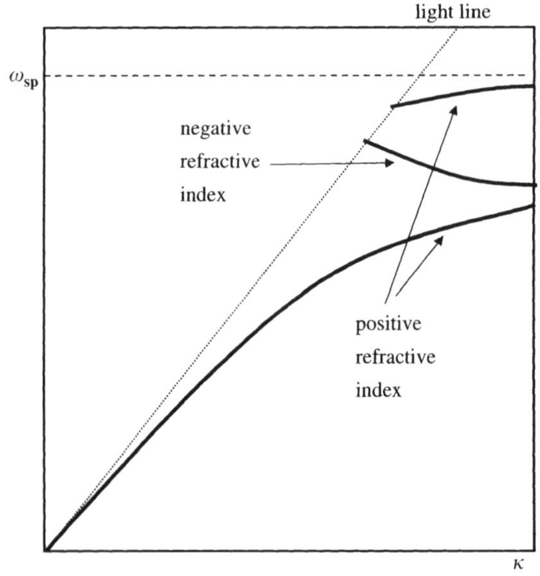

Fig. 20. Schematic view of the SPP dispersion in the first Brillouin zone.

refraction (the Snell's law) is applicable to the SPP propagation across the interface between the nanohole array region and a smooth metal film.

Let us consider the effect of this refraction on the image formation in the SPP-assisted microscope described above. As may be seen from fig. 21, positive effective refractive index of the nanohole array causes some shift in the location of the image, which is formed by the SPP crystal mirror. On the other hand, negative refractive index would produce a much more drastic effect on the imaging properties of the mirror. A real image produced by the mirror, which would be located outside the nanohole array, becomes a virtual image due to negative refraction at the SPP crystal boundary (fig. 21). However, if a real image produced by the mirror is located inside the nanohole array, negative refraction at the interface produces a second real image over the unperturbed metal film (fig. 21(b)). The character of refraction at the nanohole array boundary is clearly identifiable in the experiment. While positive refraction produces image magnification which grows with distance from the mirror along the optical axis of the system (this behaviour has been observed in experiments shown in figs. 14–18), negative refraction produces an opposite behaviour of magnification: image magnification is the highest in the immediate vicinity of the nanohole array boundary, and becomes smaller at larger distances along the optical axis. This may be seen from fig. 22(b), which has been calculated in the case of a triplet nanohole array, which

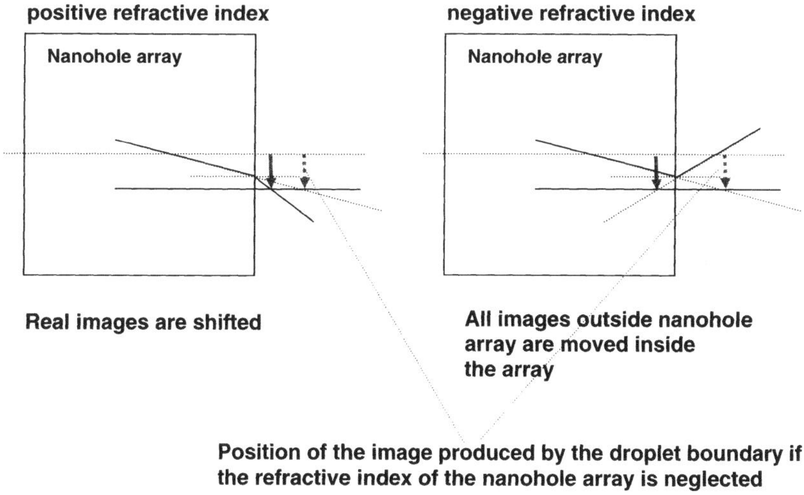

Fig. 21. The effect of the effective refractive index sign of the SPP crystal on the image formation: positive refractive index causes small shift in the image location, while negative refractive index converts real images outside the SPP crystal into virtual ones.

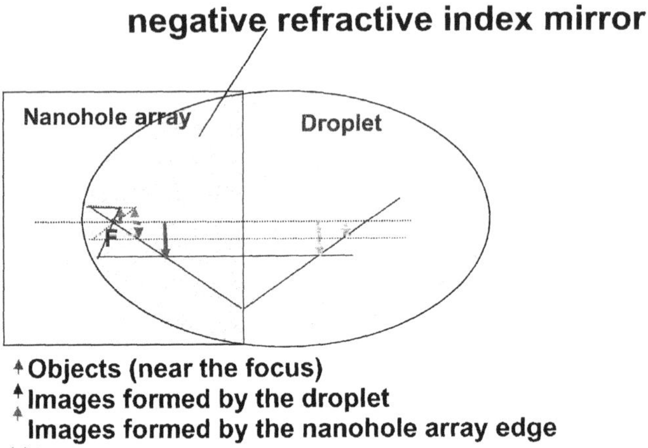

(a)

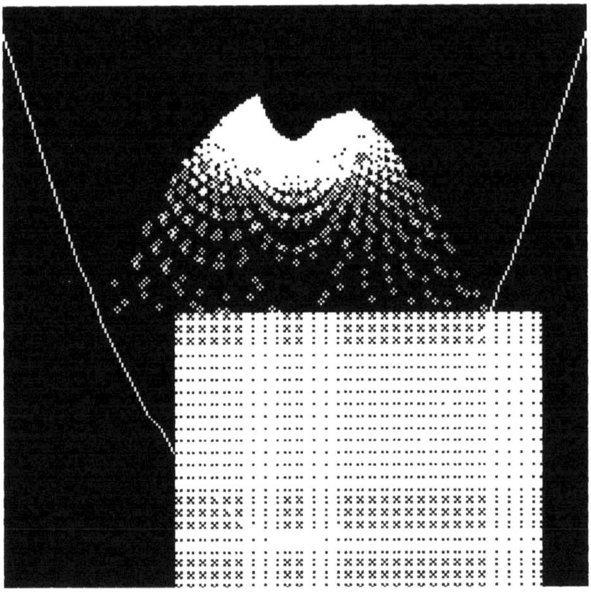

(b)

Fig. 22. (a) The magnification of the images produced by the SPP crystal mirror with negative refractive index for various distance from the mirror edge. (b) Theoretical image of a triplet array of nanoholes (rectangular area at the bottom of the image) in the case of negative effective refractive index of the SPP crystal.

has a negative effective refractive index (compare this figure with figs. 15 and 17). "Negative" behaviour of the image magnification for some nanohole arrays has been indeed observed in our experiments, as described in the next section.

§ 6. Experimental observation of negative refraction

Both positive and negative refractions have been observed in the experiments with surface polaritonic crystals. All the test patterns shown in fig. 23 had 100 nm hole diameter with 40 nm distance between the hole edges in the doublet and triplet structures, and 500 nm lattice period. Being covered with glycerine, these structures exhibited positive refraction as perceived by SPPs. All the nanohole structures were resolved in the optical images obtained using SPP-assisted microscopy with 502 nm and 515 nm light. However, when the illuminating light wavelength is such that no SPP Bloch modes can be excited on the periodic arrays ($\lambda = 488$ nm), the resolution is lost as should be expected (Smolyaninov et al., 2005b).

In the next series of experiments, negative refractive index of SPP crystals was observed in SPP-assisted imaging of aperiodic test samples (fig. 24(a)). Similar to earlier experiments, the boundary of the SPP-mirror was positioned over the array of nanoscale holes used as objects in the experiment. SPPs excited in such structure form magnified images of individual nanoholes over the unmodified area of the gold film (fig. 24(b)). Zoom of the image area, which is adjacent to the square array of nanoholes (fig. 24(c)), indicates that individual elements of the aperiodic array have been imaged with various degree of distortion (image quality appears to be the best on the right side of fig. 24(c) where the shapes of individual nanoholes are clearly recognisable). The images of the elements of the array are somewhat distorted due to their position with respect to the mirror (the same as in a conventional parabolic or elliptical mirror) and the degree of distortion is different for different elements depending on their position and orientation. Using the known mirror geometry, the shapes of the test pattern (fig. 24(a)) and the position of the array with respect to the mirror, the distorted images of the array elements formed by the SPP mirror can be modelled and compared with the experiment (fig. 24(d)). The images show good agreement with each other. The SPP-formed images are rotated and stretched/compressed compared to the respective objects, but the shapes of the individual elements of the array can be clearly recognised. The additional broadening in the experimental image is related to the finite resolution of the microscope (the test object sizes of the order of 50 nm are comparable to the optical resolution of the apparatus), aberrations due to the imperfect mirror shape and glycerine boundary quality.

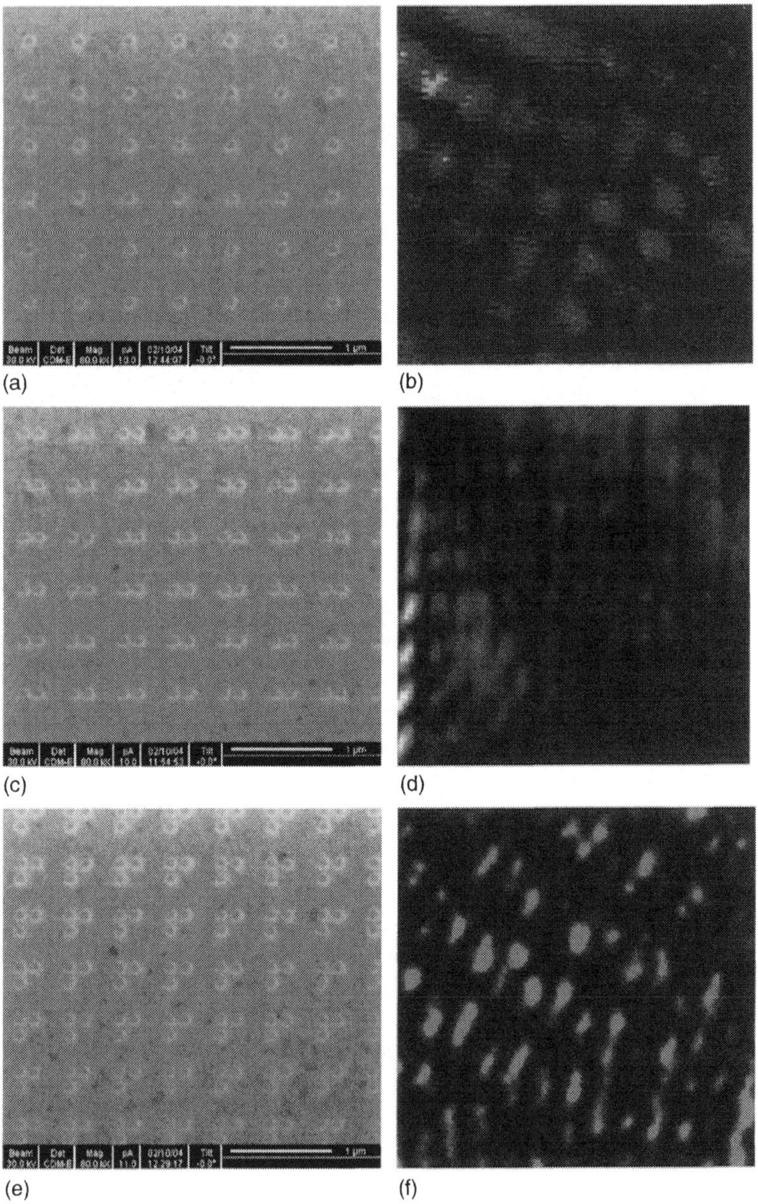

Fig. 23. The $30 \times 30\,\mu m^2$ arrays of singlet, doublet and triplet nanoholes (100 nm hole diameter, 40 nm distance between the hole edges in the doublet and triplet, 500 nm period) shown in the left column are imaged using a SPP photonic crystal mirror formed by a glycerine droplet. The optical images in the right column are obtained at $\lambda_0 = 502$ nm (singlets and doublets) and at $\lambda_0 = 515$ nm (triplets).

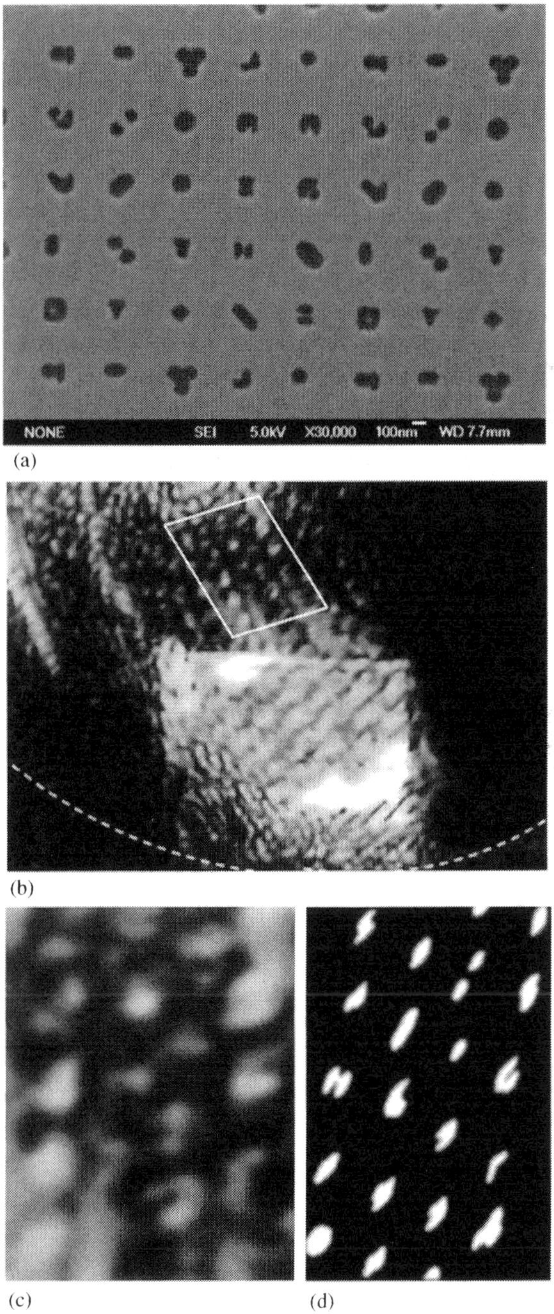

Fig. 24. (a) SEM image of the aperiodic nanohole array in the metal film. (b) Large-scale image of the array obtained with SPP-assisted microscopy. Total size of the array is $20 \times 20\,\mu m^2$. Droplet edge position is shown by the dashed line. (c) Zoom of the area marked in (b). (d) The calculated image of the array using known position of the SPP mirror.

A remarkable feature of the SPP-formed image in fig. 24(b) is the apparent inverse character of magnification in this image. This behaviour is clear from fig. 25 in which the comparison of fig. 24(b) with fig. 22(a, b), and also from fig. 25 in which the magnification in image is compared with a previously observed SPP-formed image of the triplet nanohole pattern the postive refractive index. While in the SPP-formed image of the triplet array, magnification grows with distance along the optical axis (which is consistent with a positive effective refractive index of the nanohole array), in the image of the aperiodic array magnification distribution is reversed. This behaviour is consistent with negative sign of the effective refractive index of the nanohole array in fig. 24(a). Thus, both signs of the refractive index may be realised in a magnifying SPP crystal mirror.

§ 7. SPP microscopy application in biological imaging

SPP-assisted microscopy has the potential to become an invaluable tool in medical and biological imaging, where far-field optical imaging of individual viruses and DNA molecules may become a reality. Water droplets on a metal surface can be used as elements of 2D SPP optics in measurements where aqueous environment is essential for biological studies. The application of SPP-assisted microscopy to biological imaging has been illustrated using two types of objects: nanoscale polysterene spheres and the T4 phage viruses.

The 200 nm diameter polystyrene spheres deposited on a metal surface were studied under the illumination with 502 nm light. The images taken in reflection (fig. 26(b)) and transmission (fig. 26(c)) show that individual spheres, which have attached to the nanohole array surface as a result of the deposition process, are clearly visible as standard luminous features. They appear to be bright in the transmission image because they efficiently scatter SPP waves into photons. On the other hand, for the same reason, they must efficiently scatter SPPs into light, they must appear as dark features in the 2D SPP-formed images. Dark features of appropriate size indeed appeared in the images taken using SPPs with the polysterene spheres. The size of the scatterers deposited onto a nanohole array can be estimated from the SPP-induced image via comparison with the periodicity of the nanohole array in the image. However, the quality of the SPP image of the nanoholes in the array appears to be somewhat worse in this case, compared to the images in fig. 23 because of the increased scattering by the polystyrene spheres in glycerine.

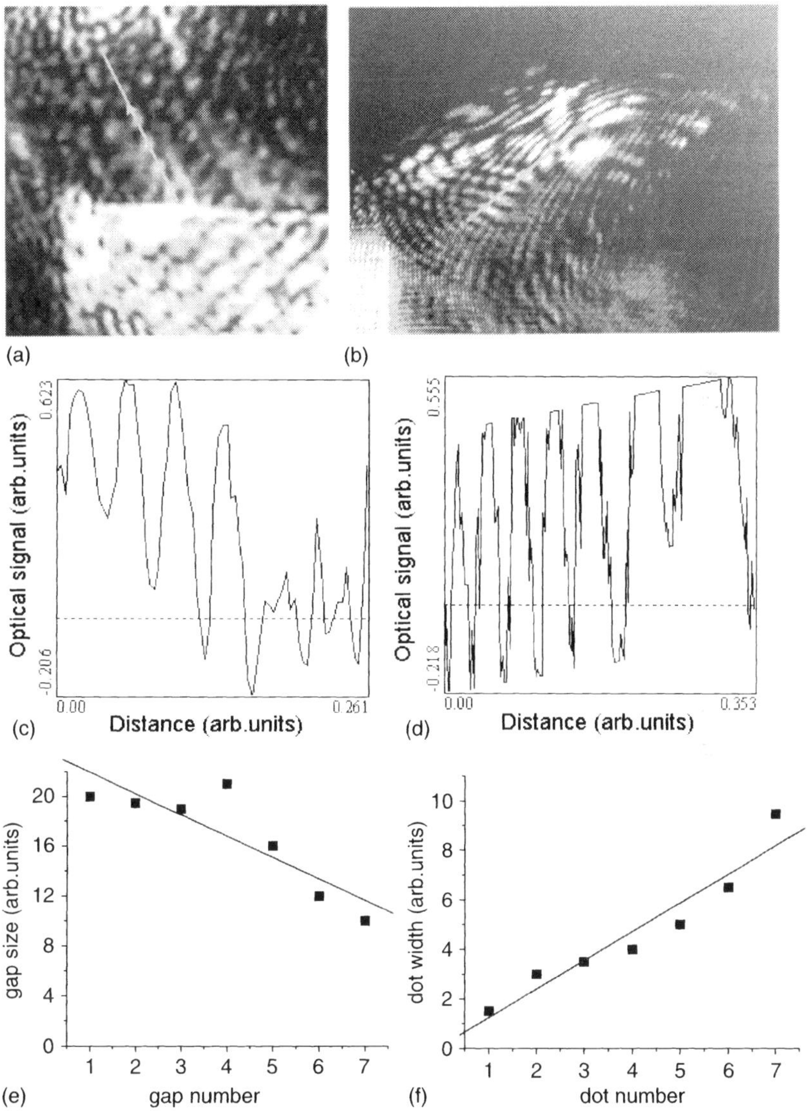

Fig. 25. Comparison of magnification dependencies on the distance from the mirror edge in the SPP-induced image of the aperiodic array (a) and in the image of the triplet array (b). (c) and (d) show cross sections of the SPP-induced images (a) and (b), respectively, along the indicated directions. The corresponding dependencies of the magnification are plotted in (e) and (f).

Similar technique has been implemented in the experiments with T4 phage viruses. Direct visualisation of viruses by using far-field microscopy techniques would constitute an important development in biosensing. A typical T4 virus is around 200 nm long and 80–100 nm wide (fig. 26(a)

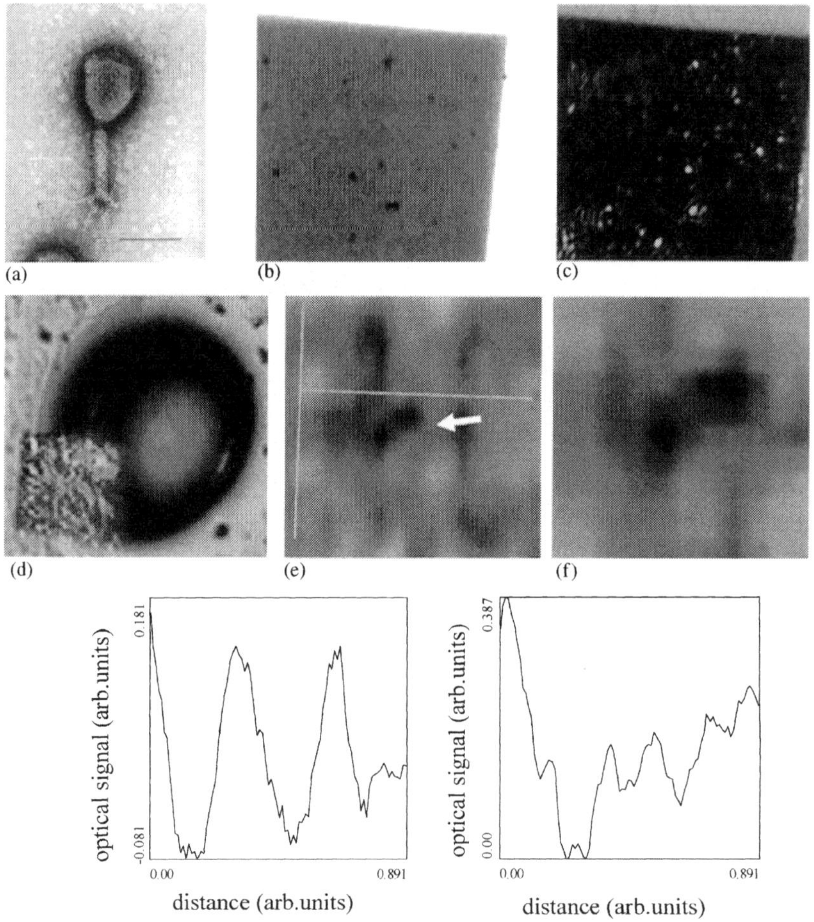

Fig. 26. (a) Electron microscope image of the T4 phage virus. (b, c) Optical images of the nanohole array with polystyrene spheres in reflection (b) and transmission (c) under illumination with the 502 nm light. (d) The droplet used to form the SPP-mirror for imaging the T4 phage. (e, f) The images of the T4 phage viruses visible as dark features (indicated by an arrow) in the SPP-formed image (g,h) Orthogonal cross sections plotted from the image (e) as indicated by the lines.

copied from the Universal Virus Database of the International Committee on Taxonomy of Viruses). In our experiments, individual T4 viruses were deposited onto an array of doublet nanoholes (fig. 23). After the glycerine droplet has been placed over the array (fig. 26(d)), the SPP image demonstrates resolution of the individual 100 nm nanoholes separated by 40 nm gaps (see the cross sections in fig. 26(e, f) in the two

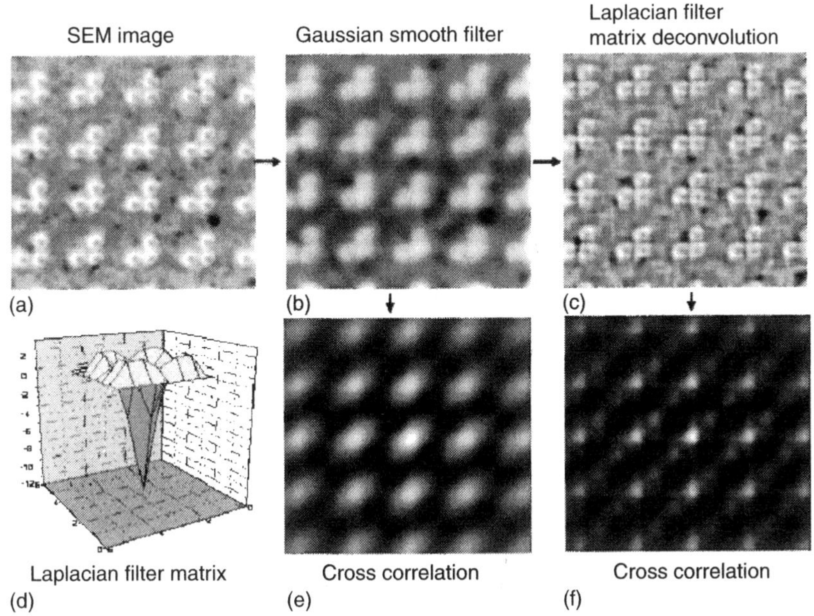

Fig. 27. Modelling of the image recovery using the Laplacian filter matrix deconvolution: the Laplacian filter allows to recover image deterioration due to the Gaussian blur, which is evidenced via calculation of the cross correlation of the original SEM image and the image recovered using Laplacian matrix deconvolution method.

orthogonal directions, as shown in fig. 26(e)). In addition, the SPP-formed image contains dark features that are similar to the one shown in figs. 26(e, f). While the surface density of these features was consistent with the T4 phage concentration in the deposited solution, the size of these features and their appearance were consistent in the way a T4 phage should look under a microscope with 50 nm resolution. Thus, the size, image contrast (T4 phages appear as dark features, similar to polystyrene spheres) and image resolution in these experiments are consistent with the known geometry of the T4 phage viruses and the resolution of the SPP-assisted microscopy.

§ 8. Digital resolution enhancement

Being quite an improvement as compared to a regular optical microscope, the 70 nm resolution is not sufficient to achieve clear visibility of many nanoholes in the test pattern in fig. 24. Even though recognisable, most nanoholes appear somewhat fuzzy. However, the blurring of optical

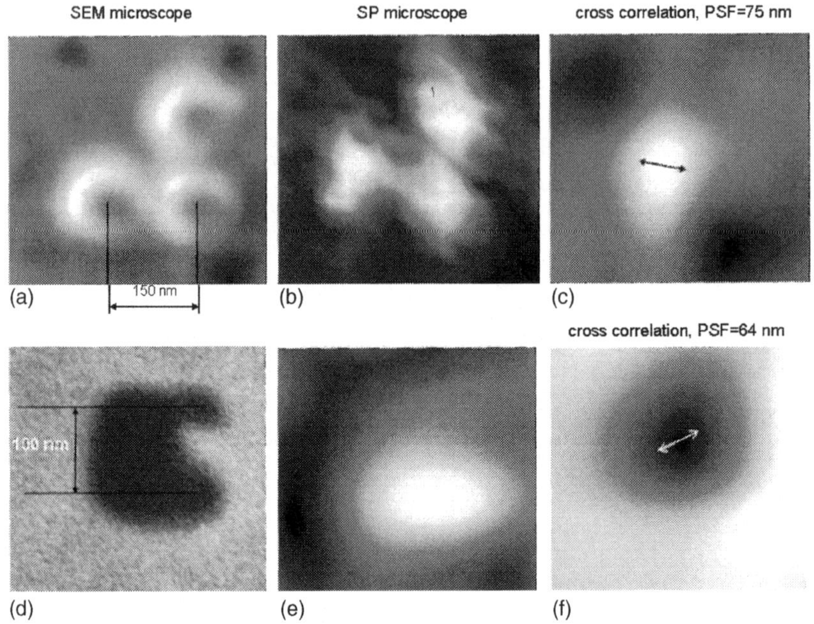

Fig. 28. Calculated cross correlation functions between the SEM- and SPP-induced images of a triplet nanohole from fig. 15 (top row) and U-shaped nanohole from fig. 24 (bottom row). These calculations indicate the point spread function of the optical microscope of the order of 70 nm.

images at the limits of optical device resolution is a very old problem (one may recall the well-publicised recent problem of Hubble telescope repair). Possible solutions of this problem are also well known. There exist a wide variety of image recovery techniques which successfully reduce image blur based on the known point-spread function (PSF) of the optical system. One of such techniques is the matrix deconvolution based on the Laplacian filter. However, precise knowledge of the PSF of the microscope in a given location in the image is absolutely essential for this technique to work, since it involves matrix convolution of the experimental image with a rapidly oscillating Laplacian filter is shown matrix (an example of such 5×5 matrix is shown in fig. 27). In the test experiments, the PSF of the SPP-based microscope was measured directly in some particular location of the SPP-formed image, as is shown in fig. 28. It can be measured directly by calculating the cross correlation between the SPP-formed image and the scanning electron microscope image of the same object. The results of these calculations in the case of triplet and U-shaped nanoholes from figs. 15 and 24 are presented in fig. 28. The calculations demonstrate the PSF of ~70 nm or $\sim\lambda/8$ achieved in these imaging experiments. The

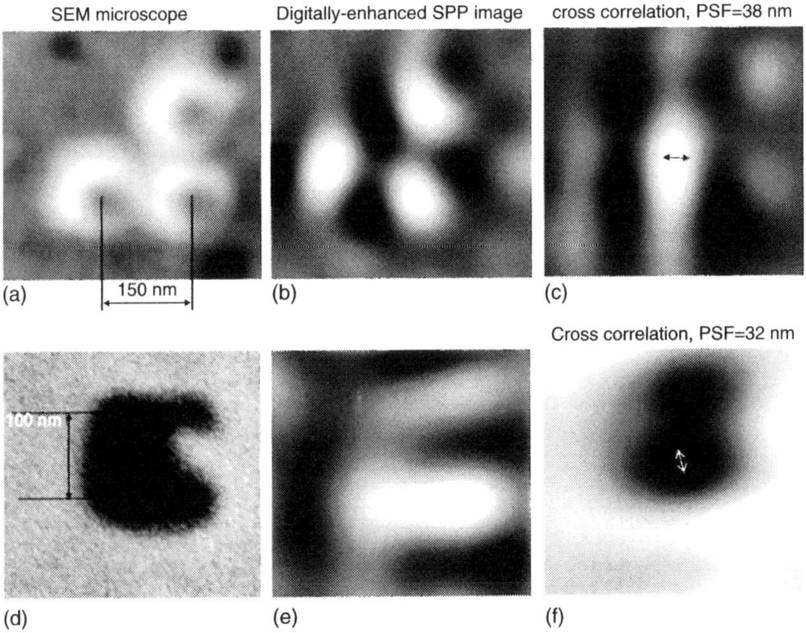

Fig. 29. Calculated cross correlation functions between the SEM and the digitally enhanced optical images of a triplet nanohole from fig. 15 (top row) and U-shaped nanohole from fig. 24 (bottom row). Comparison of these images with fig. 28 indicates approximately twofold improvement of the image resolution. The calculated PSF of the digitally enhanced optical images is of the order of 30 nm.

measured PSF was used to digitally enhance images of the nanohole arrays. Similar technique may be used to enhance resolution in the SPP-formed images of biological objects, which are measured using the nanohole array background.

The use of such digital filters led to approximately twofold improvement of resolution in the images formed by the SPP crystal mirror in both positive and negative effective refractive index cases. This twofold improvement is demonstrated in fig. 29 for both the triplet and the U-shaped nanoholes shown in fig. 28. The PSF measured as the cross correlation between the digitally processed optical image and the corresponding SEM image appears to fall into the 30 nm range, which represents improvement of resolution of the SPP-assisted optical microscopy down to $\sim\lambda/20$. This result may bring about direct optical visualisation of many important biological systems.

§ 9. Conclusion

We have discussed the principles of operation, experimental realisation and various imaging tests of far-field optical microscopy based on two-stage image magnification. The first stage must use 2D surface waves or 3D photonic crystal mirrors to produce magnified images of the object which can be then viewed with a conventional optical microscope. The development of such imaging technique becomes possible with the progress in the SPP optics which allowed creating magnifying SPP mirrors and SPP crystals with both positive and negative refractive index for SPP waves crossing the SPP crystal boundary. The resolution of up to 30 nm has been experimentally demonstrated with the illumination light wavelength around 500 nm. This technique is useful for imaging and projection applications where high, sub-diffraction limited optical resolution and fast image acquisition times are required.

Acknowledgements

This work has been supported in part by the NSF grants ECS-0304046, CCF-0508213 and EPSRC (UK). The authors are indebted to all their colleagues who contributed to the progress of this work; among them C. C. Davis, Y. J. Hung, W. Dickson, J. Elliott, G. A. Wurtz.

References

Agranovich, V.M., Shen, Y.R., Baughman, R.H., Zakhidov, A.A., 2004, Phys. Rev. B **69**, 165112.
Born, M., Wolf, E., 1999, Principles of Optics, Cambridge Univ. Press, Cambridge, UK.
Burke, J.J., Stegeman, G.I., Tamir, T., 1986, Phys. Rev. B **33**, 5186.
Challener, W.A., Mihalcea, C., Peng, C., Pelhos, K., 2005, Opt. Express **13**, 7189.
Chung, K.B., Hong, S., 2002, Appl. Phys. Lett. **81**, 1549.
Darmanyan, S.A., Zayats, A.V., 2003, Phys. Rev. B **67**, 035424.
Ditlbacher, H., Krenn, J.R., Schider, G., Leitner, A., Aussenegg, F.R., 2002, Appl. Phys. Lett. **81**, 1762.
Fang, N., Lee, H., Sun, C., Zhang, X., 2005, Science **308**, 534.
Kingslake, R., 1983, Optical System Design, Academic Press, London.
Kretschmann, M., Leskova, T.A., Maradudin, A.A., 2003, Opt. Commun. **215**, 205.
Lee, T.W., Gray, S.K., 2005, Appl. Phys. Lett. **86**, 141105.
Melville, D.O.S., Blaikie, R.J., 2005, Opt. Express **13**, 2127.
Pendry, J.B., 2000, Phys. Rev. Lett. **85**, 3966.
Pohl, D.W., Courjon, D. (Eds.), 1993. Near Field Optics, Kluwer Academic Publishers, Dordrecht.
Raether, H., 1988, Surface Plasmons, Springer, Berlin.

Ramakrishna, S.A., Pendry, J.B., 2004, Phys. Rev. B **69**, 115115.
Richards, D., Zayats, A.V. (Eds.), 2004., Nano-optics and near-field microscopy, Phil. Trans. R. Soc. Lond., Ser. A **362**, 699–919.
Smolyaninov, I.I., 2003, New J. Phys. **5**, 147.
Smolyaninov, I.I., Davis, C.C., Elliott, J., Wurtz, G.A., Zayats, A.V., 2005a, Phys. Rev. B **72**, 085442.
Smolyaninov, I.I., Davis, C.C., Zayats, A.V., 2005b, New J. Phys. **7**, 175.
Smolyaninov, I.I., Davis, C.C., Elliott, J., Zayats, A.V., 2005c, Opt. Lett. **30**, 382.
Smolyaninov, I.I., Elliott, J., Zayats, A.V., Davis, C.C., 2005d, Phys. Rev. Lett. **94**, 057401.
Smolyaninov, I.I., Mazzoni, D.L., Davis, C.C., 1996, Phys. Rev. Lett. **77**, 3877.
Weast, R.C. (Ed.), 1987. CRC Handbook of Chemistry and Physics, CRC Press, Boca Raton.
Westphal, V., Hell, S.W., 2005, Phys. Rev. Lett. **94**, 143903.
Zayats, A.V., Elliott, J., Smolyaninov, I.I., Davis, C.C., 2005a, Appl. Phys. Lett. **86**, 151114.
Zayats, A.V., Smolyaninov, I.I., 2003, J. Opt. Pure Appl. Opt. **5**, S16–S50.
Zayats, A.V., Smolyaninov, I.I., Dickson, W., Davis, C.C., 2003b, Appl. Phys. Letters **82**, 4438.
Zayats, A.V., Smolyaninov, I.I., Maradudin, A.A., 2005b, Phys. Rep. **408**, 131.

Chapter 4

Active plasmonics

by

Alexey V. Krasavin, Kevin F. MacDonald, Nikolay I. Zheludev

EPSRC Nanophotonics Portfolio Centre, School of Physics and Astronomy, University of Southampton, Highfield, Southampton SO17 1BJ, UK.

www.nanophotonics.org.uk

Nanophotonics with Surface Plasmons
Advances in Nano-Optics and Nano-Photonics
ISSN: 1871-0018

V.M. Shalaev & S. Kawata (Editors)

DOI: 10.1016/S1871-0018(06)02004-8

Contents

§ 1. Introduction

We are entering the age of integrated photonic devices for signal and information processing. Planar waveguides and photonic crystal structures are being intensively investigated as the primary solutions for guiding light in such devices; however, it may also be possible to make highly integrated optical devices with structural elements smaller than the wavelength by using metallic and metal/dielectric nanostructures to achieve strong guidance and manipulation of light. In this case the information carriers will be surface plasmon-polariton (SPP) waves, i.e. optical excitations coupled with collective electronic excitations (Agranovich and Mills, 1982; Boardman, 1982; Raether, 1988; Barnes et al., 2003; Zayats et al., 2005).

A range of very promising nanostructures capable of generating, guiding and manipulating plasmonic signals, such as SPP sources, mirrors, lenses, prisms and resonators, have been demonstrated (Smolyaninov et al., 1997; Krenn et al., 2003; Zayats and Smolyaninov, 2003). For example, a grating on a metallic surface can convert free-space optical radiation into a narrowly directed SPP wave, and a concave ridge or a row of point SPP scatterers on such a surface can focus an SPP wave. Alternatively, SPP lens design can exploit the dependence of an SPP's wave vector on the thickness of the metal film. SPP focusing may also be achieved by positioning an appropriately shaped thin dielectric layer on top of the metal layer, and a triangular dielectric area can act as an SPP prism (Agranovich, 1975; Zayats and Smolyaninov, 2003). Plasmon wave interactions with step-changes in dielectric constants and Fabry–Perot-type SPP resonators have also been extensively researched (Schlesinger and Sievers, 1980; Stegeman et al., 1981, 1984; Agranovich et al., 1981a, b, 1983; Maradudin et al., 1983; Leskova, 1984; Leskova and Gapotchenko, 1985). Periodic arrays of scatterers make efficient Bragg reflectors and thereby enable the design of polaritonic band gap materials, which can be used guide and route SPP waves (Bozhevolnyi et al., 2001; Ditlbacher et al., 2002; Krenn et al., 2003; Volkov et al., 2003). Plasmonic signals can also be guided along rows of metal nanoparticles (Quinten et al., 1998; Brongersma et al., 2000; Krenn et al., 2001; Maier et al., 2001, 2002, 2003; Weber and Ford, 2004; Maier and Atwater, 2005;

Viitanen and Tretyakov, 2005), metal strips (Yatsui et al., 2001; Nikolajsen et al., 2003; Onuki et al., 2003; Krenn and Weeber, 2004) and nanowires (Takahara et al., 1997; Takahara and Kobayashi, 2004) and gaps between metal walls (Wang and Wang, 2004; Tanaka et al., 2005; Bozhevolnyi et al., 2006). SPPs on gold/dielectric interfaces can propagate for tens of microns while symmetric waveguide configurations, comprising a thin metal film sandwiched between two dielectric layers, can support long-range SPPs with propagation lengths extending into the centimetre range (Sarid, 1981; Stegeman et al., 1983; Burke et al., 1986; Wendler and Haupt, 1986; Berini, 1999; Charbonneau et al., 2000; Nikolajsen et al., 2003). Intense effort is also being directed towards the creation of a plasmonic analogue to the laser, a coherent source of plasmons, now widely dubbed the 'spaser' (Sirtori et al., 1998; Bergman and Stockman, 2003; Nezhad et al., 2004).

Propagating plasmon-polariton excitations in metal nanostructures are thus clearly emerging as a new type of information carrier. However, we will only be able to speak about 'plasmonics' in the same way that we speak about 'photonics' when efficient techniques for active manipulation of SPP signals are identified. This chapter presents a concept for nanoscale functional elements to actively switch SPP signals.

§ 2. The concept of active plasmonics

The approach to active control of SPP waves on a metal/dielectric interface suggested here takes advantage of one of their most characteristic features, specifically the fact that their propagation depends strongly on the properties of the metal closest to (within a few tens of nanometres of) the interface. This feature can be exploited by adapting the recently developed concept of using structural phase transitions in polymorphic metals, and in particular gallium, to achieve nanoscale photonic functionality – an idea that has already been shown to facilitate all-optical switching at milliwatt power levels in thin films and nanoparticle monolayers (Bennett et al., 1998; Albanis et al., 1999, 2001; Petropoulos et al., 1999, 2001; MacDonald et al., 2001a, b, 2003, 2004, in press; Zheludev, 2002; Fedotov et al., 2003; Pochon et al., 2004), and to present opportunities for the development of a new type of photodetector (Fedotov et al., 2002) and single nanoparticle optical gates and memory elements (Soares et al., 2005; Zheludev, 2006).

It will be shown that SPP signals in metal/dielectric waveguides containing a gallium section can be effectively controlled by switching the gallium from one structural phase to another. Figure 1 shows a generic

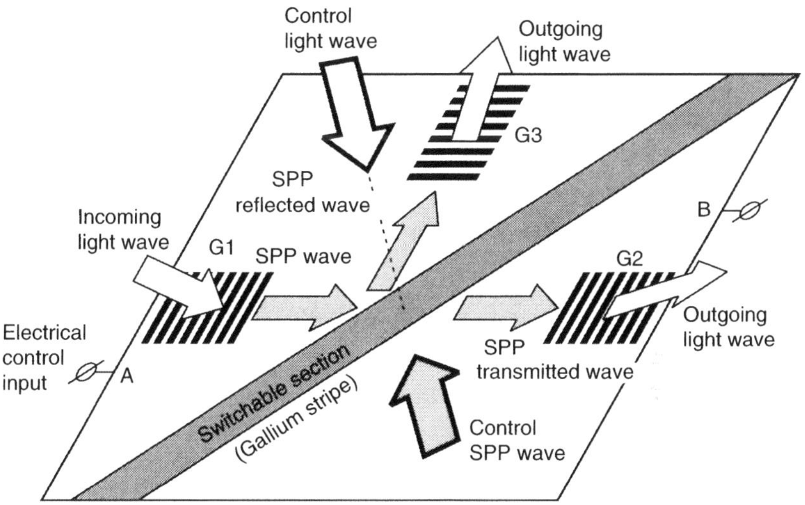

Fig. 1. Generic arrangement for optical, electrical or plasmonic control of SPP transmission through and reflection from a switchable insert in a metal/dielectric interface SPP waveguide. Following Krasavin et al. (2005).

design for an active SPP switch. Its basic elements are a metal/dielectric waveguide containing a switchable gallium section, and gratings for coupling and decoupling optical signals. In real applications several switches could be cascaded and interconnected, and more complex and sophisticated waveguide arrangements could be used. However, the generic switch presented in fig. 1 possesses all the crucial elements of the concept and is sufficient to illustrate its potential. An incoming optical signal generates an SPP wave at the coupling grating *G*1. The plasmonic signal propagates across the waveguide, passes through or is reflected by the switchable section, and is decoupled at grating *G*2 or *G*3. The intensity of the throughput signal depends on the reflectivity of and losses within the switchable section, both of which depend on its structural phase, which may be controlled in a reversible fashion by, for example, external excitations, intense SPP waves or simple Joule heating induced by a current running across the waveguide.

Any material that supports SPP waves and can be reversibly transformed between two structural phases with markedly different plasmonic properties might be used as the basis of an active plasmonic switching device, but this chapter will focus on the use of gallium, which is known for its polymorphism (Defrain, 1977; Bosio, 1978) and is particularly suited to this application. In α-gallium, the stable solid bulk phase, molecular and metallic properties coexist (Gong et al., 1991; Zuger and

Durig, 1992): some interatomic bonds are strong covalent bonds, forming well-defined Ga_2 dimers (molecules), and the rest are metallic. The structure is highly anisotropic, with much better thermal and electrical conductivity in the 'metallic planes' than along the covalent bonds. The covalent bonding leads to energy gaps at the Fermi level, and gives rise to a strong optical absorption peak centred at 2.3 eV and extending from ~0.68 eV (~1.82 μm) to ~4 eV (~310 nm). The pronounced difference between the electronic structure of α-gallium and those of the more 'metallic' (liquid and metastable solid) phases, which are akin to a free-electron gas, manifests itself as a marked difference in dielectric coefficients: e.g. $|\varepsilon_{liquid}|/|\varepsilon_{\alpha}| \sim 7$ at a wavelength of 1.55 μm. This leads to a significant difference between the SPP damping lengths of the α and metallic phases of gallium, and thereby to considerable switching contrast in the device shown in fig. 1. Mirror-like α-gallium interfaces with silica can be formed using a variety of techniques, from simply squeezing molten gallium against a substrate to ultrafast pulsed laser deposition (Gamaly et al., 1999; Rode et al., 1999, 2001; MacDonald et al., 2001a, b). To achieve optical switching and control functionality at such interfaces, one needs to induce a reversible transition from the α-phase to a metastable solid phase or the liquid phase. Such a transition may be achieved with very little input of laser, electron-beam or thermal energy (Albanis et al., 2001; MacDonald et al., 2001a; Fedotov et al., 2003; Pochon et al., 2004) because α-gallium has a remarkably low melting point of just 29.8 °C and its latent heat of fusion is relatively small, ~60 meV/atom.

In what follows, the performance of the different components of the generic active plasmonic switch will be numerically analysed, beginning with a detailed examination of the light-to-SPP coupling and SPP-to-light decoupling efficiencies of gratings fabricated on metal/dielectric SPP waveguides. Numerical models of SPP signal modulation by a gallium strip that can be switched between the 'ground-state' α-phase and a metallic phase will then be described, and the role of SPP reflection at the boundaries between switchable and non-switchable sections of the waveguide considered. The first experimental results relating to the control of SPP waves in gallium films and novel gallium/aluminium nanocomposites will be presented in the concluding part of the chapter.

§ 3. Coupling light to and from SPP waves with gratings

This section will discuss the coupling and decoupling of light to and from SPP waves using gratings, describing the relationships between the

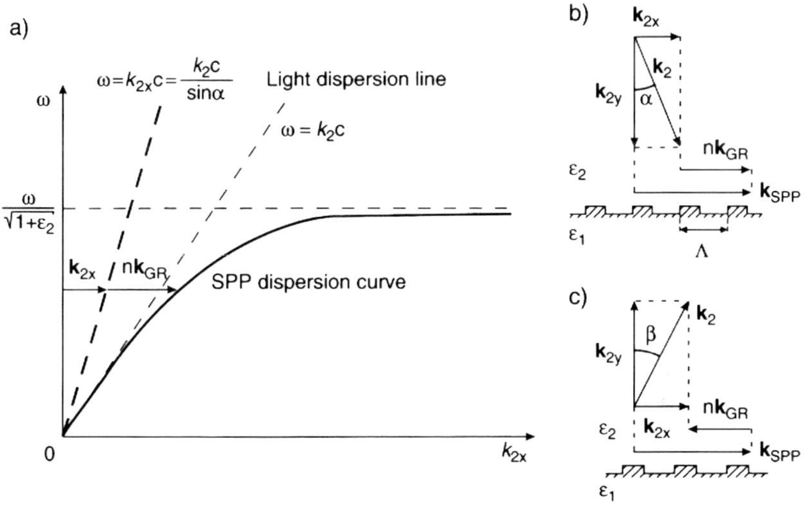

Fig. 2. Coupling light (with wave vector k_2) to and from an SPP wave (with wave vector k_{SPP}) using a grating (with vector $k_{GR} = 2\pi/\Lambda$) at the interface between a metal (with dielectric parameter ε_1) and a dielectric (ε_2). (a) Dispersion characteristics of light and SPP waves. Coupling/decoupling is possible when the mismatch between k_{SPP} and the x-component of k_2 is equal to an integer multiple of k_{GR}. (b) Wave vector diagram of light-to-SPP coupling. (c) Wave vector diagram of SPP-to-light decoupling.

efficiency of these processes, the direction of input/output light beams and the geometrical profile of the grating. Although simple momentum conservation laws govern these processes, their full complexity is only addressed by detailed numerical simulations.

The direct coupling of light to an SPP wave is forbidden by the momentum conservation law, due to the mismatch between their wave vectors (see fig. 2a) (Raether, 1988). However, a grating fabricated on a plasmon waveguide can facilitate coupling and decoupling by providing an additional k-vector equal to a multiple of the grating vector $k_{GR} = 2\pi/\Lambda$ (where Λ is the grating period). Figure 2b shows a vector diagram of the coupling process for a waveguide interface formed between a metal (with dielectric parameter ε_1) and a dielectric (ε_2). If light with a wave vector $k_2 = 2\pi/\lambda_2$ (where λ_2 is the wavelength of the light in the dielectric adjacent to the metal) illuminates a grating at an angle of incidence α, then the wave vector balance in the x direction is

$$(2\pi/\lambda_2)\sin\alpha + n2\pi/\Lambda = k_{SPP}. \tag{3.1}$$

where n is an integer.

There is a similar equation for the decoupling process, but in this case the vector supplied by the grating is subtracted from the SPP wave vector k_{SPP} (see fig. 2c):

$$k_{SPP} - n2\pi/\Lambda = (2\pi/\lambda_2)\sin\beta, \tag{3.2}$$

where β is a decoupling angle.

These equations for wave vector matching establish strict relationships between the wavelength λ_2 of light in the dielectric, the angle of incidence α or decoupling β, and the grating period Λ. However, they do not take into account the grating profile, realistic absorption losses or scattering associated with the finite size of gratings in micron-scale devices, all of which define the efficiency of the processes. A thorough investigation of the coupling and decoupling processes, taking into account all of these factors, can be performed using numerical simulations.

By numerically solving the Maxwell equations for the electromagnetic fields around gratings using the finite element method (implemented in the Comsol Multiphysics software package, see www.comsol.com), coupling and decoupling gratings on gold/silica interfaces have been analysed. This analysis concentrates on rectangular gratings with period Λ, line width s and height h (see fig. 3) as these are perhaps most suitable for microfabrication. All of the results presented in this section were calculated for p-polarized light with a free-space excitation wavelength $\lambda = 1.31$ μm (corresponding to a wavelength $\lambda_2 = 0.908$ μm in silica) – a

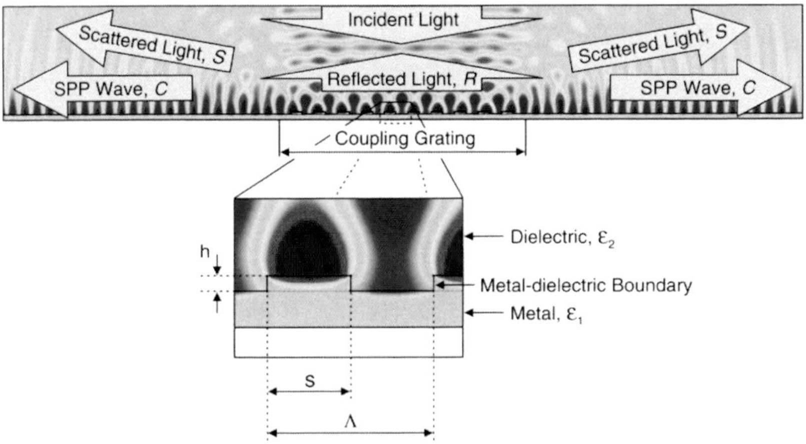

Fig. 3. Numerical simulation (mapping the magnitude of the z-component of magnetic field) of light-to-SPP coupling by a grating on a metal/dielectric waveguide. The metallic film is on the bottom surface of the silica substrate. Light is incident normally from above on the coupling grating, detail of which is shown in the zoomed section.

wavelength commonly used in current telecommunications technologies. Dielectric parameters for gold were derived from Palik (1984).

When a grating is illuminated by an electromagnetic wave, some of the incident light is coupled to an SPP wave propagating on the SPP waveguide; the rest is either reflected from the grating, scattered (and diffracted if $\Lambda > \lambda_2$) or absorbed due to ohmic damping of SPPs (see fig. 3). The coupling efficiency C, reflectivity R and scattering S were determined from numerical simulations by comparing the magnitudes of the relevant Poynting vectors to that of the incident light wave. When determining the coupling or decoupling (see below) efficiency of a grating from numerical simulations, care must be taken to isolate the desired light and SPP fields, in particular from scattered light fields. In the simulations described below, the geometry of the simulation domains was carefully designed and power-flow integration boundaries carefully chosen to achieve such isolation (Krasavin, 2006), and thereby to reduce the absolute error in coupling and decoupling efficiencies and other numerically determined values to about 3%. (Note that fig. 3, and others below depicting numerical simulations, does not necessarily show the full extent of simulation domains and may not therefore include the locations of power-flow integration boundaries.)

Figures 4 and 5 show the results of an analysis of normally incident light to SPP wave coupling by a 10-line grating on a silica/gold plasmonic waveguide interface. Note that due to the symmetry of the grating, equal portions of a normally incident light wave's energy are scattered and coupled to SPP waves propagating to either side of the grating (see fig. 3) – the efficiencies quoted below and plotted in figs. 4 and 5 are for one direction only. It is found that a grating with a fixed optimal (see below) period $\Lambda = 0.965\lambda_2$ may, depending on its height h and line width s, reflect as little as 5% of incident light, and its total scattering losses do not exceed 8% (see figs. 4a and b). For grating periods ranging from $0.86\lambda_2$ to $1.03\lambda_2$, the coupling efficiency surface in s–h parameter space has a single maximum (as exemplified in figs. 4c and d) at s and h coordinates that depend on the period as shown in fig. 5a. The optimal grating height varies significantly across the range of periods considered and it is interesting to note that with increasing period, the optimal line width does not increase to maintain a constant value of s/Λ, but in fact it actually decreases. For a 10-line grating, coupling efficiencies of ~40% can be achieved with grating periods between $0.93\lambda_2$ and $0.99\lambda_2$ (see fig. 5b), given optimal values of s and h. Figure 5c illustrates the fact that this efficiency cannot be improved by either increasing or decreasing the number of grating lines (and changing the dimensions of the light-input boundary accordingly). If fewer lines

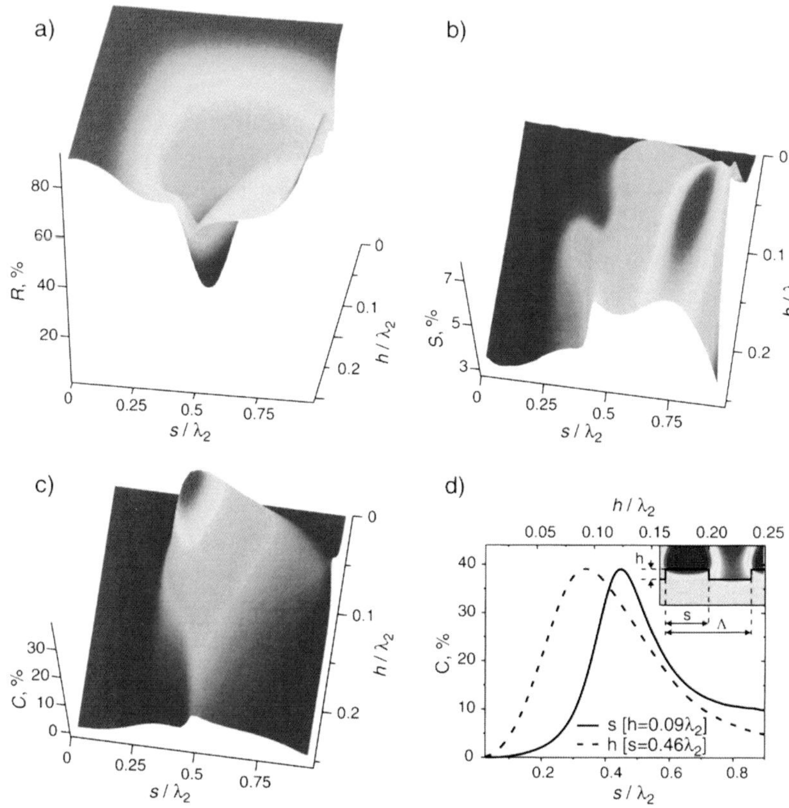

Fig. 4. Optimization of grating parameters for light-to-SPP wave coupling: (a) Reflectivity R; (b) Scattering S; and (c) Coupling efficiency C as functions of grating height h and linewidth s for a 10-line gold/silica grating with a period $\Lambda = 0.965\lambda_2$ illuminated normally by light with a free-space wavelength of 1.31 μm (wavelength in silica $\lambda_2 = 0.908$ μm). (d) Cross-sections through the maximum in part (c) in the s and h planes at $h = 0.09\lambda_2$ and $s = 0.46\lambda_2$, respectively.

are used, the coupling efficiency decreases because the grating pattern is less well defined. If more lines are used, the coupling efficiency decreases because much of the energy coupled to an SPP wave in the central part of the grating is decoupled again by another part of the same grating as the wave propagates towards the edge.

Analytically, light-to-SPP coupling at a grating can be considered using a theory for the diffraction of light by a weakly periodically modulated surface (Heitmann, 1977a), which is valid for small height modulations ($h << \lambda_2$) and slowly varying surface profiles ($h << \Lambda$), and which assumes an infinite number of grating lines. Numerical simulations provide a more powerful tool for the investigation of grating coupling but a

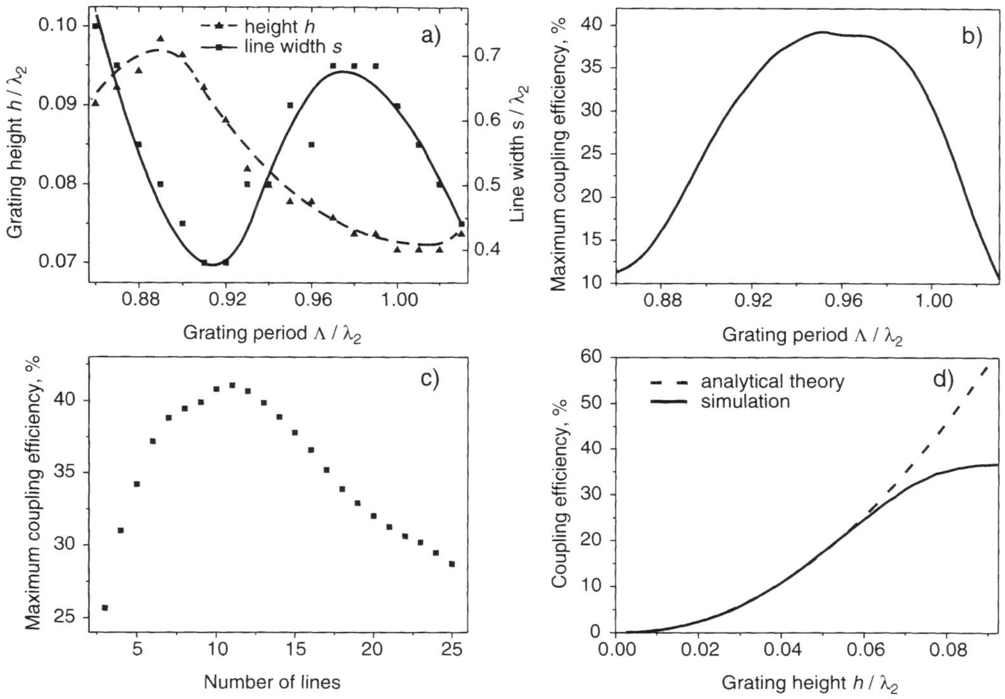

Fig. 5. Optimization of grating parameters for light-to-SPP wave coupling (normally incident light, $\lambda_2 = 0.908\,\mu\text{m}$): (a) Optimal values of grating height h and line width s as a function of grating period Λ for a 10-line grating. (b) Maximum coupling efficiency of a 10-line grating as a function of grating period Λ, assuming optimal values of s and h. (c) Maximum coupling efficiency as a function of the number of grating lines, for a grating with period $\Lambda = 0.965\lambda_2$, line width $s = 0.45\lambda_2$ and height $h = 0.095\lambda_2$. (d) Comparison of analytical and numerically simulated dependences of SPP coupling efficiency on grating height h ($\Lambda = 0.965\lambda_2$, $s = 0.46\lambda_2$).

comparison with analytical theory serves both to validate results and to determine the range of parameters over which the analytical theory is applicable. For small periodic surface modulations the energy transfer to an SPP wave is, analytically, proportional to the square of the modulation amplitude. Numerical simulations confirm this relationship, and thereby the validity of the analytical approach, even for a limited number of grating lines, for modulation amplitudes (grating heights) up to $\sim 0.06\lambda_2$ (see fig. 5d). At higher values, the efficiency determined by numerical simulation increases less rapidly, reaches a maximum at an optimal value of h and eventually decreases (as shown in fig. 4d).

Optimal dimensions for decoupling gratings can also be determined by numerical simulation. When an SPP wave propagating on a metal/dielectric waveguide encounters a grating, its energy is distributed (see fig. 6a) between transmitted (T') and reflected (R') SPP waves, a directional decoupled light wave (D') at an angle given by eq. (3.2), backward and forward scattered light waves (S'_1 and S'_2 respectively) and ohmic losses. Figure 6 illustrates how T', R', S' ($= S'_1 + S'_2$) and D' depend on the height h and line width s of a six-line grating designed to decouple light in the direction normal to the plane of the waveguide. SPP transmission (fig. 6b) is very high for gratings where the height h is small, and generally increases with line width s (i.e. as the grating profile tends towards a flat surface). In the reflection plot (fig. 6c) there are two distinct ridges. At low values of h these are simply due to Bragg reflections an SPP with a wave vector k_{SPP} is reflected strongly by a grating if the Fourier transform of the grating profile contains a significant component at $2k_{\text{SPP}}$. Figure 7 shows how the magnitude of this component $H(2k_{\text{SPP}})$ varies as a function of line width s and illustrates the correlation between the magnitude of $H(2k_{\text{SPP}})$ and the plasmonic reflectivity of the grating. At larger values of h, SPP waves scatter significantly on the first line (ridge) of the grating and the Bragg mechanism no longer applies; however, the reflection efficiency at certain line-widths in this region is still much higher than would be expected from a single scatterer (Sánchez-Gil, 1998). The low line width ($s \to 0$) ridge can be understood to result from reflection by a one-dimensional array of narrow scatterers (grating lines) separated by the grating period Λ ($\approx \lambda_2$), and the ridge at $s \sim \lambda_2/2$ to result from scattering at both the front and back edges of each grating line, i.e. from an array of scatterers separated by $\lambda_2/2$. In both cases the path difference between waves reflected from neighbouring scatterers is a multiple of $\lambda_2 \approx \lambda_{\text{SPP}}$, thus giving rise to resonant reflections. The high-reflection ridge at $s \sim \lambda_2/2$ corresponds to a significant dip in scattering (fig. 6c) and a shallower feature in the decoupling efficiency plot (fig. 6e).

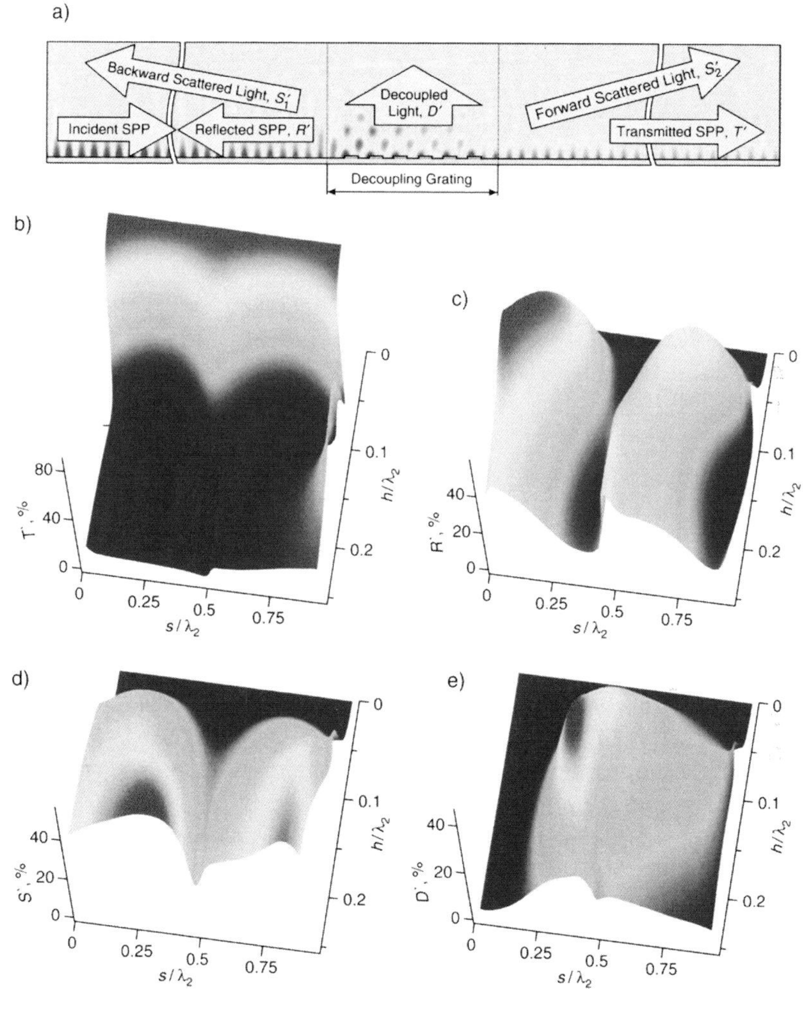

Fig. 6. Optimization of grating parameters for SPP-to-light decoupling: (a) Numerical simulation (mapping the magnitude of the z-component of magnetic field) of SPP wave decoupling, to a light beam ($\lambda_2 = 0.908\,\mu m$) propagating in the direction normal to the waveguide, by a six-line gold/silica grating with a period $\Lambda = 0.965\lambda_2$. (b) Transmission T'; (c) Reflectivity R'; (d) Scattering S' ($= S_1' + S_2'$); and (e) Decoupling efficiency D' as functions of grating height h and line-width s for the process illustrated in part (a).

Figure 6d indicates that for a six-line grating there are optimal grating parameters ($h = 0.14\lambda_2$, $s = 0.4\lambda_2$) that can decouple almost half of an SPP wave's energy into a directed light wave normal to the waveguide. Taking into account reflection and scattering losses, this level of efficiency is consistent with previous studies (Heitmann and Raether, 1976;

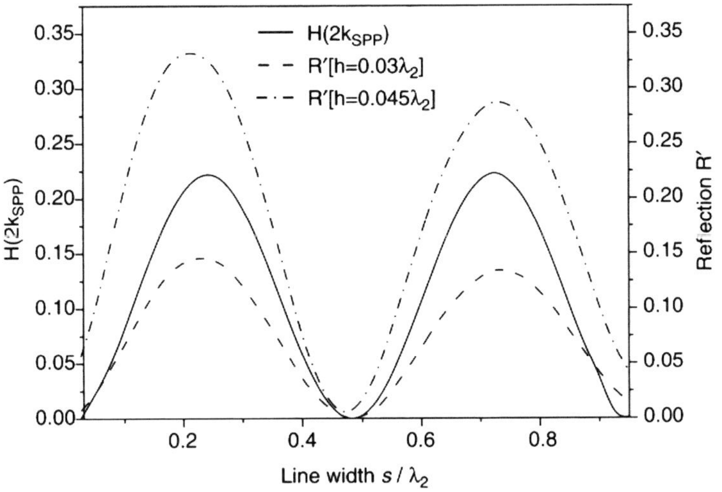

Fig. 7. Magnitude H of the $2k_{SPP}$ Fourier component of the profile of a grating with period $\Lambda = 0.965\lambda_2$ as a function of line width s. Also shown are corresponding dependences of the grating's plasmonic reflectivity R' for grating heights $h = 0.03\lambda_2$ and $0.045\lambda_2$.

Moreland et al., 1982; Worthing and Barnes, 2002). Using a longer grating (i.e. more lines) with a lower optimal height can increase the decoupling efficiency slightly by reducing scattering and reflection losses, however the improvement is small – a fourfold increase in length only increases decoupling efficiency by ~5%.

The angle β (defined in fig. 2) at which an output light beam is decoupled is determined by the period of the decoupling grating. Figure 8 compares decoupling angles calculated using eq. (3.2) with angles determined from numerical simulations, for a range of grating periods, and illustrates a good agreement between the two. The small discrepancy between the simulation data points and the theoretical curve (i.e. the fact that the points all lie slightly above the line) is related to the fact that the wave vector of an SPP on a grating differs slightly from its wave vector on a smooth surface (Pockrand and Raether, 1976; Heitmann, 1977b).

To summarize, the efficiency with which light can be coupled to and decoupled from SPP waves by gratings on metal/dielectric waveguides depends strongly on the geometric profiles of the gratings. A thorough numerical investigation of the coupling and decoupling of light with a free-space wavelength of 1.31 μm by gratings on gold/silica waveguides has shown that there are optimal grating parameters for both processes. For incident and output light beams normal to the plane of coupling and decoupling gratings, coupling efficiencies of up to 40% for SPP waves

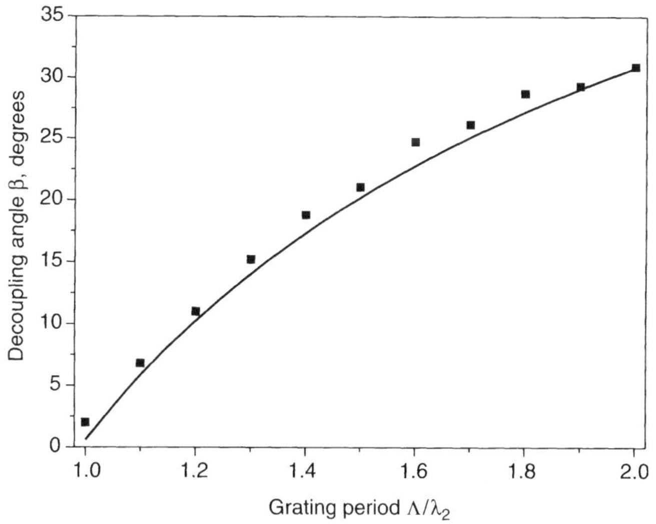

Fig. 8. Dependence of decoupling angle β on grating period Λ for a 10-line grating on a gold/silica SPP waveguide. Data points were obtained by numerical simulation, the solid line is given by eq. (3.2).

propagating to each side of the grating and decoupling efficiencies approaching 50% can be achieved.

§ 4. Modelling SPP propagation in an active plasmonic device

This section will consider the switching characteristics of an active plasmonic device such as that illustrated in fig. 1, designed to modulate SPP waves by controlling the plasmonic transmission losses of a switchable gallium section via an excitation-induced phase transition. Numerical simulations of the interaction between SPP waves and the switchable section (taking into account the full complexity of the phenomenon, including reflection, refraction and scattering of SPP waves at the boundaries between the different metals) are essential because even when an SPP wave is incident normally on the boundary between two metals, a comprehensive analytical treatment is rather complicated: reflection and transmission coefficients are expressed in terms of contour integrals, which do not reduce to any simple elementary or special functions (Agranovich et al., 1981a; Leskova, 1984). Nevertheless, the understanding of numerical results can be aided by comparison with simple relationships derived from general wave behaviour considerations.

There are two mechanisms by which an SPP wave may be modulated as the result of a phase transition in the gallium section of a device: Firstly via an associated change in the plasmonic transmission of the gallium (i.e. a change in the strength of SPP wave damping), and secondly via associated changes in the intensity of SPP waves reflected at the boundaries of the switchable gallium section. Analytically, the propagation coefficient $\xi(\varepsilon_{\mathrm{Ga}})$ of a gallium/silica SPP waveguide (the SPP intensity multiplication factor) is given by the equation

$$\xi(\varepsilon_{\mathrm{Ga}}) = \exp\left[-L/L_{\mathrm{d}}(\varepsilon_{\mathrm{Ga}})\right], \tag{4.1}$$

where L is the distance over which the SPP propagates and L_{d} is the $1/e$ energy damping length of an SPP wave propagating on the waveguide interface, which is a function of the gallium's phase. Equation (4.1) assumes isotropic media but α-gallium displays considerable anisotropy (its crystal structure belongs to the orthorhombic dipyramidal class, space group *Cmca*). However, because the SPP-related movement of electrons in a metal is defined by an electric field parallel to the direction of SPP propagation, it is only the dielectric constant in that direction which influences transmission (this fact is confirmed by the numerical results detailed below). Thus, eq. (4.1) can legitimately be used to calculate propagation coefficients for solid gallium by assuming the metal to be isotropic with a dielectric coefficient equal to that along the SPP propagation direction. Such calculations illustrate (see table 1) that for a broad range of wavelengths, plasmonic propagation through a gallium insert in an SPP waveguide depends strongly on the structural phase of the gallium (regardless of the solid-state crystalline orientation) and therefore that a phase change can indeed be used to actively control transmission.

The reflection and refraction of SPP waves at the boundary between two metals (illustrated schematically in fig. 9) has been considered previously. Assuming only that the y component of the SPP wave vector is conserved at the interface between the two metals, it has been determined that laws of reflection and refraction are analogous to those for the reflection and transmission of light at an interface between two dielectrics: the angle of reflection is equal to the angle of incidence φ and the angle of refraction θ is defined by the equation (Agranovich, 1975)

$$n_{1,2} \sin\varphi = n_{3,2} \sin\theta,$$

where $n_{i,j}$ is a refractive coefficient for the SPP waveguide formed between a metal with dielectric coefficient ε_i and a dielectric medium with

Table 1. Dielectric parameters (ε) for solid and liquid gallium with corresponding plasmonic propagation coefficients (ξ2.5) for a gallium section 2.5 μm long at free-space excitation wavelengths of 860, 1310 and 1550 nm. Complex dielectric coefficients were derived from literature (Kofman et al., 1977; Teshev and Shebzukhov, 1988)· Polycrystalline α-gallium is assumed to be isotropic with a dielectric constant equal to an average over the three primary axes..

Wavelength (nm)	Parameter	α-gallium				'Metallic' (liquid) gallium
		a-axis	*b*-axis	*c*-axis	Polycrystalline	
860	ε	$-9.1-17.7i$	$-13.3-8.4i$	$-5.6-14.1i$	$-9.3-13.4i$	$-74.7-43.2i$
	$\xi_{2.5}$	0·07	0·11	0·03	0·05	0·72
1310	ε	$-5.1-27.0i$	$-27.0-17.0i$	$1.3-21.1i$	$-10.3-21.7i$	$-115.6-98.7i$
	$\xi_{2.5}$	0·27	0·52	0·19	0·24	0·86
1550	ε	$-3.0-23.8i$	$-36.5-21.8i$	$2.8-21.4i$	$-12.2-22.3i$	$-133.0-134.4i$
	$\xi_{2.5}$	0·28	0·68	0·26	0·33	0·89

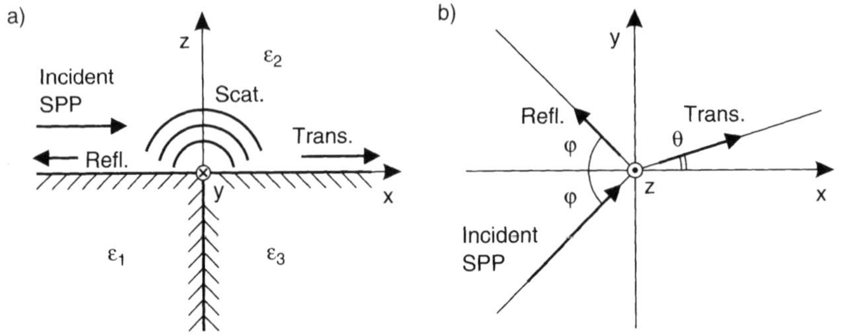

Fig. 9. Reflection, transmission and scattering of SPP waves at the boundary between two different metals ε_1 and ε_3 (ε_2 is the dielectric waveguide component). (a) Waveguide cross-section perpendicular to the boundary between the metals. (b) Plan view of the metal/dielectric interface plane. Following Agranovich (1975).

coefficient ε_j, given by the formula (Raether, 1988):

$$n_{i,j} = \mathrm{Re}\left\{\sqrt{\frac{\varepsilon_i\varepsilon_j}{\varepsilon_i + \varepsilon_j}}\right\}.$$

The SPP refractive coefficients $n_{i,j}$ for silica/α-gallium and silica/metallic-gallium waveguides (1.470 and 1.450, respectively, for an excitation wavelength of 1.55 μm, assuming polycrystalline α-gallium) are both very close to the corresponding coefficient for a silica/gold waveguide (1.457). Thus, just as there is minimal deflection of a light beam as it crosses the boundary between two dielectrics with similar refractive indices, there should be almost no refraction of an SPP wave as it crosses the boundary between silica/gold and silica/gallium waveguide sections, regardless of the angle of incidence or the structural phase of the gallium. Furthermore, in the same way that the optical reflectivity of an interface between dielectrics whose refractive indices have similar real parts is very low (except at high angles of incidence, $>70°$) even if one is moderately absorbing, the plasmonic reflectivity of gold/gallium boundaries should be low (except at high angles of incidence) regardless of the structural phase of the gallium. Three-dimensional numerical simulations (fig. 10) of SPP wave interaction with a gallium insert in a gold/silica waveguide illustrate the lack of refraction and reflection at the boundaries between the different metals. In figs. 10a and b the SPP waves are incident normally on the gold/gallium boundary. In fig. 10a the gallium is in the polycrystalline α phase and in fig. 10b it is in the metallic state. In figs. 10c and d the angle of incidence is 45°, again for the two different phases of

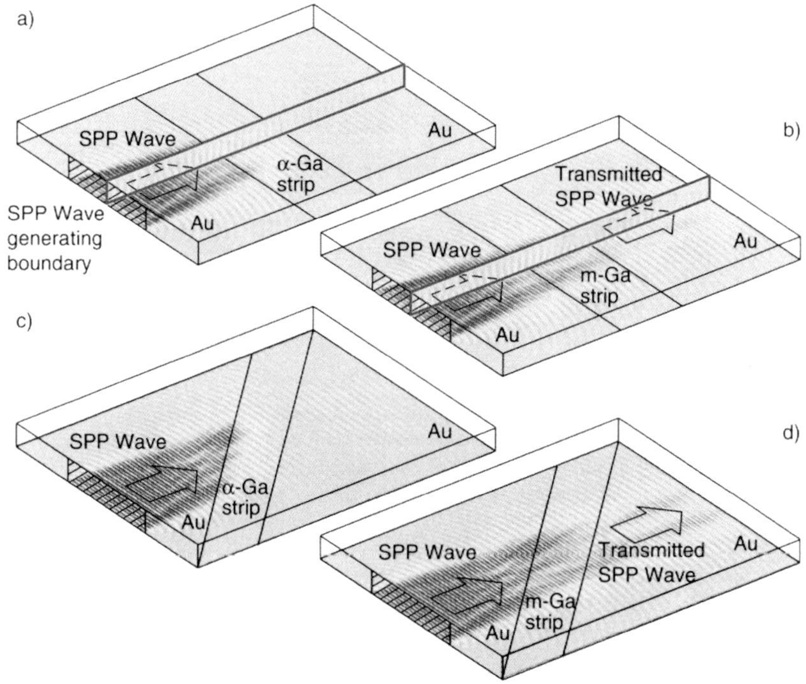

Fig. 10. Three-dimensional numerical simulations (mapping the magnitude of the z-component of magnetic field) of gold/silica/gallium active plasmonic switches with SPP waves incident (a and b) normally, and (c and d) at an angle of 45° on the gallium/gold boundaries. In cases (a) and (c) the gallium is in the polycrystalline α phase, in (b) and (d) it is in the metallic phase. The hatched border regions are the sources of SPP waves with a wavelength $\lambda_{\rm SPP} = 1.071\lambda_2$ (free-space excitation wavelength $\lambda = 1.55\,\mu$m).

gallium. The lack of refraction is clearly seen in all cases. It is also clear that there are no reflected SPP waves in figs. 10c and d, but power-flow analysis reveals that they are also absent from cases (a) and (b) of fig. 10. Thus, for the conditions presently under consideration, the functionality of the switching section is essentially independent of the angle of incidence of SPP waves – compare part (a) with (c) and part (b) with (d).

Detailed numerical investigations of the plasmonic switching device have been conducted for the case where SPP waves are incident normally on the gold/gallium boundary, in two-dimensional simulations exemplified by the vertical cross-sections outlined in figs. 10a and b, and illustrated explicitly in fig. 11. These simulations were used to calculate the transmission (ignoring coupling losses) of a waveguide structure containing a homogenous solid or isotropic liquid gallium section of length $L = 2.5\,\mu$m, as a function of free-space excitation wavelength between 0.9 and 2.0 μm. The transmission efficiency was obtained by integrating the

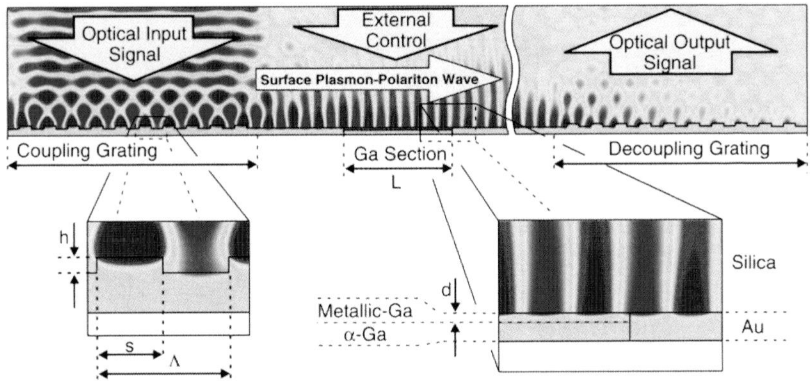

Fig. 11. Numerical simulation (mapping the magnitude of the z-component of magnetic field) of an SPP gold/silica waveguide containing a gallium switching section. The metallic film is on the bottom surface of the silica substrate. Light is incident from above on the coupling grating, detail of which is shown in the zoomed section on the left. An external excitation is used to control the transmission of the device by changing the phase of the gallium section (detail of which is shown in the zoomed section on the right). Following Krasavin and Zheludev (2004).

power flow over the region above the output coupling grating and dividing it by the integrated power flow over the same region in the absence of the gallium section. The results of these calculations, which were performed for each of the main crystallographic orientations of gallium at the silica interface, are shown in fig. 12, where the following notation is used: curve AB corresponds to a gallium crystalline structure with the A-axis parallel to the direction of SPP propagation and the B-axis perpendicular to the plane of the interface. The same convention applies to curves AC, BA, BC, CA and CB. It is clear that curves corresponding to crystalline orientations with the same axis parallel to the direction of SPP propagation (e.g. AB and AC) are essentially identical, which confirms the fact that it is the value of the dielectric constant along that axis which determines the plasmonic transmission of the gallium section.

These numerical results are plotted alongside analytical predictions (curves labelled A*, B* and C*) derived from eq. (4.1). While the results obtained by numerical simulation and analytical calculation show similar spectral trends, the actual transmission levels differ somewhat. This discrepancy probably reflects the limitations of the analytical theory, which ignores the interaction of SPPs with the gold/gallium boundaries: As described above, there is negligible reflection of SPP waves at such boundaries, but due to the dielectric coefficient mismatch, some amount of an SPP wave's energy can be decoupled at the first gold/gallium

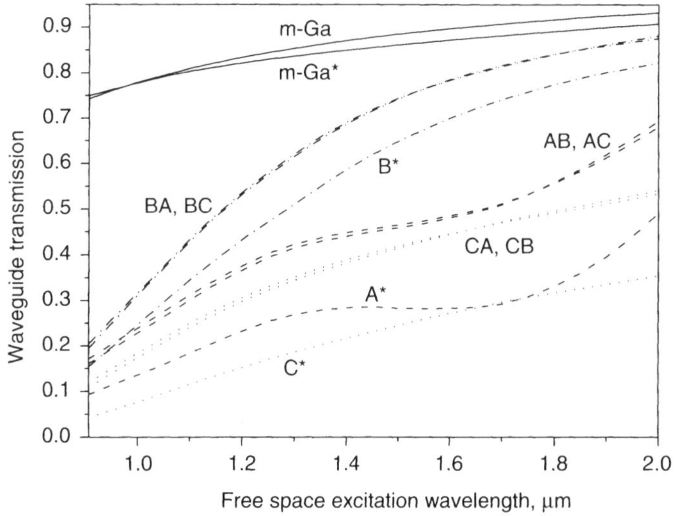

Fig. 12. Normalized transmission, as a function of free-space excitation wavelength, of a gold/silica waveguide with a 2.5 μm gallium insert for different phases and crystalline orientations of the gallium. Curves labelled with an asterisk are obtained analytically, assuming an isotropic material with a dielectric coefficient equal to that along the α-gallium axis indicated. The other curves are derived from numerical simulations and are labelled as follows: the first letter denotes the α-gallium crystalline axis parallel to the direction of SPP propagation, the second letter denotes the axis perpendicular to the plane of the metal/dielectric interface. Following Krasavin and Zheludev (2004).

boundary (Agranovich et al., 1981a) to a light wave propagating almost parallel to the metal/dielectric interface. This can then re-couple to the SPP wave after the gallium section, via diffraction at the second gallium/gold boundary, thus enabling part of the SPP signal to 'hop' over the gallium section (such a process has previously been predicted (Leskova, 1984) and an analogous phenomenon has been reported (Leskova and Gapotchenko, 1985)). This effect is observed in the numerical simulations and increases the derived transmission values relative to those obtained using the analytical theory. It is interesting to note that the discrepancy between analytical and numerical results is higher when the mismatch between the gallium and gold dielectric coefficients is larger (A and C), because in these cases the deflected light wave is stronger, and that the discrepancy tends to be larger at longer wavelengths, because the phase difference between the deflected light wave and the transmitted SPP wave after the gallium section is smaller so there is more constructive interference between the transmitted and re-coupled SPP waves (Leskova and Gapotchenko, 1985).

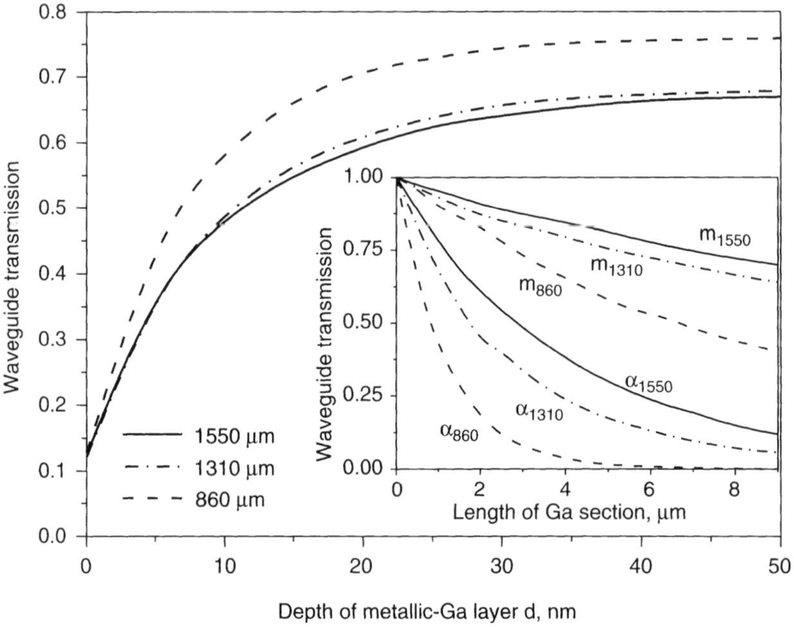

Fig. 13. Transmission (evaluated by numerical simulation) of a gold/silica waveguide containing a gallium insert as a function of metallic gallium layer thickness d (with the underlying α-gallium in polycrystalline form). Gallium section lengths were chosen so that waveguide transmission at $d = 0$ was the same for each free-space excitation wavelength: 860 nm–2.2 μm; 1310 nm–5.8 μm; 1550 nm–9 μm. The inset shows waveguide transmission as a function of gallium section length for the α and metallic phases at the same three wavelengths. Following Krasavin and Zheludev (2004).

The modulation contrast of the switch depends on the length of the gallium section as shown in the inset in fig. 13 (where the gallium section is assumed to be polycrystalline, as defined in Table 1, in the α phase). A longer section will give higher contrast, however, its absolute transmission for the metallic phases will be lower. Because the structural transformation in gallium is a surface-driven effect (see section 5), the α and metallic phases can co-exist near the interface, with a thin layer of the metallic phase sandwiched between the silica and the main body of solid α-gallium (see the zoomed section in fig. 11). This makes continuous 'analogue' control of waveguide transmission possible. Figure 13 shows waveguide transmission as a function of metallic layer thickness d. One can see that the transmission saturates as d increases and that the presence of a metallic gallium layer just a few tens of nanometres deep dramatically changes the plasmonic transmission.

In summary, theoretical estimates suggest that significant modulation of transmitted SPP wave intensity can be achieved by changing the phase composition of a short gallium insert in a gold/silica plasmon waveguide, and that an insert of this type will not refract or (at low angles of incidence) reflect incident SPP waves. A rigorous quantitative analysis of gallium's plasmonic switching characteristics, conducted using numerical simulations, confirms these inferences, and furthermore indicates that it should be possible to achieve the full level of SPP transmission switching contrast by changing the phase of a gallium layer just a few tens of nanometres thick at the interface with silica.

§ 5. Active plasmonics: experimental tests

This section will describe experimental tests designed to validate the active plasmonics concept and to determine the energy requirements and response characteristics for plasmonic switching in gallium via a nanoscale light-induced structural transformation.

The switching of structural phases in gallium can be achieved by external optical excitation, through simple laser-induced heating (MacDonald et al., 2001b; Fedotov et al., 2003). However, gallium presents an additional non-thermal metallization mechanism (Albanis et al., 2001; MacDonald et al., 2001b). Through the localization of photo-generated electron-hole pairs on the Ga_2 dimers, light at wavelengths within the dimers' absorption line can excite them from the bonding to the antibonding state, reducing the stability of surrounding crystalline cells. The α-gallium cells subsequently undergo a transition to a new configuration without necessarily achieving the melting temperature. This non-thermal mechanism is important for continuous and quasi-continuous excitation of gallium at intensities up to a few kW/cm^2. However, when the excitation takes the form of short, intense optical pulses, thermal diffusion does not have enough time to remove heat from the skin layer during the pulse, and the temperature at the excitation point increases rapidly, inevitably leading to thermal melting of the excited gallium. Whatever the mechanism behind the phase transition, it is a surface-driven effect – the metallic phase forms first at the surface of the metal and propagates into the bulk of the crystal to a depth that depends on the level of excitation and the background temperature. Following withdrawal of the excitation, the metallized layer re-crystallizes (in a time that depends on layer thickness and temperature), thus restoring the optical and electronic properties of the metal to their pre-excitation levels. This switching technique is

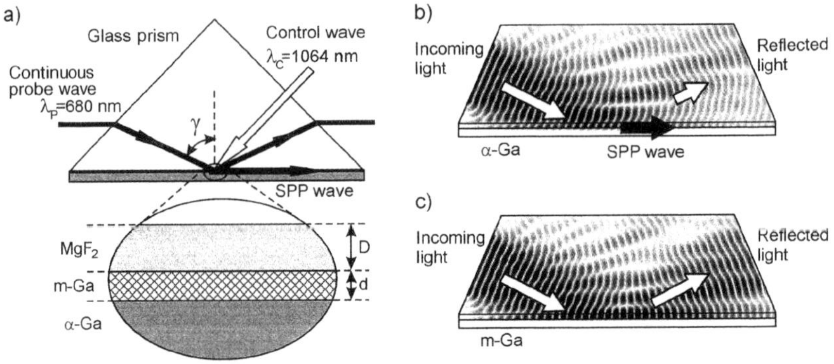

Fig. 14. Experimental test of the active plasmonics concept. (a) Arrangement for optical modulation of light-to-SPP wave coupling in a glass/MgF_2/gallium structure using the Otto configuration. (b) and (c) numerically simulated field distributions around the point at which the probe beam interacts with the glass/MgF_2/gallium structure for (b) the α phase and (c) the metallic phase of gallium. Following Krasavin et al. (2004).

inherently optically broadband (Albanis et al., 1999; MacDonald et al., 2001b; Zheludev, 2002) and the transition process is highly reproducible because it only involves a few tens of nanometres of gallium at an interface.

The potential for optical control of SPP propagation using gallium has been evaluated in reflective pump-probe experiments. These employed an attenuated total internal reflection matching scheme known as the Otto configuration (Raether, 1988), rather than a grating, to couple light to an SPP wave on a gallium/dielectric interface, and used the structural transition in gallium to modulate the coupling efficiency rather than the waveguide's propagation losses (Krasavin et al., 2004). Although this configuration would not be suited for the construction of compact or cascaded devices, it is adequate to determine the energy requirements for controlling plasmonic signals with light and as a proof of principle.

For this experiment (depicted in fig. 14a), gallium was interfaced with a BK7 glass prism previously coated with an MgF_2 film of thickness $D = 185\,\text{nm}$, simply by squeezing a bead of the liquid metal ($T_m = 29.8$ °C) against the surface of the prism then solidifying to obtain the α phase. At angles of incidence greater than 53°, p-polarized probe light from a laser diode ($\lambda_p = 680\,\text{nm}$) is totally internally reflected at the glass/ MgF_2 interface, producing an evanescent wave in the MgF_2 layer. At an incident angle $\gamma = 66°$ the interface projection of the wave vector of the evanescent wave is equal to the SPP wave vector for the glass/MgF_2/ α-gallium structure and the energy of the incident beam is efficiently (resonantly) coupled to an SPP wave. Under these conditions, probe reflectivity is low (fig. 14b) but when a pump (control) laser (an Nd:YAG laser generating 6 ns pulses at $\lambda_c = 1064$ nm with a repetition rate of 20 Hz) initiates a structural transformation from the α-phase to the metallic phase in the gallium at the probe spot, it drives the system away from resonance, decreasing the SPP coupling efficiency (which is highly sensitive to the dielectric parameters at the interface) and thereby increasing the probe reflectivity (fig. 14c). The coupling efficiency changes continuously with the thickness of the metallized layer at the interface so probe reflectivity measurements provide an effective means of monitoring that thickness.

The inset in fig. 15 shows the time dynamic of pump-induced reflectivity modulation, defined as $(R_{on} - R_{off})/R_{off}$, where R is probe reflectivity and the subscripts denote the state of the control laser. Control laser excitation of the interface leads to an immediate increase in the reflected probe intensity, and when the excitation is terminated the molten layer rapidly recrystallizes, restoring the reflectivity to its pre-excitation level. The magnitude of the effect increases with pump fluence up to ~15 mJ/cm^2 where it saturates, as illustrated by the data points in fig. 15. The solid curve in fig. 15 shows the theoretical dependence of reflectivity modulation on the

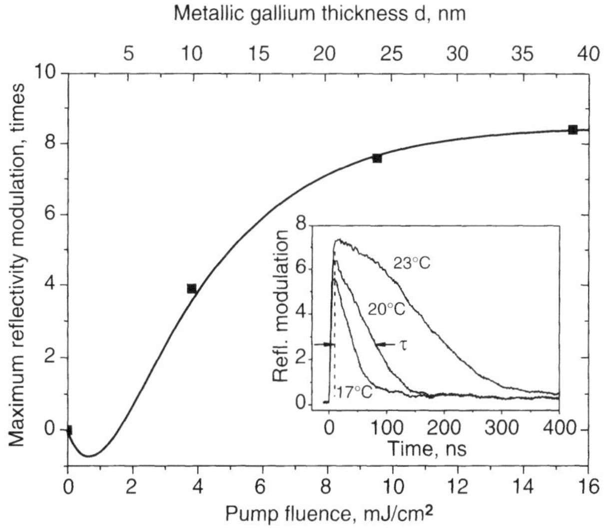

Fig. 15. Dependence of peak reflectivity modulation on pump fluence at a sample temperature $T = 28^{\circ}$C (data points), and the theoretical dependence of reflectivity on metallic gallium layer thickness d (solid line). The inset shows reflectivity dynamics following 6 ns excitation pulses at a wavelength of 1.064 μm ($Q = 15\,\mathrm{mW\,cm^{-2}}$) at a number of sample temperatures. Following Krasavin et al. (2005).

depth d of the metallized layer. The theoretical plot was scaled vertically and its extension in the horizontal direction adjusted to achieve a good fit with the experimental points. On the basis of this fitting it can be estimated that a fluence Q of $\sim$12 mJ/cm^2 produces a metallized layer with a depth d of $\sim$30 nm, which is sufficient to modulate SPP transmission by $\sim$80% in the scheme presented in fig. 11. Thus, high-contrast switching could be achieved using a 2.5 $\times$ 2.5 μm gallium insert in a silica/gold SPP waveguide with an optical excitation energy of the order of just 1 nJ.

The overall bandwidth of the data acquisition system was 100 MHz, so the transient response time was not resolved in this experiment. It may be as short as 4 ps, which is the intrinsic electronic response time of the α-phase (Rode et al., 2001). For a given excitation level, there is a steep increase in the relaxation time τ following withdrawal of the control excitation as the temperature T of the structure approaches gallium's melting temperature $T_0 = 29.8\,°C$, but relaxation times as short as 20 ns are observed at temperatures below 14 °C (see fig. 16). At the same time, the magnitude of the induced reflectivity change increases gradually as T approaches T_0. These behaviours are explained by the fact that the thickness of the metallized layer produced by a given fluence increases with proximity to T_0, and the fact that the recrystallization velocity v is inversely proportional to $[T - T_0]$ (Peteves and Abbaschian, 1991).

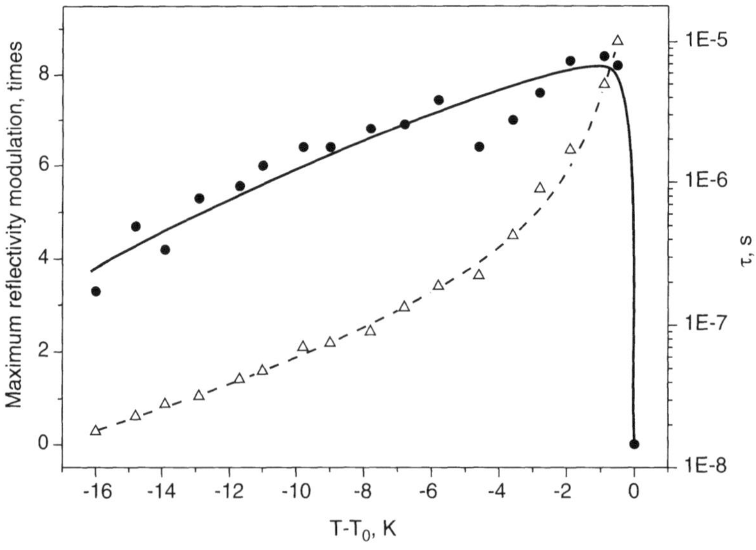

Fig. 16. Temperature dependences of the maximum pump-induced reflectivity modulation (●) and corresponding relaxation time τ (Δ) for a pump fluence $Q = 15$ mJ cm^{-2}. Following Krasavin et al. (2004).

Simply pressing liquid gallium against a dielectric as described above is not a particularly reliable method of manufacturing high-quality gallium/dielectric interfaces, and those that are produced tend to deteriorate quite rapidly because the adhesion between metal and dielectric is poor. Ultrafast pulsed laser deposition can reliably produce interfaces that retain a mirror-like quality over several years (Gamaly et al., 1999; Rode et al., 1999, 2001; MacDonald et al., 2001a, b), but this technique is complex and time consuming. It has recently been found, however, that a balance between the simplicity and reliability of the production technique might be struck by using a gallium/aluminium nanocomposite, instead of pure gallium, as a switchable medium (Krasavin et al., 2006). These composite structures comprise polycrystalline aluminium films on silica substrates wherein the grain boundaries between aluminium domains and the interface between the aluminium and the silica are infiltrated with nanolayers of gallium. The composite material retains the switching functionality provided by the externally controllable phase equilibrium between solid and liquid gallium, while benefiting, particularly during the formation process, from the uniform quality of the aluminium film and its adhesion to the substrate.

Such composites are formed simply by applying a drop of liquid gallium to the exposed surface of a thin (~250 nm) polycrystalline aluminium film on a silica substrate held at a temperature a few degrees above gallium's melting point. Once the oxide layer on the aluminium surface is broken and after a short incubation period, the liquid gallium begins to spread itself across the aluminium surface and penetrates the aluminium film through to the silica interface, where it can solidify (as the sample is cooled) with uniform crystalline orientation across the interface. This grain boundary penetration process (Tanaka et al., 2001; Pereiro-Lopez et al., 2004) produces a characteristic micron-scale pattern of 'arachnoid' lines on the metal surface (fig. 17a).

The plasmonic properties of a nanocomposite can be controlled in the same way as those of pure gallium: excitation fluences of the order of 1 mJ/cm^2 are sufficient to induce a reversible structural transition in the gallium component of the composite material and thereby to substantially modulate the properties of a composite/silica interface (fig. 17b) with sub-nanosecond response and sub-microsecond relaxation times.

§ 6. Summary and conclusions

A new concept for active plasmonics that exploits nanoscale structural transformations in the SPP waveguide materials has been described. It

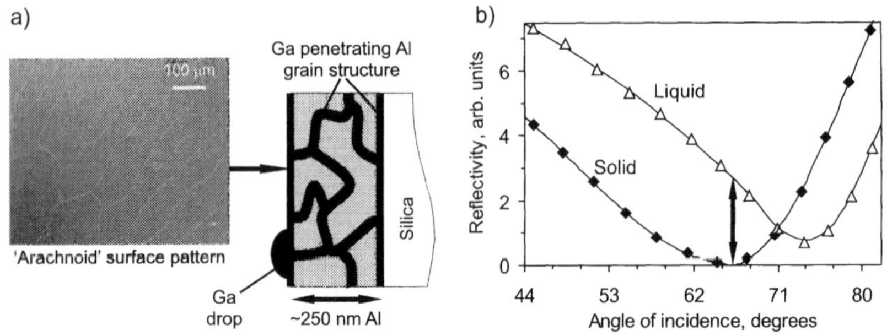

Fig. 17. Gallium/aluminium composite – a prospective material for active plasmonic applications. (a) Gallium/aluminium composite structure formed by penetration of gallium into the grain structure of an aluminium film. The scanning electron microscope image (backscattered electron detection mode) shows the arachnoid line pattern formed on the surface of the aluminium around the point at which the gallium drop is applied. (b) Angular dependence of glass-MgF_2-Gallium/Aluminium reflectivity for p-polarized light at 633 nm with the gallium component of the metal composite in the solid and liquid states. The arrow shows the significant switching contrast that can be achieved via an excitation-induced structural transition.

has been shown numerically that SPP signals, efficiently coupled/decoupled to/from a metal/dielectric waveguide using gratings, can be effectively switched by changing the structural phase of a gallium waveguide section a few microns long. Moreover, it has been demonstrated that complete switching of waveguide transmission can be achieved by inducing a structural transformation of a gallium layer just a few tens in nanometres thick adjacent to the dielectric interface.

Experimentally, it has been shown that such transformation depths can be achieved by optical pulses with fluences of just 12 mJ/cm^2. Response and relaxation times may be as short as ~4 ps and ~20 ns, respectively. Furthermore, it has been demonstrated that a gallium/aluminium composite material can provide the same switching functionality, with the added benefit that it reliably forms good quality interfaces with silica via a relatively simple production technique.

The proposed active plasmonic concept may be applied in a range of configurations as illustrated in fig. 1. Control functionality, for example, may be achieved using an electron beam instead of a light beam (Pochon et al., 2004). Alternatively, the external excitation could be replaced by an electric current in the waveguide, which would induce a structural transformation in the gallium via Joule heating, to produce an 'electro-plasmonic' modulation device analogous to current electro-optic modulators.

It has recently been found that there is a photoconductive effect associated with the light-induced phase transition in gallium: the induced change in the optical properties of the metal is accompanied by a change in its electrical conductivity, and this facilitates a new mechanism for the detection of light (Fedotov et al., 2002). A similar principle could be used to detect SPP waves: if energy dissipated by an SSP wave in a gallium film changes the structural phase, and therefore the conductivity, of even a small part of the metal, it may be detected as a change in the resistance of the gallium film.

If indeed the energy dissipated by a strong SPP wave can induce a phase transition in gallium, then 'all-plasmonic' devices should be possible. In this case two SPP waves would interact at the gallium section of a waveguide – a 'control' SPP wave would switch the phase of the gallium to modulate a transmitted 'signal' SPP wave. Various self-action effects, such as self-focusing/de-focusing, self-phase-modulation and an SPP analogue of optical bleaching (whereby propagation becomes less lossy as a thicker metallized layer is formed), may also occur as single intense SPP waves propagate across a gallium section.

Acknowledgements

The authors would like to acknowledge the financial support of the Engineering and Physical Sciences Research Council (UK) and the 'Phoremost' EU Network of Excellence.

References

Agranovich, V.M., 1975, Usp. Fiz. Nauk. **115**, 199–237.

Agranovich, V.M., Kravtsov, V.E., Leskova, T.A., 1981a, Sov. Phys. JETP **54**, 968.

Agranovich, V.M., Kravtsov, V.E., Leskova, T.A., 1981b, Solid State Commun. **40**, 687.

Agranovich, V.M., Kravtsov, V.E., Leskova, T.A., 1983, Sov. Phys. JETP **57**, 60.

Agranovich, V.M., Mills, D.L. (Eds.), 1982, Surface Polaritons: Electromagnetic Waves at Surfaces and Interfaces, North-Holland, Oxford.

Albanis, V., Dhanjal, S., Emel'yanov, V.I., Fedotov, V.A., MacDonald, K.F., Petropoulos, P., Richardson, D.J., Zheludev, N.I., 2001, Phys. Rev. B **63**, 165207.

Albanis, V., Dhanjal, S., Zheludev, N.I., Petropoulos, P., Richardson, D.J., 1999, Opt. Exp. **5**, 157.

Barnes, W.L., Dereux, A., Ebbesen, T.W., 2003, Nature (London) **424**, 824.

Bennett, P.J., Dhanjal, S., Petropoulos, P., Richardson, D.J., Zheludev, N.I., Emel'yanov, V.I., 1998, Appl. Phys. Lett. **73**, 1787.

Bergman, D.J., Stockman, M.I., 2003, Phys. Rev. Lett. **90**, 027402.

Berini, P., 1999, Opt. Lett. **24**, 1011.

Boardman, A.D. (Ed.), 1982, Electromagnetic Surface Modes, Wiley, Chichester

Bosio, L., 1978, J. Chem. Phys. **68**, 1221.

Bozhevolnyi, S.I., Erland, J., Leosson, K., Skovgaars, P.M.W., Hvam, M., 2001, Phys. Rev. Lett. **86**, 3008–3011.

Bozhevolnyi, S.I., Volkov, V.S., Devaux, E., Laluet, J.-Y., Ebbesen, T.W., 2006, Nature (London) **440**, 508.
Brongersma, M.L., Hartman, J.W., Atwater, H.A., 2000, Phys. Rev. B **62**, R16356.
Burke, J.J., Stegeman, G.I., Tamir, T., 1986, Phys. Rev. B **33**, 5186.
Charbonneau, R., Berini, P., Berolo, E., Lisicka-Shrzek, E., 2000, Opt. Lett. **25**, 844.
Defrain, A., 1977, J. Chim. Phys. **74**, 851.
Ditlbacher, H., Krenn, J.R., Schider, G., Leitner, A., Aussenegg, F.R., 2002, Appl. Phys. Lett. **81**, 1762.
Fedotov, V.A., MacDonald, K.F., Stevens, G.C., Zheludev, N.I., 2003, Nonlin. Opt., Quant. Opt. **30**, 53.
Fedotov, V.A., Woodford, M., Jean, I., Zheludev, N.I., 2002, Appl. Phys. Lett. **80**, 1297.
Gamaly, E.G., Rode, A.V., Luther-Davies, B., 1999, J. Appl. Phys. **85**, 4213.
Gong, X.G., Chiarotti, G.L., Parrinello, M., Tosatti, E., 1991, Phys. Rev. B **43**, 14277.
Heitmann, D., 1977a, Opt. Commun. **20**, 292.
Heitmann, D., 1977b, J. Phys. C Solid State **10**, 397.
Heitmann, D., Raether, H., 1976, Surf. Sci. **59**, 17.
Kofman, R., Cheyssac, P., Richard, J., 1977, Phys. Rev. B **16**, 5216.
Krasavin, A.V., 2006, Photonics of metallic nanostructures: Active plasmonics and chiral effects, Thesis. University of Southampton.
Krasavin, A.V., MacDonald, K.F., Schwanecke, A.S., Zheludev, N.I., 2006, Appl. Phys. Lett. **98**, 031118.
Krasavin, A.V., MacDonald, K.F., Zheludev, N.I., Zayats, A.V., 2004, Appl. Phys. Lett. **85**, 3369.
Krasavin, A.V., Zayats, A.V., Zheludev, N.I., 2005. J. Opt. A: Pure Appl. Opt. **7**, S85.
Krasavin, A.V., Zheludev, N.I., 2004, Appl. Phys. Lett. **84**, 1416.
Krenn, J.R., Ditlbacher, H., Schider, G., Hohenau, A., Leitner, A., Aussenegg, F.R., 2003, J. Micrbiol. **209**, 167–172.
Krenn, J.R., Salerno, M., Felidj, N., Lamprecht, B., Schider, G., Leitner, A., Aussenegg, F.R., Weeber, J.-C., Dereux, A., Goudonnet, J.P., 2001, J. Micrbiol. **202**, 122.
Krenn, J.R., Weeber, J.-C., 2004, Phil. Trans. R. Soc. Lond. A **362**, 739.
Leskova, T.A., 1984, Solid State Commun **50**, 869.
Leskova, T.A., Gapotchenko, N.I., 1985, Solid State Commun **53**, 351.
MacDonald, K.F., Fedotov, V.A., Eason, R.W., Zheludev, N.I., Rode, A.V., Luther-Davies, B., Emel'yanov, V.I., 2001, J. Opt. Soc. Am. B **18**, 331.
MacDonald, K.F., Fedotov, V.A., Pochon, S., Stevens, G.C., Kusmartsev, F.V., Emel'yanov, V.I., Zheludev, N.I., 2004, Europhys. Lett. **67**, 614.
MacDonald, K.F., Fedotov, V.A., Zheludev, N.I., 2003, Appl. Phys. Lett. **82**, 1087.
MacDonald, K.F., Fedotov, V.A., Zheludev, N.I., Zhdanov, B.V., Knize, R.J., 2001, Appl. Phys. Lett **79**, 2375.
MacDonald, K.F., Soares, B.F., Bashevoyo, M.V., Zheludev, N.I., 2006, JIEEE J. Sel. Top. Quant **12**, 371.
Maier, S.A., Atwater, H.A., 2005, J. Appl. Phys. **98**, 011101.
Maier, S.A., Brongersma, M.L., Kik, P.G., Meltzer, S., Requicha, A.A.G., Atwater, H.A., 2001, Adv. Mater. **13**, 1501.
Maier, S.A., Kik, P.G., Atwater, H.A., 2002, Appl. Phys. Lett. **81**, 1714.
Maier, S.A., Kik, P.G., Atwater, H.A., 2003, Phys. Rev. B **67**, 205402.
Maradudin, A.A., Wallis, R.F., Stegeman, G.I., 1983, Solid State Commun. **46**, 481.
Moreland, J., Adams, A., Hansma, P.K., 1982, Phys. Rev. B **25**, 2297.
Nezhad, M., Tetz, K., Fainman, Y., 2004, Opt. Exp. **12**, 4072.
Nikolajsen, T., Leosson, K., Salakhutdinov, I., Bozhevolnyi, S.I., 2003, Appl. Phys. Lett. **82**, 668.
Onuki, T., Watanabe, Y., Nishio, K., Tsuchiya, T., Tani, T., Tokizaki, T., 2003, J. Micrbiol. **210**, 284.

Palik, E.D. (Ed.), 1984, Handbook of Optical Constants of Solids, Academic Press, Orlando.
Pereiro-Lopez, E., Ludwig, W., Bellet, D., 2004, Acta Mater. **52**, 321.
Peteves, S.D., Abbaschian, R., 1991, Metall. Trans. **22A**, 1259.
Petropoulos, P., Kim, H.S., Richardson, D.J., Fedotov, V.A., Zheludev, N.I., 2001, Phys. Rev. B **64**, 193312.
Petropoulos, P., Offerhaus, H.L., Richardson, D.J., Dhanjal, S., Zheludev, N.I., 1999, Appl. Phys. Lett. **74**, 3619.
Pochon, S., MacDonald, K.F., Knize, R.J., Zheludev, N.I., 2004, Phys. Rev. Lett. **92**, 145702.
Pockrand, I., Raether, H., 1976, Opt. Commun. **18**, 395.
Quinten, M., Leitner, A., Krenn, J.R., Aussenegg, F.R., 1998, Opt. Lett. **23**, 1331.
Raether, H., 1988, Surface Plasmons on Smooth and Rough Surfaces and on Gratings, Springer-Verlag, Berlin.
Rode, A.V., Luther-Davies, B., Gamaly, E.G., 1999, J. Appl. Phys. **85**, 4222.
Rode, A.V., Samoc, M., Luther-Davies, B., Gamaly, E.G., MacDonald, K.F., Zheludev, N.I., 2001, Opt. Lett. **26**, 441.
Sánchez-Gil, J.A., 1998, Appl. Phys. Lett. **73**, 3509.
Sarid, D., 1981, Phys. Rev. Lett. **47**, 1927.
Schlesinger, Z., Sievers, A.J., 1980, Appl. Phys. Lett. **36**, 409.
Sirtori, C., Gmachl, C., Capasso, F., Faist, J., Sivco, D.L., Hutchinson, A.L., Cho, A.Y., 1998, Opt. Lett. **23**, 1366.
Smolyaninov, I.I., Mazzoni, D.L., Mait, J., Davis, C.C., 1997, Phys. Rev. B **56**, 1601.
Soares, B.F., MacDonald, K.F., Fedotov, V.A., Zheludev, N.I., 2005, Nano Lett. **5**, 2104.
Stegeman, G.I., Burke, J.J., Hall, D.G., 1983, Opt. Lett. **8**, 383.
Stegeman, G.I., Maradudin, A.A., Rahman, T.S., 1981, Phys. Rev. B **23**, 2576–2585.
Stegeman, G.I., Maradudin, A.A., Shen, T.P., Wallis, R.F., 1984, Phys. Rev. B **29**, 6530–6539.
Takahara, J., Kobayashi, T., 2004, Opt. Phot. News **15**, 54.
Takahara, J., Yamagishi, S., Taki, H., Morimoto, A., Kobayashi, T., 1997, Opt. Lett. **22**, 475.
Tanaka, K., Tanaka, M., Sugiyama, T., 2005, Opt. Exp. **13**, 256.
Tanaka, R., Choi, P.-K., Koizumi, H., Hyodo, S., 2001, Mater. Trans. **42**, 138.
Teshev, R.S., Shebzukhov, A.A., 1988, Opt. Spectrosc. (USSR) **65**, 693.
Viitanen, A.J., Tretyakov, S.A., 2005, J. Opt. A: Pure Appl. Opt. **7**, S133.
Volkov, V.S., Bozhevolnyi, S.I., Leosson, K., Boltasseva, A., 2003, J. Micrbiol. **210**, 324–329.
Wang, B., Wang, G.P., 2004, Opt. Lett. **29**, 1992.
Weber, W.H., Ford, G.W., 2004, Phys. Rev. B **70**, 125429.
Wendler, L., Haupt, R., 1986, J. Appl. Phys. **59**, 3289.
Worthing, P.T., Barnes, W.L., 2002, J. Mod. Opt. **49**, 1453.
Yatsui, T., Kourogi, M., Ohtsu, M., 2001, Appl. Phys. Lett. **79**, 4583.
Zayats, A.V., Smolyaninov, I.I., 2003, J. Opt. A: Pure Appl. Opt. **5**, S16.
Zayats, A.V., Smolyaninov, I.I., Maradudin, A.A., 2005, Phys. Rep. **408**, 131.
Zheludev, N.I., 2002, Contemp. Phys. **43**, 365.
Zheludev, N.I., 2006, J. Opt. A: Pure Appl. Opt. **8**, S1.
Zuger, O., Durig, U., 1992, Phys. Rev. B **46**, 7319.

Chapter 5

Surface plasmons and gain media

by

M.A. Noginov, G. Zhu

Center for Materials Research, Norfolk State University, Norfolk, VA, 23504

V.P. Drachev, V.M. Shalaev

School of Electrical & Computer Engineering and Birck Nanotechnology Center, Purdue University, West Lafayette, IN 47907

Nanophotonics with Surface Plasmons
Advances in Nano-Optics and Nano-Photonics
ISSN: 1871-0018

V.M. Shalaev & S. Kawata (Editors)

DOI: 10.1016/S1871-0018(06)02005-X

Contents

§ 1. Introduction

The technique of coloring stain glasses by gold and silver nanoparticles was known to Romans. The British Museum has a famous Lycurgus Cup (4th Century A.D.), which changes its color depending on the illumination (fig. 1). When viewed in reflected light, for example, in daylight, it appears green. However, when a light is shone into the cup and transmitted through the glass, it appears red. Nowadays it is known that the coloration of the Cup is determined by the frequency of localized surface plasmon (SP) resonance in metallic nanoparticles embedded into the glass.

Localized SP is the oscillation of free electrons in a metallic particle (driven by an external electromagnetic wave), whose resonance frequency is the plasma frequency adjusted by the size and, mainly, the shape of the particle. A phenomenon relevant to localized SPs is a surface plasmon polariton (SPP) or a surface electromagnetic wave propagating along the interface between two media possessing permittivities with opposite signs, such as metal–dielectric interface. The plasmon's electromagnetic field is concentrated in the close vicinity to the surface of the particle or metal–dielectric boundary.

Fig. 1. Lycurgus Cup (4th Century A.D.) at different illuminations.

Localized plasmons are found on rough surfaces (Ritchie, 1973; Fleischmann et al., 1974; Moskovits, 1985), in engineered nanostructures (Quinten et al., 1998; Averitt et al., 1999; Brongersma et al., 2000; Mock et al., 2002; Pham et al., 2002), as well as in clusters and aggregates of nanoparticles (Kreibig and Vollmer, 1995; Quinten, 1999; Su et al., 2003). In the spots where local fields are concentrated, both linear and nonlinear optical responses of molecules and atoms are largely enhanced. This leads to a number of important applications, the most matured of which is the surface enhanced Raman scattering (SERS) (Fleischmann et al., 1974).

SPs result in enhanced local fields and, thus, can dramatically enhance the Raman signal, which depends linearly on the local field intensity. Among other advantages, SERS makes possible rapid molecular assays for detection of biological and chemical substances (Kneipp et al., 2002). A very high sensitivity of SERS enables observation of Raman scattering from a single molecule attached to a metal colloidal particle (Kneipp et al., 1997; Nie and Emory, 1997).

Raman scattering sensing techniques can not only detect the presence of a biomolecular analyte, but also provide a great deal of information on exactly what specific molecules are being detected. SERS enables molecular "fingerprinting," which is of particular interest for molecule sensing and bio-applications. A detailed analysis of this very powerful sensing technique is outside the scope of this chapter. We mention here only some recently developed, particularly efficient and sensitive SERS substrates. Those include nanoshells developed by the Halas group (Prodan et al., 2003), substrates fabricated with nanosphere lithography developed in the Van Duyne group (Hayes and Van Duyne, 2003), and adaptive silver films developed by Drachev et al. (2004).

Fractal aggregates of metallic nanoparticles supporting localized SPs can lead to extremely large enhancements of local field amplitudes exceeding those of single metallic particles (Markel et al., 1996; Shalaev et al., 1996; Shalaev, 2000). A number of interesting optical phenomena (such as highly efficient harmonic generation, SERS, Kerr effect, etc.) caused by dramatic field enhancement in hot spots of fractal aggregates of silver nanoparticles have been theoretically predicted and experimentally demonstrated by Shalaev and co-workers (Shalaev, 1996; Shalaev, 2000; Shalaev, 2002). We note that a fractal aggregate can be roughly thought of as a collection of spheroids, with different aspect ratio, formed by various chains of nanoparticles in the aggregate (Shalaev, 2000).

Brus and Nitzan (1983) proposed to use gigantic localized fields to influence photochemistry of reactions. Later this phenomenon was studied in application to photocells, detectors, and other processes including vision (Hutson, 2005). The enhancement of the response of a p-n junction by localized SPs has been studied recently (Schaadt et al., 2005). A strong field enhancement in the vicinity of metallic tip enables linear and nonlinear near-field scanning microscopy, spectroscopy, and photomodification with nanoscale resolution (Stockman, 1989; Ferrell, 1994; S'anchez et al., 1999). An extraordinary high transmission of light through periodic arrays of subwavelength holes in metallic films has been explained in terms of resonant excitation of SPs by Ghaemi et al. (1998). An enhancement of surface magneto-optical interaction by SPs has been discussed by Bonod et al. (2004).

Negative-index materials (NIMs) have a negative refractive index, so that the phase velocity is directed against the flow of energy. There are no known naturally occurring NIMs in the optical range. However, artificially designed materials (metamaterials) can act as NIMs. Metamaterials can open new avenues to achieving unprecedented physical properties and functionality unattainable with naturally existing materials, as was first described by Veselago in his seminal paper (Veselago, 1968). Optical NIMs (ONIMs) promise to create entirely new prospects for controlling and manipulating light, optical sensing, and nanoscale imaging and photolithography.

Proof-of-principle experiments (Smith et al., 2000; Shelby et al., 2001) have shown that metamaterials can act as NIMs at *microwave* wavelengths. NIMs drew a large amount of attention after Pendry predicted that a NIM can act as a superlens, allowing an imaging resolution which is limited not by the wavelength but rather by material quality (Pendry, 2000). The near-field version of the superlens has recently been reported by the Zhang and Blaikie groups (Fang et al., 2005; Melville, and Blaikie, 2005).

While negative permittivity $\varepsilon' < 0$ $(\varepsilon = \varepsilon' + i\varepsilon'')$ in the optical range is easy to attain for metals, there is no magnetic response for naturally occurring materials at such high frequencies. Recent experiments showed that a magnetic response and negative permeability $\mu' < 0$ $(\mu = \mu' + i\mu'')$ can be accomplished at terahertz frequencies (Linden et al., 2004; Yen et al., 2004; Zhang et al., 2005). These studies showed the feasibility of ONIMs because a magnetic response is a precursor for negative refraction. The metamaterial with a negative refractive index at 1.5 μm, based on paired metal nanorods embedded in a dielectric, was designed (Podolskiy et al., 2002, 2003, 2005) and experimentally demonstrated by the Shalaev group (Shalaev et al., 2005; Drachev et al., 2006; Kildishev et al., 2006).

Metallic surfaces and nanoparticles can influence not only nonlinear but also linear optical responses. Thus, the dependence of emission spectra and emission lifetimes of luminescent centers on (submicron) distance from a metallic mirror has been discussed in detail in review by Drexhage (1974) and references therein. An increase of luminescence intensity of dye molecules *adsorbed* onto islands and films of metal nanoparticles, when the plasma resonances are coupled to the absorption spectra of the molecules, has been observed in the early 80s by Glass et al. (1980, 1981) and Ritchie and Burstein (1981). An enhancement and modification of the emission of dye molecules and trivalent rare-earth ions adsorbed onto rough metallic surfaces, metallic islands, engineered structures, etc. have been studied more recently (Weitz et al., 1983; Denisenko et al., 1996; Kikteva et al., 1999; Selvan et al., 1999; Lakowicz et al., 2001). An increase of optical absorption of CdS quantum dots by gold nanospheres has been demonstrated by Biteen et al. (2004).

If a mixture of dye solution and metallic nanoparticles is used as a laser medium, then the SP-induced enhancement of absorption and emission of dye can significantly improve the laser performance. Thus, according to the studies by Kim et al. (1999) and Drachev et al. (2002), in a mixture of rhodamine 6G (R6G) dye with aggregated silver nanoparticles placed in an *optical micro-cavity* (glass capillary tube), local electromagnetic fields are enhanced by many orders of magnitude, and the laser action can be obtained at the pumping power, which is not enough to reach the lasing threshold in a pure dye solution of the same concentration. The experiments in the studies by Kim et al. (1999) and Drachev et al. (2002) were not highly reproducible and the mechanisms of the dramatic reduction of the lasing threshold were not clearly understood. This motivated us to carry out a more systematic study of an analogous system, investigating separately the effects of spontaneous emission, the stimulated emission in a pump-probe setup, and the laser emission in an easy-to-tune two-mirror laser cavity (see Sections 4.2, 4.5, and 4.6).

Most of the existing and potential future applications of nanoplasmonics suffer from damping caused by metal absorption and radiation losses. Over the years, several proposals have been made on how to conquer the plasmon loss. Thus, in 1989 Sudarkin and Demkovich suggested to increase the propagation length of SPP by creating the population inversion in the dielectric medium adjacent to the metallic film (Sudarkin and Demkovich, 1989). The proposed experimental test was based on the observation of increased reflectivity of a metallic film in the frustrated total internal reflection setup. Sudarkin and Demkovich (1989) also briefly discussed the possibility of creating a SP-based laser. This work was preceded by the observation of super-luminescence and light generation by a dye solution under the condition of internal reflection (Kogan et al., 1972) and the study of gain-enhanced total internal reflection in the presence of metallic film through the mediation of SPPs on the metal surface (Plotz et al., 1979).

The ideas above have been further developed in more recent years. Thus, gain-assisted propagation of SPPs in planar metal waveguides of different geometries has been studied by Nezhad et al. (2004). SPPs at the interface between metal and a dielectric with strong optical gain have been analyzed theoretically by Avrutsky (2004). In particular, it has been shown (Avrutsky, 2004) that the proper choice of optical indices of the metal and dielectric can result in an infinitely large effective refractive index of surface waves. Such resonant plasmons have extremely low group velocity and are localized in a very close vicinity to the interface. The amplification of SPPs at the interface between silver film and dielectric medium with optical gain (laser dye) has been recently demonstrated by Seidel et al. (2005). The observation was done in an experimental setup similar to that proposed by Sudarkin and Demkovich (1989) and the experimentally observed change in the metal reflection was as small as 0.001%.

In a similar way, Lawandy (2004) has predicted the *localized* SP resonance in metallic nanospheres to exhibit a singularity when the surrounding dielectric medium has a critical value of optical gain. This singularity, resulting from canceling *both real* and *imaginary* terms in the denominator of the field enhancement factor in metal nanospheres $\propto (\varepsilon_d - \varepsilon_m)/(2\,\varepsilon_d + \varepsilon_m)$ can be evidenced by an increase of the Rayleigh scattering within the plasmon band and lead to low-threshold random laser action, light localization effects, and enhancement of SERS (Lawandy, 2004) (here ε_d and ε_m are complex dielectric constants of dielectric and metal, respectively).

This study was continued in the work by Lawandy (2005), where a three-component system consisting of (i) metallic nanoparticle, (ii) shell of adsorbed molecules with optical gain, and (iii) surrounding dielectric (solvent) has been considered. In particular, it has been shown that depending on the thickness of the layer of an amplifying shell, the absorption of the complex can be increased or decreased with the increase of the gain in the dye shell (Lawandy, 2005).

A seemingly similar phenomenon has been described in an earlier publication (Bergman and Stockman, 2003) using a completely different set of arguments. Thus, Bergman and Stockman (2003) have proposed a new way to excite localized fields in nanosystems using SP amplification by stimulated emission of radiation (SPASER). SPASER radiation consists of SPs (bosons), which undergo stimulated emission, but, in contrast to photons, can be localized on the nanoscale. SPASER consists of an active medium with population inversion that transfers its excitation energy by radiationless transitions to a resonant nanosystem, which plays a role analogous to the laser cavity (Bergman and Stockman, 2003; Stockman, 2005). Alternatively, the major features of SPASER can be probably described in terms of Förster dipole–dipole energy transfer (Förster, 1948; Dexter et al., 1969) from an excited molecule (ion, quantum dot, etc.) to a resonant SP oscillation in a metallic nanostructure.

In this chapter, we discuss the demonstrated (i) enhancement of localized SP oscillation in the aggregate of Ag nanoparticles by optical gain in the surrounding dye and (ii) enhancement of spontaneous and stimulated emission of rhodamine 6G (R6G) laser dye by Ag aggregate. The results presented in this chapter have been published in Refs. (Noginov et al., 2005a; Noginov et al. 2006a, b, c).

§ 2. Estimation of the critical gain

Let us estimate a critical gain needed to compensate metal loss of localized SPs. The polarizability (per unit volume) for isolated metallic nanoparticles is given by $\beta = (4\pi)^{-1}[\varepsilon_m - \varepsilon_d] / [\varepsilon_d + p(\varepsilon_m - \varepsilon_d)])$, where p is the depolarization factor (Shalaev, 2000). If the dielectric is an active medium with $\varepsilon''_d = -p\varepsilon''_m/(1-p)$, then at the resonance wavelength λ_0 both the real and imaginary parts in the denominator become zero, leading to extremely large local fields limited only by saturation effects (Drachev et al., 2004; Lawandy, 2004). Thus, for the gain coefficient needed to compensate loss of *localized* SP, we find $\gamma = (2\pi/\lambda_0)\varepsilon''_d/\sqrt{\varepsilon'_d} = (2\pi/\lambda_0)[p/(1-p)]\varepsilon''_m/n = (2\pi/n\lambda_0)(\Gamma/\omega_p)[p/(1-p)][\varepsilon_b + n^2(1-p)/p]^{3/2} \sim 10^3 \text{cm}^{-1}$ at $\lambda_0 = 0.56\,\mu\text{m}$ (we used the Drude formula $\varepsilon_m = \varepsilon_b - \omega_p^2[\omega(\omega + i\Gamma)]$, $n = 1.33$, $\varepsilon_d = 1.77$, and known optical constants from Johnson and Christy (1972)). The value of the critical gain above is close to that estimated in Ref. (Lawandy, 2005).

The gain $\gamma \sim 10^3$ cm^{-1} needed to compensate loss of SPP or localized SP is within the limits of semiconducting polymers (Hide et al., 1997) or laser dyes (highly concentrated, $\sim 10^{-2}$ M (Lawandy, 2005), or adsorbed onto metallic nanoparticles). One can estimate that a single excited molecule of R6G characterized by the cross section $\sim 4 \times 10^{-16}$ cm^2 adsorbed onto metallic nanoparticle with the diameter d = 10 nm causes the gain coefficient (per the volume occupied by the nanoparticle) of the order of 10^3 cm^{-1}. If the number of adsorbed R6G molecules per nanoparticle exceeds one, the effective gain can be even higher.

§ 3. Experimental samples and setups

Experimentally, we studied mixtures of R6G dye (Rhodamine 590 Chloride from Exciton) and aggregated Ag nanoparticles. The starting solutions of R6G in methanol had concentrations of dye molecules in the range 1×10^{-6} to 2.1×10^{-4} M. Poly(vinylpyrrolidone) (PVP)-passivated silver aggregate was prepared according to the procedure described by Noginov et al. (2005b). The estimated concentration of Ag nanoparticles in the aggregate was 8.8×10^{13} cm^{-3}. In many measurements, Ag was diluted severalfold with methanol before mixing it with the dye.

Absorption spectra of dye solutions, Ag aggregate solutions, and dye–Ag aggregate mixtures were recorded with the UV-VIZ-IR Lambda 900 spectrophotometer (Perkin Elmer). In the emission measurements, the mixtures of R6G dye and Ag aggregate were excited with the frequency-doubled radiation of Q-switched Nd:YAG laser (Quanta Ray, $\lambda = 532$ nm, $t_{pulse} = 10$ ns, repetition rate 10 Hz). The same pumping was used in the pump-probe gain measurements, the laser experiment, and the pump-probe Rayleigh scattering experiment. The emission spectra were recorded using an MS257 monochromator (Oriel), a photomultiplier tube, and a boxcar integrator. A cw 594 nm He-Ne laser beam was used as a probe in the gain measurements.

In the luminescence kinetics studies, the emission of pure R6G dye solutions and dye–Ag aggregate mixtures was excited with an optical parametric amplifier Topaz (Quantronics/Light Conversion $\lambda = 530$ nm, $t_{pulse} = 2.5$ ps) pumped with the Spitfire laser system (Spectra Physics). The signal was detected and recorded using 10 GHz GaAs PIN photodetector model ET-4000 (EOTech, rise time <35 ps) and 2.5 GHz oscilloscope TDS7254 (Tektronix).

§ 4. Experimental results and discussion

4.1. Absorption spectra

The absorption spectrum of Ag aggregate has one structureless band covering the whole visible range and extending to near-infrared (fig. 2, trace 8). The major feature in the absorption spectrum of R6G is the band peaking at ≈ 528 nm (fig. 2a, trace 1 and inset of fig. 2a). The absorption spectra of the mixtures were recorded when Ag aggregate solution was added to the dye solution by small amounts (fig. 2a, traces 2–7). In different particular experiments, the "step" size varied between 1% and

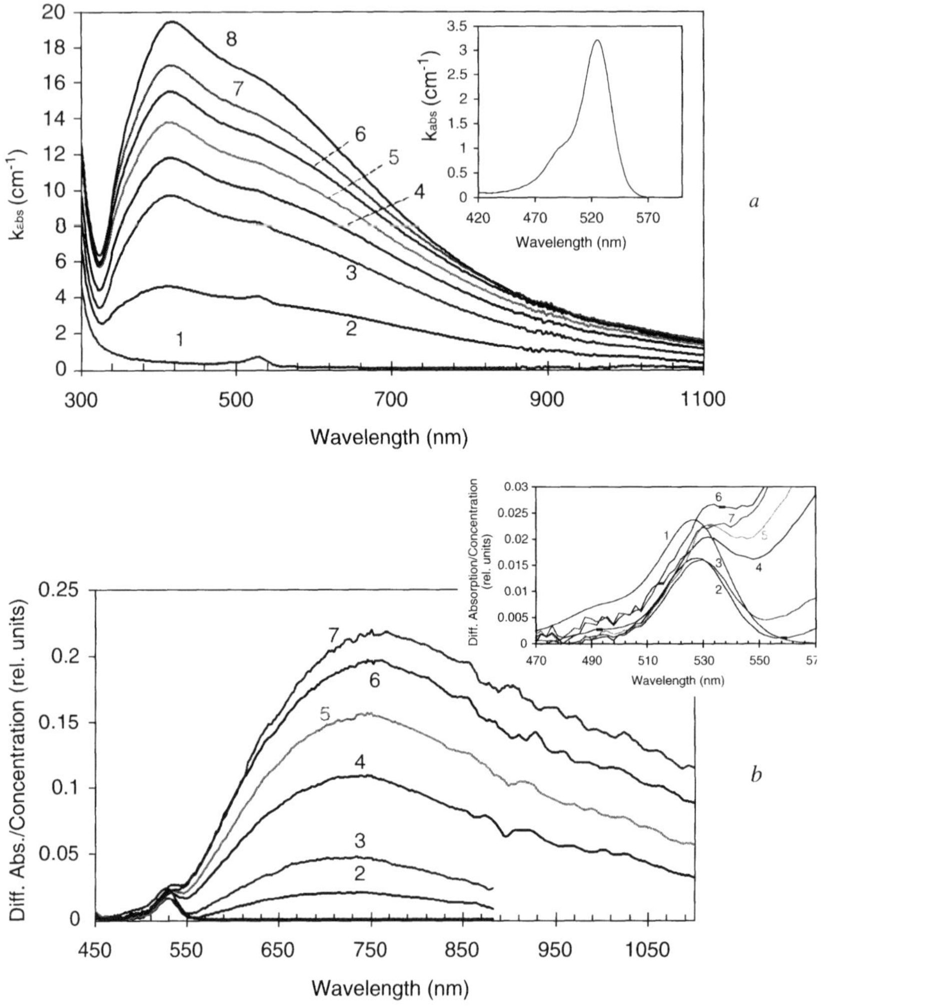

Fig. 2. (a) Trace 1 – Absorption spectrum of pure R6G dye solution (2.1×10^{-6} M); trace 8 – absorption spectrum of pure Ag aggregate solution (8.8×10^{13} cm^{-3}); traces 2–7 – absorption spectra of R6G–Ag aggregate mixtures. The ratio of Ag aggregate solution to R6G solution in the mixture is equal to 27.1/72.9 (2), 57.3/42.7 (3), 66.5/33.5 (4), 74.8/25.2 (5), 80.8/19.2 (6), and 86.8/13.2 (7). Inset: Absorption spectrum of pure R6G dye solution (1.25×10^{-5} M). (b) Difference absorption spectra (absorption spectrum of the mixture minus scaled spectrum of the aggregate, normalized to the concentration of R6G dye in the mixture). The traces in figure b correspond to the spectra with similar numbers in figure a. Inset: Enlarged fragment of the main frame. Trace 1 corresponds to pure R6G dye.

25% of the maximal Ag concentration. We then scaled the absorption spectrum of pure Ag aggregate to fit each of the spectra of the mixtures at $\leqslant$450 nm and calculated the differential spectra (*mixture–aggregate*). The differential spectra obtained this way, normalized to the concentration of the R6G dye in the mixture, reveal the regular absorption band of R6G dye, at ~0.53 μm, and a much broader new absorption band centered at 0.72–0.75 μm (fig. 2b). The latter broad band can be due to hybrid states formed by R6G molecules chemisorbed (via Cl^-) onto silver nanoparticles (Fang, 1999) or due to restructuring of the Ag aggregate in the presence of dye molecules.

With the increase of Ag aggregate concentration in the mixture, the intensity of the absorption band of R6G (calculated using the procedure described above) decreased, with the rate exceeding the reduction of the R6G concentration (fig. 2b). This reduction will be discussed in detail in Section 4.4. (The observed reduction of the R6G absorption is due to Ag aggregate but not an aggregate solvent without Ag (Noginov et al., 2005c)). The slight spectral shift of the R6G absorption band (from 528 to 531 nm) with the increase of the Ag aggregate concentration can be partially explained by its "mechanical" overlap with the strong absorption band peaking at 0.72–0.75 μm.

4.2. *Spontaneous emission*

Spontaneous emission spectra of the dye–Ag aggregate mixtures were studied in the setup schematically shown in the low left corner inset of fig. 3, when the dye or a mixture of dye with Ag aggregate was placed in a 1 mm thick cuvette. The samples were pumped and the luminescence was collected nearly normally to the surface of the cuvette. We found that while the shape of the emission band (upper right corner inset of fig. 3) was practically unaffected by the presence of Ag aggregate in the mixture, its intensity changed significantly. At the starting concentrations of R6G and Ag aggregate, equal to R6G $= 1.25 \times 10^{-5}$ M, Ag $= 8.7 \times 10^{12}$ cm^{-3}, the emission intensity of the dye (measured in the maximum at ~558 nm) increased up to 45% with the addition of small amounts of aggregated silver nanoparticles (fig. 3). At the further increase of the concentration of Ag aggregate in the mixture, the emission intensity decreased. This reduction was, in part, due to the absorption by Ag aggregate of both pumping and emission. Another possible reason for the reduction of the emission intensity was quenching of dye luminescence by silver nanoparticles. However, the relative decrease of the emission intensity was

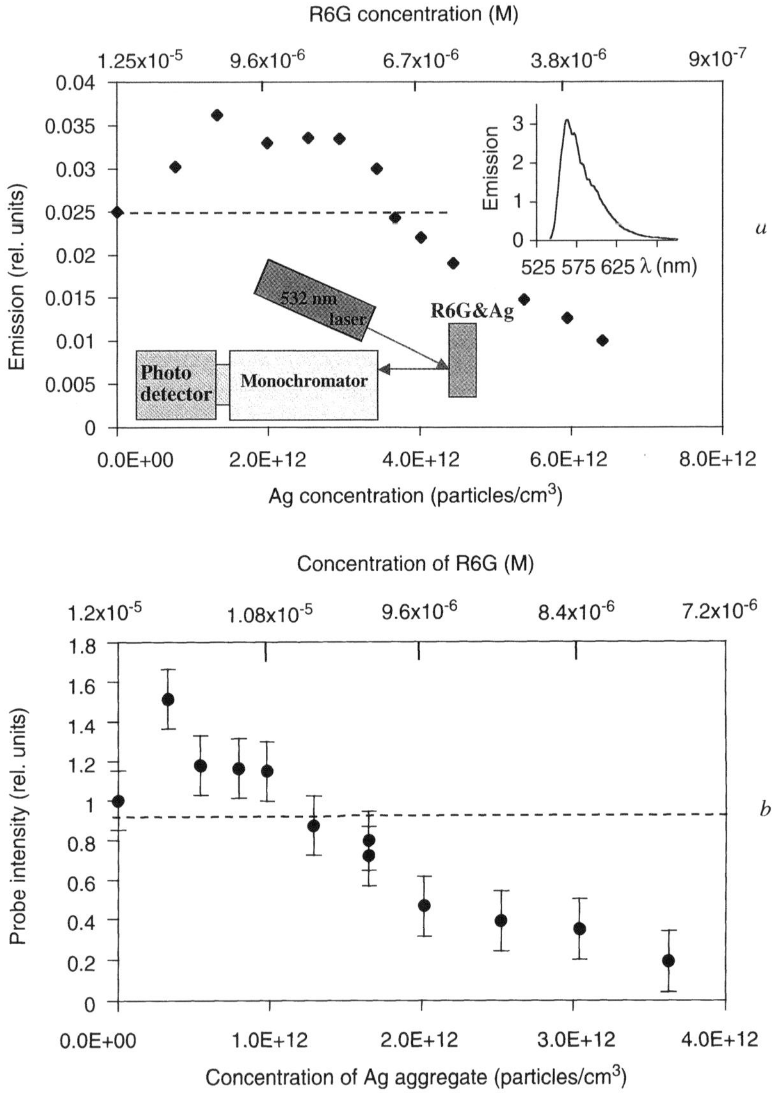

Fig. 3. (a) Emission intensity recorded at the addition of Ag aggregate to R6G dye. The starting concentrations of R6G and Ag aggregate are, respectively, 1.25×10^{-5} M and 8.7×10^{12} cm^{-3} The emission intensity corresponds to "as is" detected signal that has not been a subject to any normalization. Inset in the low left corner: Schematic of the experimental setup used at the emission intensity measurement. Inset in the upper right corner: Emission spectrum of R6G dye. (b) Probe light intensity (at $\lambda = 594$ nm) measured in a pump-probe gain experiment in 10 mm cuvette at the 532 nm pumping energy equal to 0.38 mJ (after 0.5 mm pinhole) as a function of the R6G and Ag aggregate concentrations. All detected signals are normalized to that in pure dye solution.

much smaller than the relative reduction of the R6G absorption determined according to the procedure described above.

At low concentrations, i.e., $\leqslant 2.1 \times 10^{-5}$ M, of R6G used in the majority of our experiments, the spontaneous emission kinetics recorded in pure dye solutions was nearly single exponential. The measured time constant was ~25% larger than the 3.6–3.8 ns lifetime of R6G known from the literature (Selanger et al., 1977; Müller et al., 1996; Zander et al., 1996). The elongation of the decay kinetics was probably due to the reabsorption, which could not be completely neglected, since at certain emission wavelengths (~550 nm) the absorption length of dye was comparable to the linear size of the cuvette (~1 cm). When dye ($<2 \times 10^{-5}$ M) was mixed with Ag aggregate ($<1.3 \times 10^{13}$ cm^{-3}), the effective emission decay time shortened to ~88% of its maximal value in pure dye (fig. 4). (We assume that the reabsorption was weak enough to affect the qualitative character of the dependence in fig. 4 significantly.) One can hypothesize that dye molecules, which luminescence decay-times are shortened by the presence of Ag aggregate, are situated close to metallic nanoparticles and their absorption and emission properties are affected by SP-enhanced fields.

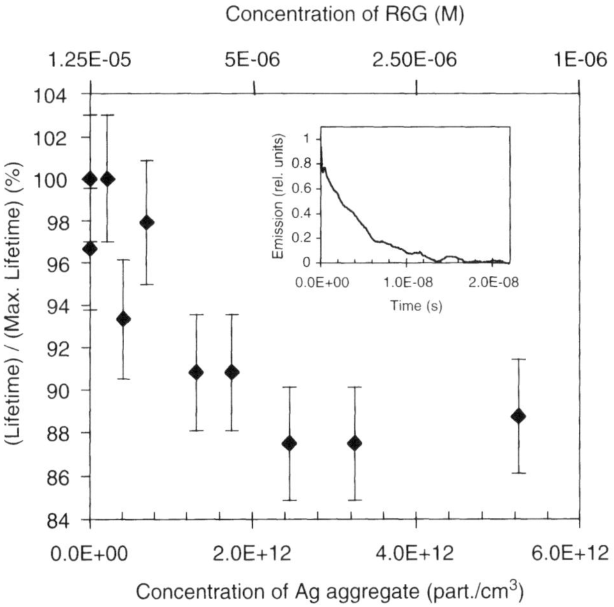

Fig. 4. Dependence of the R6G emission lifetime on concentration of Ag aggregate in dye–Ag aggregate mixtures characterized by low concentrations of R6G dye. All values are normalized to the lifetime measured in the pure R6G solution. Inset: Typical emission kinetics.

4.3. Enhanced Rayleigh scattering due to compensation of loss in metal by gain in dielectric

Following the prediction by Lawandy (2004), we sought for the enhancement of Rayleigh scattering by silver nanoparticles embedded in the dye with optical gain. In the pump-probe experiment, a fraction of the pumping beam was split off and used to pump a laser consisting of the cuvette with R6G dye placed between two mirrors (inset of fig. 5a). The emission line of the R6G laser (~558 nm) corresponded to the maximum of the gain spectrum of R6G dye in the mixtures studied. The beam of the R6G laser, which was used as a probe in the Rayleigh scattering, was aligned with the pumping beam in the beamsplitter and sent to the sample through a small (0.5 mm) pinhole (inset of fig. 5a). The pump and probe beams were collinear, and their diameters at the pinhole were larger than 0.5 mm.

The scattered light was collected by an optical fiber that was placed within several millimeters from the cuvette at the angles ranging from ~45° to ~135° relative to the direction of the beam propagation. (We did not notice that the results of the SP enhancement measurement depended on the detection angle.) The fiber collected scattered *probe* light as well as scattered *pumping* light and spontaneous emission of dye. To separate the scattered probe light, we used a monochromator and ran the emission spectrum from 540 to 650 nm.The scattered probe light was seen in the spectrum as a relatively narrow (~5 nm) line on the top of a much broader spontaneous emission band.

Experimentally, we kept the energy of the probe light constant and measured the intensity of the scattered probe light as the function of the varied pumping light energy. The sixfold increase of the Rayleigh scattering observed in the dye–Ag aggregate mixture with the increase of the pumping energy (fig. 5a, squares) is the clear experimental demonstration of the compensation of loss in metal and the enhancement of the quality factor of SP resonance by optical gain in surrounding dielectric.

The dye–Ag aggregate mixtures were placed in 1 mm thick cuvettes. At the dye concentration, 2.1×10^{-5} M, which we used in the majority of our experiments, the maximal optical amplification (of the pure dye solution or dye–Ag aggregate mixtures) at 1 mm length did not exceed ~7% (see Section 4.5). The lateral dimension of the pumped volume was smaller than 1 mm.The dye–Ag aggregate mixtures were visually clear, with the transport mean free path of the order of centimeters. Correspondingly, the probability of elongation of photon path in the *pumped* volume due to

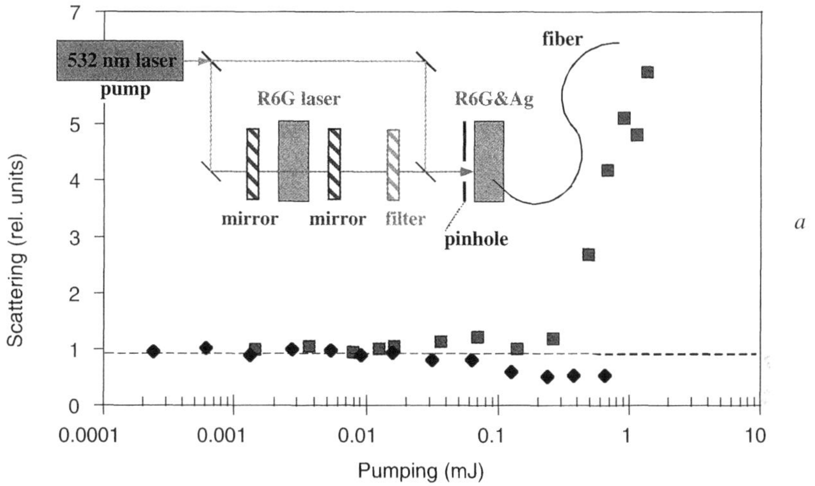

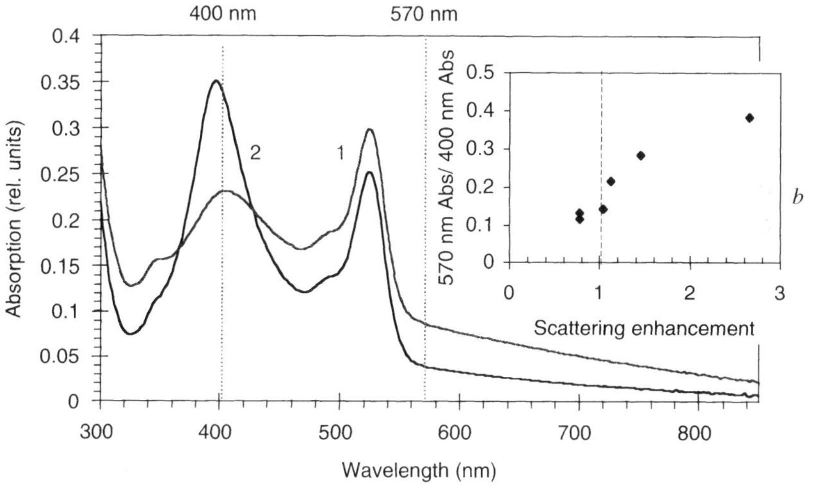

Fig. 5. (a) Intensity of the Rayleigh scattering as the function of the pumping energy. (b) Absorption spectra of the dye–Ag aggregate mixtures; R6G – 2.1×10^{-5} M, Ag aggregate – 8.7×10^{13} cm^{-3}. Squares in (a) and trace 1 in (b) correspond to one mixture, and diamonds in (a) and trace 2 in (b) correspond to another mixture. Inset of (a): Pump-probe experimental setup for the Rayleigh scattering measurements. Inset of (b): The ratio of the absorption coefficients of the dye–Ag aggregate mixtures at 570 and 400 nm plotted vs. the enhancement of the Rayleigh scattering measured at 0.46 mJ.

scattering was insignificantly small. Thus, we conclude that an increase of the intensity of scattered light in our experiment was due to an enhancement of the Rayleigh scattering cross section of metallic particles rather than a simple amplification of scattered light in a medium with gain.

(Note that low average gain in the volume of the cuvette is not inconsistent with the existence of high local gain in vicinity of silver nanoparticles caused by adsorbed molecules). Experimentally, no noticeable enhancement of scattering was observed in the pure R6G dye solution or pure Ag aggregate suspension.

Depending on the shape of the absorption spectrum of the Ag aggregate–dye mixture (fig. 5b), the intensity of the Rayleigh scattering could increase or decrease with the increase of pumping (fig. 5a). Inset of fig. 5b shows a monotonic dependence of the scattering enhancement measured at 0.46 mJ pumping energy vs. the ratio of the absorption coefficients of the mixture at 570 and 400 nm. One can see that the relatively strong absorption of the mixture at 570 nm, which is a signature of *aggregated* Ag nanoparticles, helps to observe an enhanced Rayleigh scattering. Although we do not precisely understand the relationship between the absorption spectra and the physical properties of the mixtures, which govern the results of the scattering experiments, the correlation exists with no doubts.

A plausible explanation for different scattering properties of different mixtures could be in line with the theoretical model developed by Lawandy (2005), in which a three-component system consisting of metallic nanoparticle, shell of adsorbed molecules with optical gain, and surrounding dielectric (solvent) has been studied. In particular, it has been shown that depending on the thickness of the layer of an amplifying shell, the absorption of the complex can be increased or decreased with the increase of the gain in dye (Lawandy, 2005). Similarly, we can speculate that in our experiments, in different mixtures we had different numbers of adsorbed molecules per metallic nanoparticle, which determined the enhancement or the reduction of the Rayleigh scattering.

4.4. *Discussion of the results of the absorption and emission measurements*

4.4.1. *Suppression of the SP resonance by absorption in surrounding dielectric media*

Using the same line of arguments as was used by Lawandy (2004), one can infer that by embedding metallic nanoparticle in a dielectric medium with *absorption*, one can further increase the imaginary part of the denominator in the field enhancement factor $\propto [\varepsilon_m - \varepsilon_b]/[\varepsilon_d + p(\varepsilon_m - \varepsilon_d)]$ and, correspondingly, reduce the quality factor of the plasmon resonance and the peak absorption cross section. This effect should always accompany the effect of the SP enhancement by gain and be observed in the same dye–Ag aggregate mixtures. The absorption coefficient of the dielectric medium, which is comparable to the critical value of gain $\sim 10^3$ cm^{-1}, should be adequate for the observation of a significant damping of a SP resonance.

4.4.2. *Emission intensity and absorption*

What is the reason for the increase of the spontaneous emission intensity with the addition of Ag aggregate to the R6G dye solution? The quantum yield of spontaneous emission of low-concentrated R6G dye is ~95% (Kubin and Fletcher, 1982; Magde et al., 2002); thus, the experimentally observed emission enhancement cannot be due to the increase of the quantum yield. An enhancement of emission can be explained by increased absorption of R6G in the presence of Ag aggregate. However, at the first glance, this explanation contradicts with the experimental observation: seeming reduction of the R6G absorption with the increase of Ag aggregate concentration (inset of fig. 2b).

We argue that the commonly accepted procedure of the decomposition of an absorption spectrum into its components, which we used to treat data of fig. 2, is not applicable to the mixture of two substances (dye and aggregated Ag nanoparticles), which affect each other, and that the paradox above has a clear physical explanation.

The absorption spectrum of a fractal aggregate is comprised of a continuum of homogeneous bands corresponding to metallic nanostructures with different effective form factors. The homogeneous widths of individual bands are comparable to the characteristic widths of the absorption and emission bands of R6G dye. Following our prediction of the suppression of the SP resonance by the absorption in a dielectric, we speculate that the absorption band of R6G "burns" a hole in the absorption spectrum of the aggregate in the frequency range corresponding to the absorption of dye. Thus, the conventional method that we used to extract the absorption band of R6G from the absorption spectrum of the mixture was not applicable to our system. Instead, the absorption spectrum of the mixture should be decomposed according to the method schematically shown in fig. 6. This explanation suggests that we have experimentally observed the predicted suppression of the SP resonance by absorption in the surrounding dielectric medium. The experimental observation of this effect provides additional support to our claim above of the enhancement of the SP by gain.

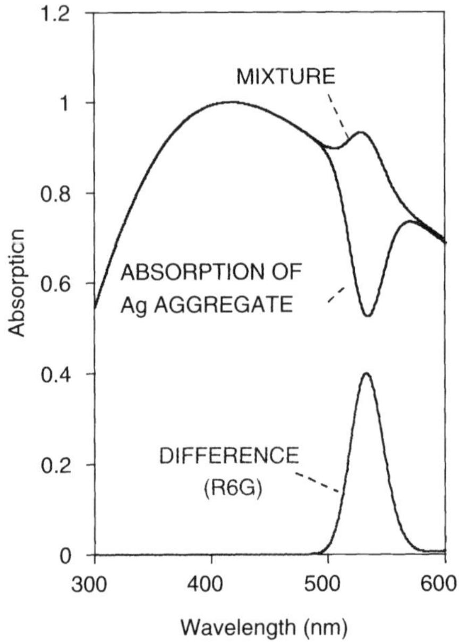

Fig. 6. Schematic of the proposed decomposition of the absorption spectrum of the mixture into the absorption spectra of the components.

4.5. Stimulated emission studied in a pump-probe experiment

The enhancement of stimulated emission of R6G dye by Ag aggregate was studied in a pump-probe experiment, which scheme is shown in inset of fig. 7a. The mixture of R6G dye and silver nanoparticles (placed in 10 mm cuvette) was pumped with ~10 ns pulses of a frequency-doubled Nd-YAG laser (532 nm) and probed with a cw 594 nm He-Ne laser. The two beams were collinear. They were centered at 0.5 mm pinhole that was attached to the front wall of the cuvette and restricted the diameters of the incoming beams. The amplification of the probe light (during the short pumping pulse) was measured using 1 GHz oscilloscope TDX 784D (Tektronix). To minimize the amount of spontaneous emission light reaching the detector, we used a monochromator, which wavelength was set at 594 nm.To subtract the residual luminescence signal from the amplified probe light, we repeated the same measurements two times, with and without the probe light.

The result of the measurements for a pure R6G dye solution (1.25×10^{-5} M) is depicted in fig. 7 (diamonds). The dependence of the amplification of the probe light (defined as the ratio of the output and input

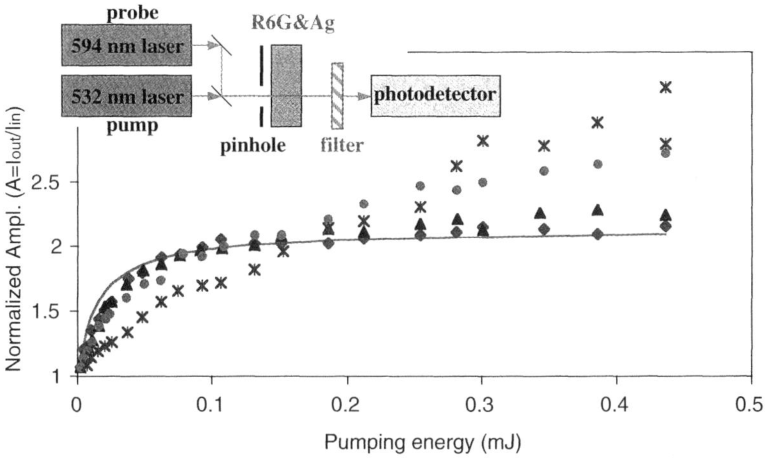

Fig. 7. Amplification $A = I_{out}/I_{in}$ (at $\lambda = 594$ nm) as a function of the 532 nm pumping energy (after 0.5 mm pinhole) measured in a series of R6G dye–Ag aggregate solutions in 10 mm cuvette. Solid line – calculation corresponding to pure dye, 1.25×10^{-5} M. Characters – experiment. Diamonds – pure dye, 1.25×10^{-5} M; triangles – dye, 1.25×10^{-5} M, Ag aggregate, 3.6×10^{11} particles/cm^3; circles – dye, 9.0×10^{-6} M, Ag aggregate, 2.5×10^{12} particles/cm^3; crosses – dye, 5.1×10^{-6} M, Ag aggregate, 5.2×10^{12} particles/cm^3. All amplification signals are normalized to the transmission in corresponding not pumped media. Inset: schematic of the experiment.

probe light intensities $I_{\text{out}}/I_{\text{in}}$, on the pumping power density $P(t)/S$) was described in terms of equations

$$0 \approx \frac{\mathrm{d}n^*}{\mathrm{d}t} = \frac{P(t)}{Sh\nu_p}\sigma^{\text{dye}}_{\text{abs }532}(N - n^*) - \frac{n^*}{\tau} \quad \text{and} \tag{4.1}$$

$$A \equiv \frac{I_{\text{out}}}{I_{\text{in}}} = \exp(n^*\sigma^{\text{dye}}_{\text{em }594}l) \tag{4.2}$$

where n^* is the population of the metastable excited state of R6G; N is the concentration of R6G molecules in the solution; $h\nu_p$ is the energy of the pumping photon; τ is the spontaneous emission lifetime; $\sigma^{\text{dye}}_{\text{em }594}$ is the emission cross section of R6G at the probe wavelength (Svelto, 1998); $\sigma^{\text{dye}}_{\text{abs }532}$ is the absorption cross section of R6G at the pumping wavelength; and l is the length of the cuvette. In the model, we assumed a four-level scheme of R6G (Svelto, 1998), with the pumping transition terminating at a short-lived state above the metastable state. Thus, we neglected any stimulated emission at the *pumping* wavelength. Since the duration of the pumping pulse was twice longer than the luminescence decay time, we assumed a quasi-cw regime of pumping ($\mathrm{d}n^*/\mathrm{d}t \approx 0$).

A good qualitative and quantitative agreement between the calculation (solid line in fig. 7) and the experiment (diamonds) confirms the adequacy of the model. The saturation in the system occurs when all dye molecules are excited and stronger pumping cannot cause a larger gain.

When Ag aggregate was added to the mixture, it caused absorption at the wavelength of the probe light. Correspondingly, in order to obtain the true value of A, all measured amplification signals in the dye–Ag aggregate mixtures were normalized to the transmission T of the probe light through the not pumped sample. The transmission of the sample was checked before and after each amplification measurement. In addition, the solution was thoroughly stirred before measuring each new data point. This helped us to minimize the possible effect of the photomodification of the mixture on the measurement results. The analysis of the probe signal kinetics revealed the existence of at least two different photoinduced lenses, both having characteristic times longer that the duration of the pumping pulse, $\sim$10 ns. Those photoinduced lenses did not interfere with the observation significantly, and the study of their nature was outside the scope of this work.

The results of the amplification measurements in the mixed samples are summarized in fig. 7. The most remarkable feature of this experiment is that in the presence of Ag aggregate, the value of the amplification is not limited by the saturation level characteristic of pure dye but grows to larger magnitudes. Since at strong pumping ($\geqslant$2 mJ) all dye molecules are already excited, the observed enhancement of the amplification cannot be due to the increased absorption efficiency. We speculate that it is rather due to the enhancement of the *stimulated emission* efficiency caused by the field enhancement in the vicinity of aggregated Ag nanoparticles. Note that the enhancement of *stimulated emission* and the enhancement of *absorption* are caused by the same physical mechanisms and are expected to accompany each other.

When we added Ag aggregate to the mixture, we diluted R6G dye. The dilution of dye, on its own, should lead to the reduction of the gain saturation level. Partial absorption of pumping by Ag aggregate should also lead to the reduction of the amplification. These two factors make the result presented in fig. 7 (enhancement of amplification) even more remarkable.

An increase of the *spontaneous* emission intensity with the addition of Ag aggregate to the mixture is shown in fig. 3a. In approximately the same range of the R6G and Ag aggregate concentrations and at very small pumping energies ($\leqslant$0.4 mJ), we have observed an *absolute* enhancement of the probe light amplification, without any renormalization to the sample transmission (fig. 3b). The observed increase of the amplified signal could be due to the combination of enhancements of stimulated emission and absorption.

It appears likely that the interplay between the enhancements of absorption and emission, which increase an amplified signal, and the dilution of dye as well as an absorption dye to Ag aggregate, which reduce amplification, may cause a complex behavior of the system, which is critically dependent on the Ag concentration and pumping intensity.

4.6. Effect of Ag aggregate on the operation of R6G dye laser

In this particular experiment, we studied the effect of Ag aggregate on the performance of R6G dye laser. We expected to reduce the lasing threshold by adding Ag aggregate to the dye solution, as it was demonstrated by (Kim et al., 1999; Drachev et al., 2002), where a glass capillary resonator was used.

The easy-to-tune laser setup consisted of the rear dichroic mirror, through which the gain element (10 mm cuvette with dye or dye–Ag aggregate mixture) was pumped, and the output mirror (fig. 8a). All mirrors were flat. The back mirror had the reflectivity coefficient $R = 99.96\%$ (at the stimulated emission wavelength $\lambda \sim 558$ nm). Several output mirrors, which we used, had reflectivities R equal, respectively, to 99.7%, 87.8%, 46.0%, and 17.5%. The distance between the mirrors was 6.8 cm, and the cuvette with the gain medium was positioned approximately in the center of the cavity. The lens with the focal length equal to 18 cm focused pumping light (Q-switched 532 nm laser radiation) into approximately 0.5 mm spot in the cuvette. The color filter placed after the output mirror helped to separate the laser emission from the residual pumping leaking through the laser cavity.

With the addition of the Ag aggregate to the R6G dye solution, instead of anticipated enhancement of the laser output, we observed an increase of the lasing threshold and a reduction of the slope efficiency. The representative series of input–output curves is shown in fig. 8b. Apparently, at the sets of the parameters, which were used in our experiments, the reduction of the stimulated emission output caused by the dilution of dye and the addition of a gray absorbing substance (Ag aggregate) to the laser cavity overcame the increase of the stimulated emission intensity caused by the SP field enhancement in metallic nanostructures.

In order to evaluate the positive impact of the SP field enhancement on laser operation, we compared the experimental laser thresholds measured

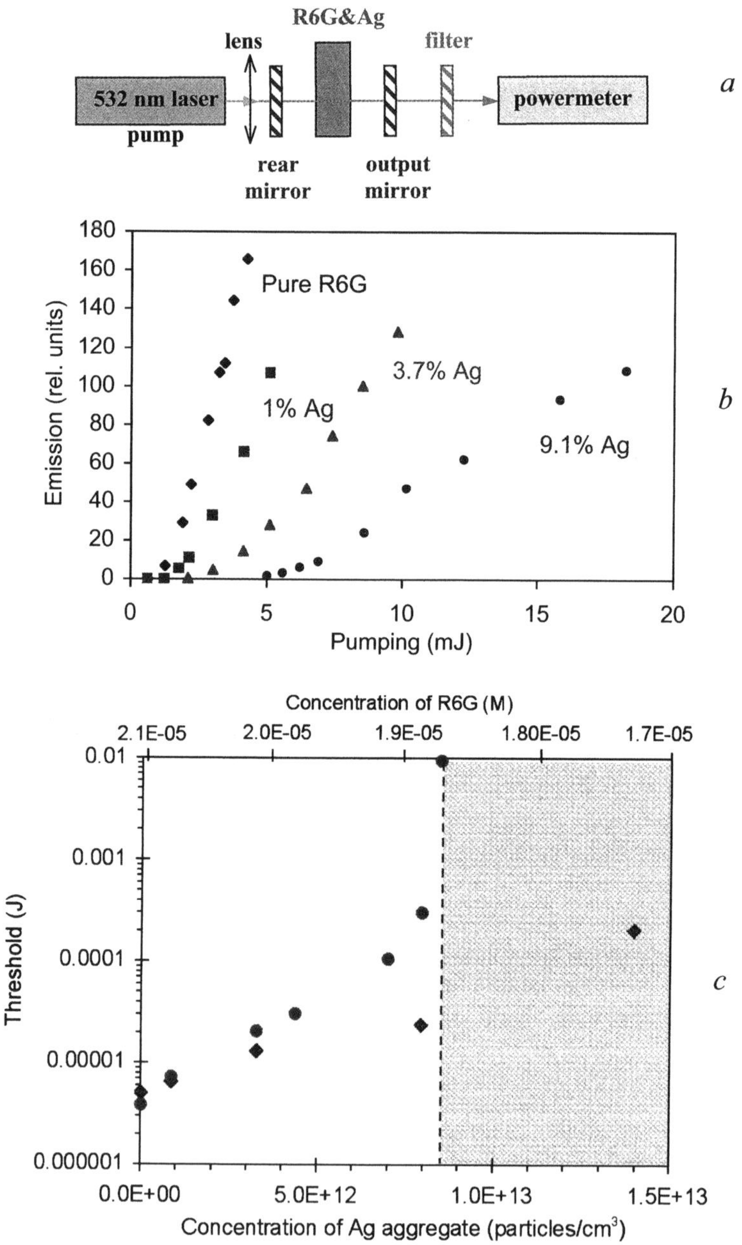

Fig. 8. (a) Laser setup; (b) the input–output laser curves (at $\lambda \sim 560$ nm) recorded in a series of samples prepared by mixing small amounts of Ag aggregate (8.8×10^{12} particles/cm^3) with R6G dye (1.1×10^{-5} M). The volume percentage of the Ag aggregate is indicated next to each curve. (c) Experimental (diamonds) and calculated (circles) dependences of the laser threshold on Ag aggregate concentration. $R_{\rm out} = 99.7\%$, $L = 15.6\%$.

in a series of dye–aggregate mixtures with the thresholds calculated under the assumption that the only two effects of the aggregate were (i) the dilution of dye and (ii) an increase of parasitic absorption at the pumping and emission wavelengths. We first evaluated the threshold population inversion n_{th} [cm^{-3}], which is required to conquer loss at the lasing wavelength, and then calculated the value of the incident pumping energy E_{th} [J], which is needed to create this population inversion.

Thus, the value n_{th} was calculated from the formula equating the gain and the loss at the lasing threshold

$$\exp(2l\sigma^{dye}_{em\ 560}n_{th})\exp\{-2l[(k^{dye}_{abs\ 560} - \sigma^{dye}_{abs\ 560}n_{th}) + k^{Ag}_{abs\ 560}]\} \times R_1R_2(1-L) = 1, \tag{4.3}$$

and the value E_{th} was calculated from the formula for the threshold rate of the pumping absorption per unit volume F_{th} [$cm^{-3}s^{-1}$]

$$F_{th} = \frac{E_{th}}{t_p h\nu_p(Sl)}\{1 - \exp[-((k^{dye}_{abs\ 532} - n_{th}\sigma^{dye}_{abs\ 532}) + k^{Ag}_{abs\ 532})l]\} \left\{\frac{(k^{dye}_{abs\ 532} - n_{th}\sigma^{dye}_{abs\ 532})}{[(k^{dye}_{abs\ 532} - n_{th}\sigma^{dye}_{abs\ 532}) + k^{Ag}_{abs\ 532}]}\right\}, \tag{4.4}$$

which, in turn (in quasi-cw approximation), is related to n_{th} as

$$F_{th} = \frac{n_{th}}{\tau}. \tag{4.5}$$

Here l is the length of the laser element, $\sigma^{dye}_{abs(em)xxx}$ is the absorption (emission) cross section of dye at the wavelength λ (nm), $k^{dye(Ag)}_{abs\ xxx}$ is the absorption coefficient of dye (Ag aggregate) at the concentration corresponding to the particular mixture at λ (nm), R_1 and R_2 are the reflection coefficients of the laser mirrors, L is the parasitic intracavity loss per roundtrip (reflections of the cuvette walls, etc.), t_p is the duration of the pumping pulse, and S is the cross-sectional area of the pumping beam. In eq. (4.4), the first term in figure brackets determines the fraction of incident pumping energy, which is absorbed in the sample, and the second term in figure brackets determines the fraction of the total absorbed pumping energy, which is consumed by *dye* but not by Ag aggregate.

The experimental and the calculated dependences of the laser threshold on the Ag aggregate concentration are depicted in fig. 8c. A good agreement between the experimental and the calculated values of the threshold in the absence of Ag aggregate justifies the accuracy of our model. One can see in fig. 8c that if the enhancements of absorption and stimulated emission by Ag aggregate are neglected and the role of the aggregate is reduced to the dilution of dye and the increase of the parasitic absorption, the calculated threshold values in dye–Ag aggregate mixtures significantly exceed the experimental ones. On the right hand side of the dashed line in fig. 8c, no lasing is predicted. (In that range of Ag aggregate concentrations, the calculated threshold value of the population inversion n_{th} exceeds the concentration of dye molecules in the solution.) This confirms that Ag aggregate enhances the stimulated emission of R6G dye. However, the overall effect of the aggregate on the laser operation is negative.

How can one make the overall effect of Ag aggregate on the laser output to be positive? As it is shown in Ref. (Noginov et al. 2006c) the improvement of the laser performance at the addition of Ag aggregate to R6G dye should be expected at large values of the output coupling and the value of the threshold population inversion close to the total concentration of dye molecules in the solution. The detailed discussion of the optimal laser parameters is outside the scope of this Chapter.

§ 5. Summary

To summarize, we have observed the compensation of loss in metal by gain in interfacing dielectric in the mixture of aggregated silver nanoparticles and rhodamine 6G laser dye. The demonstrated sixfold enhancement of the Rayleigh scattering is the evidence of the quality factor increase of the SP resonance. This paves the road for numerous applications of nanoplasmonics, which currently suffer from strong damping caused by absorption loss in metal. We have also predicted and experimentally observed the counterpart of the phenomenon above, namely, the suppression of the SP in metallic nanostructure embedded in a dielectric medium with absorption.

We have demonstrated that by adding the solution of aggregated silver nanoparticles to the solution of rhodamine 6G laser dye, one can enhance the efficiency of the spontaneous and the stimulated emission. We attribute an increase of the *spontaneous emission* intensity of dye to the increase of the *absorption* efficiency caused by the field enhancements in metallic nanostructures associated with SPs. The enhancement of the *stimulated emission* of R6G dye, which has the same nature as the enhancement of absorption (SP-induced field enhancement), was observed in the pump-probe and laser experiments. In the dye–Ag aggregate mixtures studied, the positive effect of the stimulated emission enhancement could not overcome in the laser experiment the negative effects associated with the dilution of dye by Ag aggregate and parasitic absorption of Ag aggregate.

Acknowledgments

The work was supported by the following grants: NASA #NCC-3-1035, NSF #HRD-0317722 and DMR-0611430, NSF-NIRT #ECS-0210445, ARO #W911NF-04-1-0350, MURI-ARO 50342-PH-MUR and MURI-ARO 50372-CH-MUR.

References

Averitt, R.D., Westcott, S.L., Halas, N.J., 1999, Linear optical properties of gold nanoshells, J. Opt. Soc. Am. B, **16**, 1824–1832.

Avrutsky, I., 2004, Surface plasmons at nanoscale relief gratings between a metal and a dielectric medium with optical gain, Phys. Rev. B **70**, 155416/1–155416/6.

Bergman, D.J., Stockman, M.I., 2003, Surface plasmon amplification by stimulated emission of radiation: quantum generation of coherent surface plasmons in nanosystems, Phys. Rev. Lett. **90**, 027402/1–027402/4.

Biteen, J., Garcia-Munoz, I., Lewis, N., Atwater, H., 2004, California Institute of Technology, Pasadena, California, MRS Fall Meeting.

Bonod, N., Reinisch, R., Popov, E., Neviere, M., 2004, Optimization of surface-plasmon-enhanced magneto-optical effects, J. Opt. Soc. Am. B **21**, 791–797.

Brongersma, M.L., Hartman, J.W., Atwater, H.A., 2000, Electromagnetic energy transfer and switching in nanoparticle chain arrays below the diffraction limit, Phys. Rev. B **62**, R16356–R16359.

Brus, L.E., Nitzan, A., 1983, Chemical processing using electromagnetic field enhancement", U.S. Patent No.: 4,481,091 (21 October, 1983)., US patent no.: 4,481,091, Chemical processing using electromagnetic field enhancement.

Denisenko, G.A., Malashkevich, G.E., Tziganova, T.V., Galstyan, V.G., Voitovich, A.P., Perchukevich, P.P., Kalosha, I.I., Bazilev, A.G., Mchedlishvili, B.V., Strek, W., Oleinikov, V.A., 1996, Influence of silver surface structure on luminescence characteristics of europium doped polymer films, Acta Physica Polonica A **90**, 121–126.

Dexter, D.L., Förster, Th., Knoh, R.S., 1969, The radiationless transfer of energy of electronic excitation between impurity molecules in crystals, Phys. Stat. Solidi **34**, K159–K162.

Drachev, V.P., Buin, A.K., Nakotte, H., Shalaev, V.M., 2004, Size dependent $\chi(3)$ for conduction electrons in Ag nanoparticles, Nano Lett. **4**, 1535–1539.

Drachev, V.P., Cai, W., Chettiar, U., Yuan, H.-K., Sarychev, A.K., Kildishev, A.V., Shalaev, V.M., 2006, Experimental verification of optical negative-index materials, Laser Phys. Lett. **3**, 49–55/DOI 10.1002/lapl.200510062.

Drachev, V.P., Kim, W.-T., Safonov, V.P., Podolskiy, V.A., Zakovryashin, N.S., Khaliullin, E.N., Shalaev, V.M., Armstrong, R.L., 2002, Low-threshold lasing and broad-band multiphoton-excited light emission from Ag aggregate-adsorbate complexes in microcavity, J. Mod. Opt. **49**, 645–662.

Drachev, V.P., Thoreson, M.D., Khaliullin, E.N., Davisson, V.J., Shalaev, V.M., 2004, Surface-enhanced raman difference between human insulin and insulin lispro detected with adaptive nanostructures, J. Phys. Chem. B **108**, 18046–18052.

Drexhage, K.H., 1974, Interaction of dye with monomolecular dye layers. In: Wolf, E. (Ed.), Progress in Optics XII, New York, North-Holland, 164–232.

Fang, Y., 1999, Acid with visible and near-infrared excitations, J. Raman Spectrosc. **30**, 85–89.

Fang, N., Lee, H., Zhang, X., 2005, Sub-diffraction-limited optical imaging with a silver superlens, Science **308**, 534–537.

Ferrell, T.L., 1994, Thin-foil surface-plasmon modification in scanning-probe microscopy, Phys. Rev. B **50**, 14738–14741.

Fleischmann, M., Hendra, P.J., McQuillan, A.J., 1974, Raman spectra of pyridine adsorbed at a silver electrode, Chem. Phys. Lett. **26**, 163–166.

Förster, Th., 1948, Zwischenmolekulare energiewändler und fluoreszenz, Ann. Phys. **2**, 55–75.

Ghaemi, H.F., Thio, T., Grupp, D.E., Ebbesen, T.W., Lezec, H.J., 1998, Surface plasmons enhance optical transmission through subwavelength holes, Phys. Rev. B **58**, 6779.

Glass, A.M., Liao, P.F., Bergman, J.G., Olson, D.H., 1980, Interaction of metal particles with adsorbed dye molecules: absorption and luminescence, Opt. Lett. **5**, 368–370.

Glass, A.M., Wokaun, A., Heritage, J.P., Bergman, J.P., Liao, P.F., Olson, D.H., 1981, Enhanced two-photon fluorescence of molecules adsorbed on silver particle films, Phys. Rev. B **24**, 4906–4909.

Hayes, C.L., Van Duyne, R.P., 2003, Dichroic optical properties of extended nanostructures fabricated using angle-resolved nanosphere lithography, Nano Lett. **3**, 939–943.

Hide, F., Schwartz, B.J., Díaz-Garcia, M.A., Heeger, A.J., 1997, Conjugated polymers as solid-state laser materials, Synth. Metals **91**, 35–40.

Hutson, L., 2005, Golden eye, Mater, World **13**(6), 18–19.

Johnson, P.B., Christy, R.W., 1972, Optical constants of the noble metals, Phys. Rev. B **6**, 4370–4379.

Kikteva, T., Star, D., Zhao, Z., Baisley, T.L., Leach, G.W., 1999, Molecular orientation, aggregation, and order in Rhodamine films at the fused silica/air interface, J. Phys. Chem. B **103**, 1124–1133.

Kildishev, A.V., Cai, W., Chettiar, U., Yuan, H.-K., Sarychev, A.K., Drachev, V.P. Shalaev, V.M., 2006, Negative refractive index in optics of metal-dielectric composites, JOSA B, will be published in the March issue, 2006.

Kim, W., Safonov, V.P., Shalaev, V.M., Armstrong, R.L., 1999, Fractals in microcavities: giant coupled, multiplicative enhancement of optical resonances, Phys. Rev. Lett. **82**, 4811–4814.

Kneipp, K., Kneipp, H., Itzkan, I., Dasari, R.R., Feld, M.S., 2002, Topical review: surface-enhanced Raman scattering and biophysics, J. Phys. **14**, R597–R624.

Kneipp, K., Wang, Y., Kneipp, H., Perelman, L.T., Itzkan, I., Dasari, R.R., Feld, M.S., 1997, Single molecule detection using surface-enhanced Raman scattering (SERS), Phys. Rev. Lett. **78**, 1667–1670.

Kogan, B.Ya., Volkov, V.M., Lebedev, S.A., 1972, Superluminescence and generation of stimulated radiationunder internal-reflection conditions, JETP Lett. **16**, 100.

Kreibig, U., Vollmer, M., 1995, Optical Properties of Metal Clusters, vol. 25, Springer, New York.

Kubin, R.F., Fletcher, A.N., 1982, Fluorescence quantum yields of some rhodamine dyes, J. Luminescence **27**, 455–462.

Lakowicz, J.R., Gryczynski, I., Shen, Y., Malicka, J., Gryczynski, Z., 2001, Intensified fluorescence, Photon. Spectra, 96–104.

Lawandy, N.M., 2004, Localized surface plasmon singularities in amplifying media, Appl. Phys. Lett. **85**, 540–542.

Lawandy, N.M., 2005, Nano-particle plasmonics in active media, in: McCall, M.W., Dewar, G., Noginov, M.A. (Eds.), Proceedings of SPIE Complex Mediums VI: Light and Complexity, vol. 5924, SPIE, Bellingham, WA, pp. 59240G/1–59240G/13.

Linden, S., Enkrich, C., Wegener, M., Zhou, J., Koschny, T., Soukoulis, C.M., 2004, Magnetic response of metamaterials at 100 terahertz, Science **306**, 1351–1353.

Magde, D., Wong, R., Seybold, P.G., 2002, Fluorescence quantum yields and their relation to lifetimes of rhodamine 6G and fluorescein in nine solvents: improved absolute standards for quantum yields, Photochem. Photobiol. **75**, 327–334.

Markel, V.A., Shalaev, V.M., Stechel, E.B., Kim, W., Armstrong, R.L., 1996, Small-particle composites. I. linear optical properties, Phys. Rev. B **53**, 2425–2436.

Melville, D.O.S., Blaikie, R.J., 2005, Super-resolution imaging through a planar silver layer, Opt. Expr. **13**, 2127–2134.

Mock, J.J., Barbic, M., Smith, D.R., Schultz, D.A., Schultz, S., 2002, Shape effects in plasmon resonance of individual colloidal silver nanoparticles, Chem. Phys. **116**, 6755–6759.

Moskovits, M., 1985, Surface-enhanced spectroscopy, Rev. Mod. Phys. **57**, 783–826.

Müller, R., Zander, C., Sauer, M., Deimel, M., Ko, D.-S., Siebert, S., Arden-Jacob, J., Deltau, G., Marx, N.J., Drexhage, K.H., Wolfrum, J., 1996, Time-resolved identification of single molecules in solution with a pulsed semiconductor diode laser, Chem. Phys. Lett. **262**, 716–722.

Nezhad, M.P., Tetz, K., Fainman, Y., 2004, Gain assisted propagation of surface plasmon polaritons on planar metallic waveguides, Optics Express **12**, 4072–4079.

Nie, S., Emory, S.R., 1997, Probing single molecules and single nanoparticles by surface-enhanced Raman scattering, Science **275**, 1102 1104.

Noginov, M.A., Zhu, G., Shalaev, V.M., Drachev, V.P., Bahouza, M., Adegoke, J., Small, C., Ritzo, B. A., 2005a, "Enhancement and suppression of surface plasmon resonance in Ag aggregate by optical gain and absorption in surrounding dielectric medium", arxiv, physics/0601001 (2005).

Noginov, M.A., Zhu, G., Davison, C., Pradhan, A.K., Zhang, K., Bahoura, M., Codrington, M., Drachev, V.P., Shalaev, V.M., Zolin, V.F., 2005b, Effect of Ag aggregate on spectroscopic properties of Eu:Y_2O_3 nanoparticles, J. Mod. Opt. **52**, 2331–2341.

Noginov, M.A., Vondrova, M., Williams, S.N., Bahoura, M., Gavrilenko, V.I., Black, S.M., Drachev, V.P., Shalaev, V.M., Sykes, A., 2005c, Spectroscopic studies of liquid solutions of R6G laser dye and Ag nanoparticle aggregates, J. Opt. A Pure Appl. Opt. **7**, S219–S229.

Noginov, M.A., Zhu, G., Bahoura, M., Adegoke, J., Small, C., Ritzo, B.A., Drachev, V.P., Shalaev, V.M., 2006a, Applied Physics β; published online August 2006.

Noginov, M.A., Zhu, G., Bahouza, M., Adegoke, J., Small, C., Ritzo, B.A., Drachev, V.P., Shalaev, V.M., 2006b, "Enhancement of surface plasmons in Ag aggregate by optical gain in dielectric medium", Optics Letters, in print.

Noginov, M.A., Zhu, G., Bahouza, M., Adegoke, J., Small, C., Darvison, C, Drachev, V.P., Nyga, P., Shalaev, V.M., 2006c, "Enhancement of spontaneous and stimulated emission of rhodamine 6G dye by Ag aggregate", communicated to the journal.

Pendry, J.B., 2000, Negative refraction makes a perfect lens, Phys. Rev. Lett. **85**, 3966–3969.

Pham, T., Jackson, J.B., Halas, N.J., Lee, T.R., 2002, Preparation and characterization of gold nanoshells coated with self-assembled monolayers, Langmuir **18**, 4915–4920.

Plotz, G.A., Simon, H.J., Tucciarone, J.M., 1979, Enhanced total reflection with surface plasmons, JOSA **69**, 419–421.

Podolskiy, V.A., Sarychev, A.K., Narimanov, E.E., Shalaev, V.M., 2005, Resonant light interaction with plasmonic nanowire systems, J. Opt. A Pure Appl. Opt. **7**, S32–S37.

Podolskiy, V.A., Sarychev, A.K., Shalaev, V.M., 2002, Plasmon modes in metal nanowires and left-handed materials, J. Nonlinear Opt. Phys. Mat. **11**, 65–74.

Podolskiy, V.A., Sarychev, A.K., Shalaev, V.M., 2003, Plasmon modes and negative refraction in metal nanowire composites, Opt. Expr. **11**, 735–745.

Prodan, E., Radloff, C., Halas, N.J., Nordlander, P., 2003, A hybridization model for the plasmon response of complex nanostructures, Science **302**, 419–422.

Quinten, M., 1999, Optical effects associated with aggregates of clusters, J. Cluster Sci. **10**, 319–358.

Quinten, M., Leitner, A., Krenn, J.R., Aussenegg, F.R., 1998, Electromagnetic energy-transport via linear chains of silver nanoparticles, Opt. Lett. **23**, 1331–1333.

Ritchie, R.H., 1973, Surface plasmons in solids, Surf. Sci. **34**(1), 1–19.

Ritchie, G., Burstein, E., 1981, Luminescence of dye molecules adsorbed at a Ag surface, Phys. Rev. B **24**, 4843–4846.

S'anchez, E.J., Novotny, L., Xie, X.S., 1999, Near-field fluorescence microscopy based on two-photon excitation with metal tips, Phys. Rev. Lett. **82**, 4014–4017.

Schaadt, D.M., Feng, B., Yu, E.T., 2005, Enhanced semiconductor optical absorption via surface plasmon excitation in metal nanoparticles, Appl. Phys. Lett. **86**(063106), 3.

Seidel, J., Grafstroem, S., Eng, L., 2005, Stimulated emission of surface plasmons at the interface between a silver film and an optically pumped dye solution, Phys. Rev. Lett. **94**(177401), 1–4.

Selanger, K.A., Falnes, J., Sikkeland, T., 1977, Fluorescence life-time studies of Rhodamine 6G in Methanol, J. Phys. Chem. **81**, 1960–1963.

Selvan, S.T., Hayakawa, T., Nogami, M., 1999, Remarkable influence of silver islands on the enhancement of fluorescence from Eu^{3+} ion-doped silica gels, J. Phys. Chem. B **103**, 7064–7067.

Shalaev, V.M., 1996, Electromagnetic properties of small-particle composites, Phys. Reports **272**, 61–137.

Shalaev, V.M., 2000, Nonlinear Optics of Random Media: Fractal Composites and Metal-Dielectric Films, Springer Tracts in Modern Physics, vol. 158, Springer, Berlin, Heidelberg.

Shalaev, V.M. (Ed.), 2002. Optical Properties of Random Nanostructures, Topics in Applied Physics vol. 82, Springer Verlag, Berlin, Heidelberg.

Shalaev, V.M., Cai, W., Chettiar, U., Yuan, H.-K., Sarychev, A.K., Drachev, V.P., Kildishev, A.V., 2005, Negative index of refraction in optical metamaterials, Opt. Lett. **30**, 3356–3358; arXiv:physics/0504091, Apr. 13, 2005.

Shalaev, V.M., Poliakov, E.Y., Markel, V.A., 1996, Small-particle composites. II. Linear optical properties, Phys. Rev. B **53**, 2437–2449.

Shelby, R.A., Smith, D.R., Schultz, S., 2001, Experimental verification of a negative index of refraction, Science **292**, 77–79.

Smith, D.R., Padilla, W., Vier, J.D.C., Nemat-Nasser, S.C., Schultz, S., 2000, Composite medium with simultaneously negative permeability and permittivity, Phys. Rev. Lett. **84**, 4184–4187.

Stockman, M.I., 1989, Possibility of laser nanomodification of surfaces with the use of the scanning tunneling microscope, Optoelectron. Instrument Data Process **3**, 27–37.

Stockman, M.I., 2005, Ultrafast, nonlinear, and active nanoplasmonics, in: McCall, M.W., Dewar, G., Noginov, M.A. (Eds.), Proceedings of SPIE Complex mediums VI: Light and Complexity, vol. 5924, SPIE, Bellingham, WA, pp. 59240F/1–59240F/12.

Su, K.-H., Wei, Q.-H., Zhang, X., Mock, J.J., Smith, D.R., Schultz, S., 2003, Interparticle coupling effects on plasmon resonances of nanogold particles, Nano Lett. **3**, 1087–1090.

Sudarkin, A.N., Demkovich, P.A., 1989, Excitation of surface electromagnetic waves on the boundary of a metal with an amplifying medium, Sov. Phys. Tech. Phys. **34**, 764–766.

Svelto, O., 1998, Principles of Lasers, 4th ed., Plenum Press, New York.

Veselago, V.G., 1968, The electrodynamics of substances with simultaneously negative values of ε and μ, Soviet Physics Uspekhi **10**(4), 509–514.

Weitz, D.A., Garoff, S., Gersten, J.I., Nitzan, A., 1983, The enhancement of Raman scattering, resonance Raman scattering, and fluorescence from molecules absorbed on a rough silver surface, J. Chem. Phys. **78**, 5324–5338.

Yen, T.J., Padilla, W.J., Fang, N., Vier, D.C., Smith, D.R., Pendry, J.B., Basov, D.N., Zhang, X., 2004, Terahertz magnetic response from artificial materials, Science **303**, 1494–1496.

Zander, C., Sauer, M., Drexhage, K.H., Ko, D.-S., Schulz, A., Wolfrum, J., Brand, L., Eggelin, C., Seidel, C.A.M., 1996, Detection and characterization of single molecules in aqueous solution, Appl. Phys. B Lasers Opt. **63**, 517–523.

Zhang, S., Fan, W., Minhas, B.K., Frauenglass, A., Malloy, K.J., Brueck, S.R.J., 2005, Midinfrared resonant magnetic nanostructures exhibiting a negative permeability, Phys. Rev. Lett. **94**(037402), 1–4.

Chapter 6

Optical super-resolution for ultra-high density optical data storage

by

Junji Tominaga

National Institute of Advanced Industrial Science and Technology, Center for Applied Near-Field Optics, Tsukuba Central 4, 1-1-1 Higashi, Tsukuba 305-8562, Japan
e-mail: j-tominaga@aist.go.jp

Nanophotonics with Surface Plasmons — *V.M. Shalaev & S. Kawata (Editors)*
Advances in Nano-Optics and Nano-Photonics — *© 2007 Published by Elsevier B.V.*
ISSN: 1871-0018 — DOI: 10.1016/S1871-0018(06)02006-1

Contents

§ 1. Introduction

The resolution of almost all optical microscopes is indispensably governed by the diffraction limit, which is determined by the relationship $1.22\lambda/(2NA)$, where λ and NA are wavelength of light and numerical aperture of a lens, respectively. To improve the resolution to nanometer scale, near-field optics has recently attracted much attention. It is well known that for an optical near-field generated on any surface of objects, surface plasmons give rise to an interface between a metal surface and a dielectric material, and localized plasmons especially emerging on a metallic nanoparticle's surface are nonpropagating electromagnetic fields and are localized within micrometer range (Raether, 1988). In particular, the intensity normal to the surface decays exponentially and the coupling length with the other electromagnetic field is approximately limited to less than 100 nm. Several different scanning near-field optical microscopes (SNOMs) have been developed and utilized to explore the physical properties of the localized photons using a glass pipette with a small aperture, or a sharp edge covered with a metallic film (Kawata, 2001). Although SNOMs are now one of the most popular tools in optics to observe small objects, a part of the systems to control a space between a probe and an object has prevented the use of the localized fields from many industrial applications. For example, SNOMs will never be used to observe micro- or nanocracks generated on a fuselage or wings of an aircraft as a tool for routine inspection; it will be hard to be adapted as a biomedical routine tool because of its slow scanning speed and limited observation area. It is clear that firms and hospitals need a simpler and more convenient equipment.

An alternative way to increase the resolution of optical microscopes is by inserting a high-index liquid between the objective lens and tissues, as is often done in biomedical research (Guerra, 1988; Guerra et al., 1993). Thanks to the high index the diffraction limit can be modified to $\lambda/(2n_{\rm eff}NA)$, where $n_{\rm eff}$ is the refractive index of the liquid. Although refractive index of the liquids is usually smaller than 2.0, the resolution can be improved up to twofold. Recently, another alternative to focus light on a very small spot has been examined by an uncommon optical property of so-called meta-material with a negative refractive index (Pendry, 2000). Smolyaninov et al. (2005) recently succeeded in increasing n_{eff} by surface plasmons from a gold film. Zhang and coworkers, in contrast, transferred a nanometer-sized pattern (master) into a photopolymer layer through a thin silver layer (Fang et al., 2005). Apart from such topics of light focusing, optical super-resolution data storage disks have also been reported here, which are referred to as "super-resolution near-field structure (super-RENS)" since 1998 (Tominaga et al., 1998). The most advanced super-RENS disk is now able to resolve 50-nm pit patterns with signal-to-noise ratio (SNR) of more than 100 (about 40 dB) (Kim et al., 2004). Although the optical super-resolution effect of super-RENS disk has long been excluded from the plasmon family because Sb has small plasmon frequency in visible light, it has recently been found that composite diffracted evanescent waves (CDEWs), reported by Lezec and Thio, play a major role in the resolution when an optical small aperture is opened in the active layer (Lezec and Thio, 2004). In super-RENS, the thickness of an intermediate dielectric layer inserted between an Sb thin layer and a recording layer is very important in convoluting the far-field light and CDEWs generated around the aperture edge. Therefore, the thinner the layer thickness, the more convoluted are the two electromagnetic fields: a very high signal is retrieved from the disk. In contrast, the resolution limit is determined by the aperture size. However, it has gradually been found that up-to-date super-RENS disk (named "third generation" super-RENS disk) cannot be explained using such a simple super-resolution model with an optical aperture, but that more complicated processes are related to increase the signal intensity and the small resolution, which means in turn that once the detailed mechanism is revealed further improvement on the resolution and signal intensity is expected.

In this chapter we briefly review the optical super-resolution effect and basic properties of our early super-RENS disks, and focus on the third-generation super-RENS disk.

§ 2. Features and mechanisms of super-RENS disk – types A and B

Since 1998, three different types of super-RENS optical disks have been designed in our research center: super-RENS disk using an Sb thin layer (hereafter, type A), which generates an optical window; super-RENS disk using a silver oxide (AgOx) thin layer (hereafter, type B), which generates a single light-scattering center due to the photo-thermal decomposition; and super-RENS disk using a platinum oxide (PtOx) layer (hereafter, type C), which generates a nanosize gas bubble. Each resolution and feature is depicted in fig. 1.

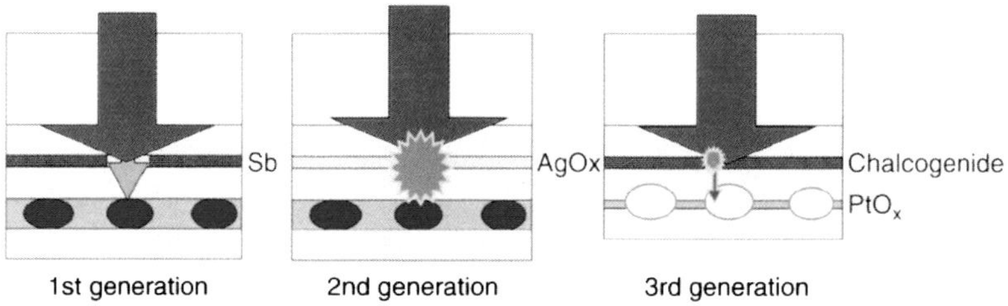

Fig. 1. Three types of super-resolution near-field structure (super-RENS) optical disks. First-generation: type A disk; second-generation: type B disk; and third-generation: type C disk. In type A, an optical aperture in an antimony thin film is created by an incident laser beam, and near-field is generated, which is scattered by adjacent recorded marks or pits through a transparent solid layer. In type B, instead of antimony, silver oxide film is used to generate a single light-scattering center. In type C, a special phase-change thin film is split into two phases and the phase boundary reads pits recorded in platinum oxide layer.

Type A disk was designed to make a 15 nm Sb thin layer generate an optical small window in a laser spot focused upon. The window size was controlled by adjusting the power of incident laser. This means that the window size determines the resolution of type A disk (see also fig. 2). In conjunction with the careful control of the thickness of the intermediate layer separating the Sb layer and recording layer ($Ge_2Sb_2Te_5$), 60 nm resolution was experimentally achieved against the diffraction limit of 540 nm, although the signal intensity of approximately threefold (~10dB) to the noise was too small to satisfy the commercially available value (Tominaga et al., 1998, 1999). This drawback is due to the fact that the number of photons that contribute to SNR relatively decreases on decreasing the window size. This property is crucial in all the near-field devices with an aperture. Many experimental and computational efforts have already been made to make clear the readout mechanism of type A disk, and all the results are in good agreement with the original aperture model (Tsai and Lin, 2000; Tsai et al., 2000; Tominaga and Tsai, 2003). Another drawback of type A disk is that the power of readout laser to obtain a maximum intensity strongly depends on each pit or mark size as shown in Fig. 2. This induces another crucial fault in case data coding used in the super-RENS disk succeeds to the similar ones in currently available optical disks including CD, DVD, and high-end disks, since information pits in CD or DVD usually consist of seven or nine different code-length pits. Even though type A super-RENS disk still holds such problems, it has provided a variety of attractive nano-optics in thin multilayers because adjusting the intermediate dielectric layer's thickness between the Sb and recording layer has enabled to understand the aperture's behavior, near-field interaction,

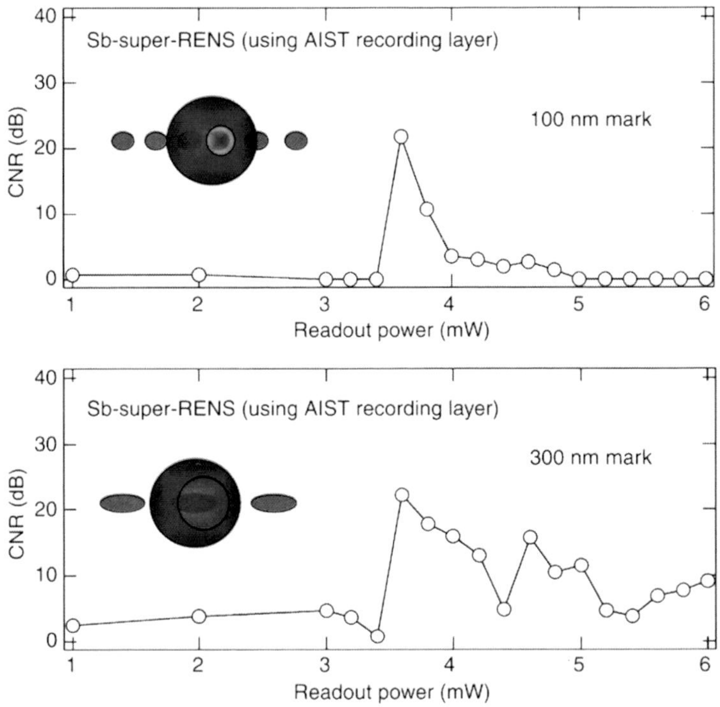

Fig. 2. Typical super-resolution properties against readout laser power. Top: 100 nm recorded mark; and bottom: 300 nm mark. Hence, AgInSbTe (AIST) phase-change film is used as a recording layer. On increasing the power, the optical aperture expands, and the signal of 100 nm mark suddenly disappears when the aperture includes plural marks.

and its focusing. In the last 7 years, many papers related to type A super-RENS disk have been published in journals worldwide (Tominaga and Tsai, 2003; Tominaga and Nakano, 2005).

In order to overcome the drawbacks of type A disk, a single light-scattering center has been investigated since 1999. For this, the Sb layer was replaced with a silver oxide (AgOx) layer because AgOx is known to decompose into Ag and oxygen at 160 °C under atmospheric pressure (Fuji et al., 2000; Tominaga et al., 2000). But at high pressure, it is expected that a Ag-rich phase is reversibly generated in the focused laser spot by decomposition and scatters light out of the disk due to the plasmonic effect. Adjusting the composition of the layer for Ag_2O enabled to enhance faint signals from a mark smaller than the diffraction limit up to the same level as that of type A disk, without the signal drawback depending on the pit size (Fuji et al., 2000). Hereafter we name the light-scattering disk as type B super-RENS disk. While the crucial issue of type A super-RENS disk was resolved, the resolution was scarified by ~100–150 nm. Although Ag nanoparticles were thought to become a strong light-scattering center by huge localized plasmons at first, the particles were soon condensed, leading to rapid signal reduction. We gradually understood through type B disk study that the signal from the plasmons was unstable and the gain was not as high as expected so far: less than threefold (~10 dB) (Tominaga and Nakano, 2005). Furthermore, when the laser power to record phase-change marks in a recording layer (GeSbTe or AgInSbTe chalcogenides) was once beyond a certain limit (~10 mW), it was noticed that the AgOx layer irreversibly left an oxygen gas bubble with Ag nanoparticles. In such a case, the mechanism of the signal enhancement became more complicated. Through dedicated efforts by our collaborated colleagues, however, it has gradually been revealed that more than half of the total signal from type B super-RENS disk is not generated by the plasmons from the metallic particles, but somewhere from the phase-change layer itself.

In order to resolve all problems in type A and B disks, in 2003 the AgOx layer was replaced with a platinum oxide (PtOx) layer (type C disk), which has a higher decomposition temperature (550°C) and leaves an oxygen gas bubble as a "recording pit," since plasmons are no longer needed for the super-RENS effect. Since then, type C disk has been called the third-generation super-RENS disk; the resolution and signal intensity were amazingly improved by 50 nm and more than 100-fold (>40 dB) (Kikukawa et al., 2002; Kim et al., 2003). By the invention of the third-generation disk, almost all issues in type A and B super-RENS disks were resolved. All characteristics of the third-generation super-RENS disk can satisfy commercially available signal values. The only exception is that nobody knows why such a huge signal enhancement and small resolution are easily obtained by using "chalcogenide" thin layers. So, what is the major mechanism for the super-RENS disk?

§ 3. Features of super-RENS disk – type C

Type C super-RENS disk exhibits completely different features in readout and in recording from the other disks (types A and B). First, even when a pit size is smaller than 100 nm, a large signal intensity of more than 40 dB is stably obtained using a present DVD optical head with 635 nm wavelength and a 0.6NA. As shown in fig. 2, the signal raising power of the type A disk is slightly different in readout of 100 and 300 nm pit trains, which means that the readout mechanism of type A disk is due to an optical aperture. In addition, the signal intensity of 100 nm pit trains suddenly falls down with increasing laser power, after reaching the maximum. In contrast, the intensity of the 300 nm pit trains still remains at more than 5 mW. These results are explained by the fact that an optical aperture is expanded with the power; when its size becomes comparable to the 100 nm pit, the intensity reaches the maximum; the aperture size is further expanded with raising power, and finally the intensity disappears when the aperture includes more than two pits. In the case of 300 nm pit train readout, however, the signal is still held at some level at more than 4 mW, where the 100 nm pit signal almost fades out. In addition, it should be noticed that the pit size (300 nm) is a little larger than the optical resolution limit in the far-field; therefore, the signal remains even at 6 mW. So, we can say that the signal obtained from 300 nm pits at high power consists of both near-field and far-field components. The behavior of modulation transfer function (MTF) derived from a computer-simulated model with a simple aperture also supports the validity of the model.

By contrast, the features of type C disk are astonishingly different from the above results and discussion. Figure 3 shows a typical feature of a relationship between readout pit size and laser power. All recorded pits, large or small, never feel the size of an aperture. In addition, the thinner the film, the better the signal intensity is. If it were concluded that the same aperture model was applied to type C disk as the mechanism, the

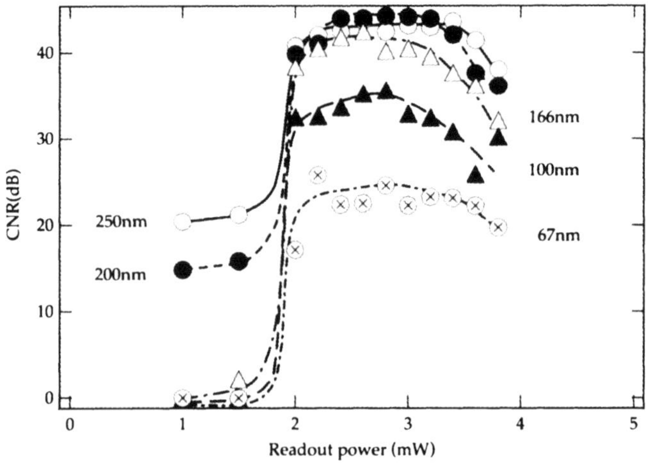

Fig. 3. Typical signal intensity of type C super-RENS disk against readout laser power. All signals from 67 to 250 nm marks have the same standing power at around 2 mW and a falling power at around 4 mW. It should be noticed in comparison to Fig. 2 that very huge and stable signals at more than 40 dB are obtained using the same optics.

higher signal intensity would be observed in the thicker film because it seals more incident beam than the thinner one. The MTF of the disk is dissimilar to that of type A disk; MTF closed to the diffraction limit is only enhanced (Nakano et al., 2005). Another result absolutely denies the simple aperture model. That is, the signal profiles against readout power do not depend on pit sizes within the near-field region, and the maximum intensities are held constant within 2.0–3.5 mW.The latter property is fortunately very much familiar and valuable for engineers designing the readout system in type C super-RENS disk because the wide power margin is inevitable for environmental tolerance, especially temperature difference between cold and tropical places, while a needle-like power adjustment is needed in type A and B super-RENS disks, and in any other near-field recording devices proposed so far. Therefore, it is concluded that a simple aperture model is no longer applied to type C super-RENS disk, even if an aperture is generated.

The potential of type C disk is thought to be limitless at the moment. For example, using the most advanced next-generation DVD drive with a 405 nm blue laser and $0.85NA$, named Blu-ray, SNR of more than 40 dB, which is thought to be required for the commercialized value at least, is obtained with a pit size of 37.5 nm that is one-tenth smaller than the laser spot size. According to a recent result (Shima et al., 2005), the resolution depends only on a laser spot size. If type C disk is once combined with solid immersion lens (SIL) technology, further small resolution will be attained in future, although nobody till now has confirmed how much small pit is recorded in the PtOx film. Therefore, we can conclude once again that the readout mechanism of type C disk is absolutely different from that of the other near-field devices and a simple aperture model can no longer be applied.

§ 4. Understanding the super-resolution mechanism of type C disk

So, what actually induces such a huge signal enhancement from type C disk? Probably, the most noticeable property to understand the super-resolution mechanism is a relationship between the threshold power and disk rotation speed (Kuwahara et al., 2004). The threshold power in readout becomes lower on decreasing the rotation speed. The relationship is somewhat linear, which simply means that temperature in the active region in the disk plays a critical role in readout. However, refractive index of a material slightly decreases in general with increasing temperature because of volume expansion (Tominaga, 1985). However, the thermal expansion ratio of metal is usually within $10^{-4} \times T(\mathrm{K})$, and therefore the index may be varied by about 5%. In addition, the change in volume, to which the refractive index is subject, usually occurs linearly. Therefore, a simple model under thermal index change cannot explain a steep threshold power observed in type C super-RENS disks. In conclusion, except for a phase transition with a threshold temperature there is no model to explain the experimental results. In 2004, we proposed a model based on a large electronic polarization change like ferroelectricity (Tominaga et al., 2004). In ferroelectrics, a dielectric constant may vary at a specific temperature called Curie temperature T_0 and it belongs to "second phase transition." Second phase transition differs from first phase transition with a latent heat at the transition. In general, second phase transition has no latent heat, but only the heat capacity changes slightly. Therefore, it is hard to observe a second phase transition without a stress-induced layer because the transition is only observed as a slightly bent point in a heat flow chart, while a transition point in first phase transition is clearly obtained with a huge latent heat. The free energy F of ferroelectrics at second phase transition may simply be expressed by Landau's model (Lines and Glass, 1997):

$$F = \frac{1}{2}sP^2 + \frac{1}{4}tP^4 + \frac{1}{6}uP^6 + \cdots, \tag{3.1}$$

where, s, t, and u are parameters, and P is polarization.

Deriving the energy minimum and using the relation

$$\frac{\partial^2 F}{\partial P^2} = \frac{\partial E}{\partial P} = \chi^{-1}$$

we can obtain the Curie temperature T_0.

$$\varepsilon \propto (T_0 - T)^{-1} \tag{3.2}$$

Hence, $P = \chi E = 4\pi\varepsilon^{-1}E$. However, this model must be modified to the condition including internal stress:

$$F\frac{1}{2}sP^2 + \frac{1}{4}tP^4 + \frac{1}{2}u(x - x_0)^2 + vxP^2. \tag{3.3}$$

The third term is of strain energy and the fourth one is of a coupling term with the deviation due to the stress, where u and v are parameters. According to eq. (3.3), we can derive an important relationship between atomic displacement and polarization:

$$\Delta x \propto -P^2. \tag{3.4}$$

It means that an atomic displacement by a strong internal stress may induce a huge polarization and vice versa. The importance of this relationship would be discussed later.

Figure 4 shows a typical heat flow curve of a $Ge_2Sb_2Te_5$ phase-change film (thickness: 500 nm) sandwiched between ZnS-SiO_2 films by differential scanning calorimetry (DSC). Mechanically, the thinner the film, the larger internal stress is applied at the interface. As mentioned already, the fact that a thinner film generates higher signal intensity and better resolution in type C disk supports that the internal stress induced in the disk plays a major role in the super-resolution besides the optical factors. In 2004, our group experimentally discovered an important fact in the switching mechanism of a $Ge_2Sb_2Te_5$ phase-change film using X-ray analysis in fine structure (XAFS) at SPring-8 in Harima, and at Photon factory in Tsukuba: the phase-change film holds two stable structures in energy potential with a different configuration of a Ge atom in a cubic unit cell of Te atoms (Kolobov et al., 2004; Weinic et al., 2006). Amazingly, in one of the structures, a Ge atom is slightly shifted off from the central position in the Te unit cell because of charge balance due to vacancies and Sb atoms in the adjacent cells. Therefore, the structure is really ferroelectric. In the structure that has generally been called "crystal" so far, a Ge atom is chemically bonded with surrounding six Te neighbors; however, the bonding energy is not equal to each other: three of them consist of a stronger bonding energy

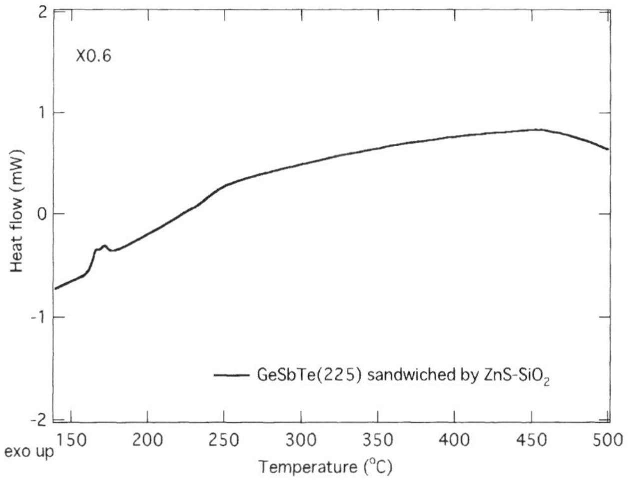

Fig. 4. DSC curve obtained from stress-induced GeSbTe film. First transition is at 160°C with a latent heat from as-deposited amorphous to crystal (distorted simple cubic, d-sc), and second transitions are allowed at around 250°C (hexagonal) and around 450°C (unidentified).

and the remaining of a weaker one. As temperature rises, the stable phase at low-temperature transits to the second stable phase along the line of chemical potential. Although the second phase has been believed to be "amorphous" so far, it was confirmed that the phase has a rigid structure to some extent in short-range order like a crystal, where a Ge atom is connected with four Te atoms by four strong bonds instead. This phase transition due to the Ge atomic dislocation in about 0.2 nm induces a huge optical property. Usually, in metals and semiconductors the reflectivity is increased with temperature because carriers (electrons and holes) are generated and the film becomes more opaque. In contrast, the chalcogenide films like Sb, Te, and Se oppositely and unusually become transparent by generating a band gap at around the Fermi level in the high-temperature phase (Weinic et al., 2006). This atomic dislocation probably induces a very large variance on polarizability as well as the band gap because refractive index depends on electronic polarizability α, and is expressed by Clausius–Mossotti equation (3.5) (Ibach and Luth, 1995; Atkins, 1998):

$$\alpha = \frac{3}{4\pi N_A}\frac{n^2-1}{n^2+2}V, \tag{3.5}$$

where α and n are electronic polarizability and refractive index at infinitely long wavelengths, respectively. N_A is Avagadro number and V volume in a unit cell. In actual, α is also expressed by quantum mechanics, and it consists of the summation of electron dipole transitions in each atom, which is stimulated by an incident light wavelength (Davydov, 1965). Now, we change eq. (3.5) with a simple form using a constant $3V/(4\pi N_A) = \zeta$.

$$n^2 = \frac{2\alpha+\zeta}{\zeta-\alpha}. \tag{3.6}$$

Hence, ζ is thought as the free volume of an atom in a unit cell. If ζ attains the same value as α, refractive index can be diverged. Figure 5 shows electronic polarizabilities of several ions experimentally obtained at a high temperature. There are no data for Ge ions unfortunately, but α would not exceed $3A^3$. Therefore, we can fully expect that ζ exceeds α at the phase transition boundary, since ζ of Ge in $Ge_2Sb_2Te_5$ is roughly estimated as $8A^3$ from the displacement at the transition. For example, if ζ is smaller than α in a low-temperature phase because of the low mobility, n^2 (ε) becomes negative: Ge behaves like a metal. However, at the transition, the large Ge dislocation makes it positive beyond the singular point, $\zeta = \alpha$: it in turn becomes dielectric and behaves like a transparent aperture in a high-temperature phase. This means that the refractive index is greatly changed in the vicinity of the phase boundary. At the boundary, especially, hot carriers

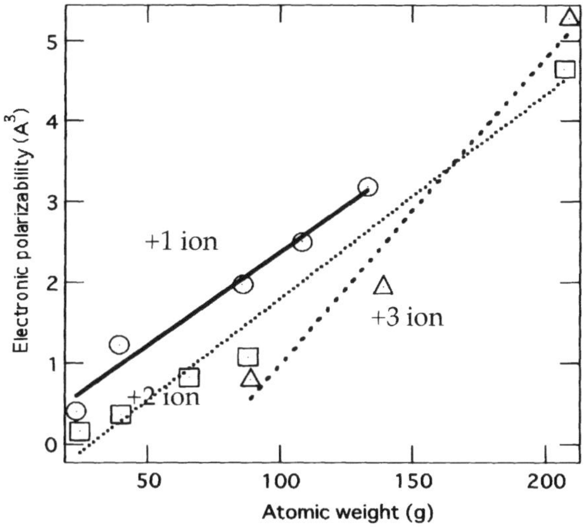

Fig. 5. Experimentally estimated electronic polarizabilities of several ions at a high temperature.

generated in the low-temperature phase are trapped and some are reflected back to the phase because they cannot diffuse into the high-temperature phase beyond the energy barrier. Therefore, the refractive index probably gives rise to a large dispersion and may become "negative" within a very narrow region. Once the phase transition occurs by laser heating with a Gaussian beam profile during the disk rotation, in turn, the phase transition boundary appears at a singular point or ring with a huge refractive index gradient, where a huge number of CDEWs are generated and scattered by pit trains closely placed under the layer. Therefore, the thinner layer can produce more signal and smaller resolution because of the sharper boundary edge. In this model, the aperture size in readout is not taken into account, since the boundary edge is only active for super-resolution, but the two phases at low and high-temperature are not. In conclusion, the threshold power in readout in fig. 3 is independent of the pit size.

§ 5. Combination of plasmonic enhancement and type C super-RENS disk

In the previous section, we discussed the super-resolution power of CDEWs due to the refractive index diversion. Here, let us go back and focus on plasmonic enhancement once again. This is very exciting and the most fascinating science in near-field optics. As introduced, metallic hole-array device is one of the up-to-date topics in near-field optics (Ebbesen et al., 1998; Barnes et al., 2003). In super-RENS technology, we have once examined plasmonic enhancement in type B disk. Through the study of type B disk one convenient method to transform AgOx films into Ag nanostructured layers in a very large area by reduction was discovered (Tominaga, 2003; Arai et al., 2006). One of the typical examples is shown in fig. 6. The main factor to transform AgOx films into the layer consisting of Ag nanoparticles is the nucleation process leading to nanoparticles. This is achieved by the following procedures: First, CF_4 gas is introduced into the vacuum chamber to generate AgF on the AgOx film surface; Second, a gas mixture of H_2 and O_2 is subsequently introduced to reduce AgOx into Ag nanoparticles around AgF nano-seeds. The whole reduction is completed in 3–5 min. This simple method is very useful because a uniform nanostructure with Ag nanoparticles is fabricated on any shaped surface on which a AgOx film can be deposited. The main advantage of the layer is that the nanoparticle's plasmonic absorption peak appears at around 400 nm wavelength, which is almost the same as that used in the next-generation DVD systems: Blu-ray and HD-DVD. Now, we are ready for combining type C disk and the plasmonic layer on the top.

Plasmonic signal enhancement is evaluated by the following method. First, typical type C super-RENS disks are prepared with thickness of each layer as shown in fig. 7. Hence, thickness of the top dielectric layer ($ZnS–SiO_2$: $n = 2.25$), d, is varied between 5 and 120 nm.It should be noticed that the top layer thickness is not sensitive to recording signals because pit recording is only active between the phase-change layer and the PtOx layer through the intermediate dielectric layers. Second, pit trains of 100 nm diameter (20 MHz with a duty of 50%) are recorded on several tracks at a constant linear velocity of 4.0 m/s, which is two-third less than that of DVD. The recording power is set at 7 mW with a 405 nm wavelength and 0.60NA optics. After evaluating the super-resolution signal intensities against the readout laser power, a Ag_2O film is deposited on the top of the disks and subsequently is deoxidized by the above method. Hence, the Ag_2O thickness was set at 100 nm. The signal intensities are evaluated once again on the same tracks evaluated previously. The difference in the signal intensities before and after fabricating the Ag nanostructured layer becomes the plasmonic enhancement. Figure 8 shows a cross-sectional image of a recorded track with 100 nm pit trains taken by a transmission electron microscope (TEM), and the result is shown in fig. 9. The real TEM image of a prepared disk sample may be very similar to that shown in fig. 7. In addition, on the top round Ag nanoparticles of diameter 49 ± 8 nm are confirmed (Arai et al., in press). The signal intensity from the 100 nm pit trains has the maximum

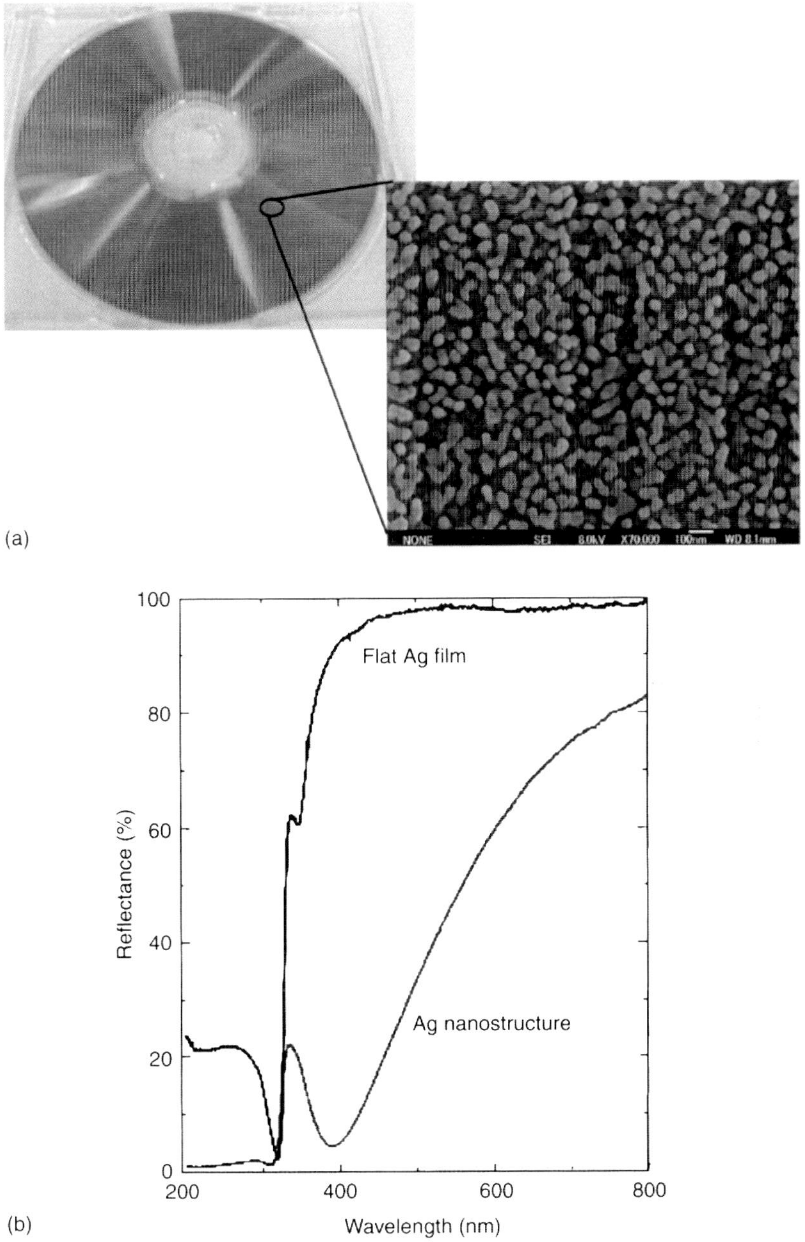

Fig. 6. (a) Nanoparticled structure consisting of silver on 12 cm optical disk surface. The averaged diameter of silver nanoparticles is 49 ± 8 nm. The structure was fabricated through the reduction of silver oxide film. Vertical steps in the SEM picture are grooves to guide a laser beam for recording. (b) The reflectance spectrum from the surface with a spectrum from a flat silver film. A specific drop in the spectrum at 400 nm is the plasmonic absorption due to the nanostructure.

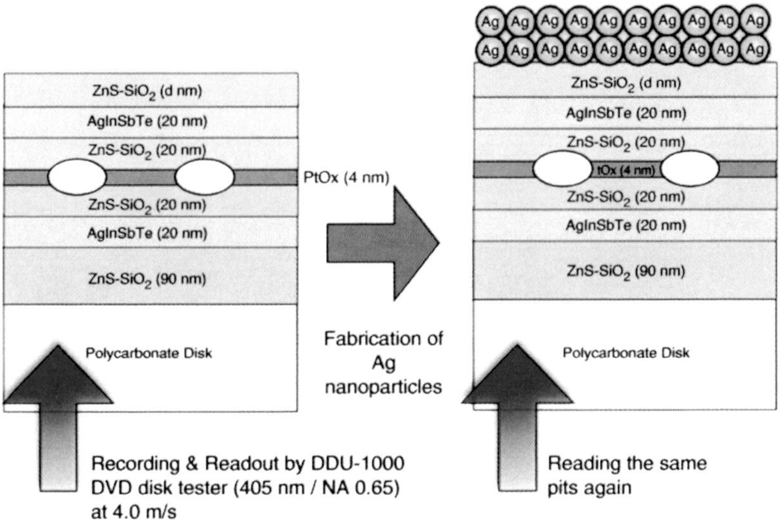

Fig. 7. Experimental procedures to estimate plasmonic enhancement. First super-RENS disks (type C disk with double AIST layers) are prepared and recorded with several patterns. After signal intensities are measured, a Ag nanostructure is fabricated. The prerecorded patterns are remeasured and compared.

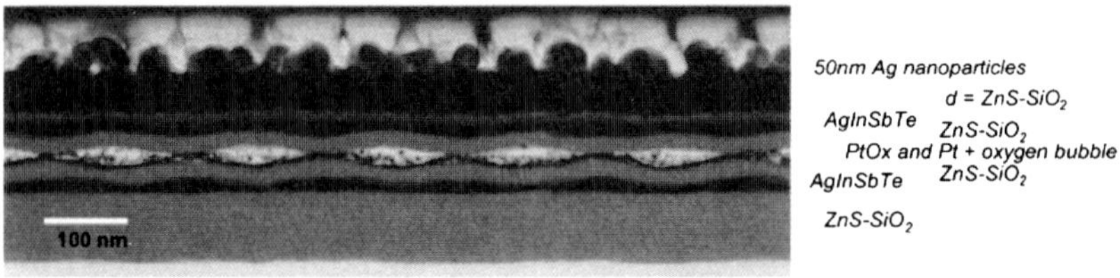

Fig. 8. TEM image of the cross-section of the super-RENS disk with Ag nanoparticles. The real image is very similar to the right-hand side picture of Fig. 7.

at the top dielectric layer thickness of 20 nm (see solid triangles in fig. 9), and it gradually decreases with the layer thickness. Without Ag nanoparticles, this super-RENS disk can attain a signal intensity nearly 100-fold (40dB) to the noise at the 20 nm layer thickness. In contrast, the intensity curve from the disks with Ag nanoparticles is shown as solid circles. The intensity curve becomes almost flat at a layer thickness of more than 40 nm.The signal intensity is improved about threefold (~10dB) at the 120 nm thickness. This is because the distractive CDEW interferences are suppressed by the strong light absorption of the Ag nanoparticle's layer at 400 nm wavelength. On the contrary, in the disks

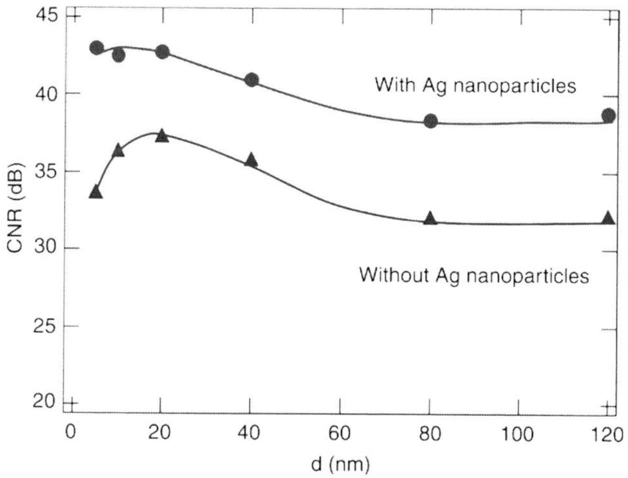

Fig. 9. Signal enhancement of the super-RENS disks with and without Ag nanoparticles. Triangles and circles are original signals, and signals on the same track after Ag nanoparticles are fabricated, respectively.

with a dielectric layer thinner than 40 nm the signal intensity gradually increases with the inverse of the thickness, and reaches the maximum (130-fold: 42 dB) at the thinnest layer thickness of 5 nm. As a result, it can be estimated and concluded that the increment of the signal gain of 5–6 dB (twofold) is the contribution from the plasmonic enhancement.

From the above experiment and discussion it is clear that the plasmonic enhancement is limited within a very short range, less than 40 nm.In addition, the gain obtained is not as large as expected, but only two- or threefold to the noise level as compared to the contribution from CDEWs due to the phase transition. Although a lot of studies have been carried out so far on plasmonic devices consisting of metallic hole arrays, most results are computer simulations with nice color pictures but a very few experimental evidences, since the device fabrication in precise is still difficult in nanometer scale. Although the plasmonic enhancement and gain of our results are not enough, it probably suggests that plasmonic control is not yet mature and we are still at the entrance.

§ 6. Summary

Three different types of super-resolution near-field disks and their features were introduced. Especially, it was discussed that type C super-RENS disk differs from the other disks and indicates extremely high signal intensity and small resolution. Although the detailed mechanism has not yet been clearly understood, it was addressed that a simple aperture model is not applicable, but the super-resolution readout depends on the optical property at the edge of a phase-change boundary, and the generation of a huge number of CDEWs plays a role in the high signal intensity and resolution, rather than plasmonic enhancement.

Acknowledgement

The author thanks T. Nakano, M. Kuwahara, T. Shima, A. Kolobov, P. Fons, S. Petit, C. Rockstuhl, and T. Arai for their helpful discussion.

References

Arai, T., Kurihara, K., Nakano, T., Rockstuhl, C., Tominaga, J., Appl. Phys. Lett. (in press).
Arai, T., Rockstuhl, C., Fons, P., Kurihara, K., Nakano, T., Awazu, K., Tominaga, J., 2006, Nanotechnology **17**, 79.
Atkins, P.W., 1998, Physical Chemistry, Oxford Univ. Press, Oxford.
Barnes, W.L., Dereux, A., Ebbesen, T.W., 2003, Nature **424**, 824.
Davydov, A.S., 1965, Quantum Mechanics, Pregamon, Oxford.
Ebbesen, T.W., Lezec, H.J., Ghaemi, H.F., Thio, T., Wolff, P.A., 1998, Nature **391**, 667.
Fang, N., Lee, H., Sun, C., Zhang, X., 2005, Science **308**, 534.
Fuji, H., Katayama, H., Tominaga, J., Men, L., Nakano, T., Atoda, N., 2000, Jpn. J. Appl. Phys. **39**, 980.
Guerra, J.M., 1988, J. Appl. Opt. **29**, 3741.
Guerra, J.M., Srinivasarao, M., Stein, R.S., 1993, Science **262**, 1395.
Ibach, H., Luth, H., 1995, Solid-State Physics, Springer, Heidelberg, Berlin.
Kawata, S. (Ed.), 2001, Near-Field Optics and Surface Plasmons Polaritons (Topics in Applied Physics, vol. 81), Springer, Heidelberg, Berlin.
Kikukawa, T., Nakano, T., Shima, T., Tominaga, J., 2002, Appl. Phys. Lett. **81**, 4697.
Kim, J., Hwang, I., Kim, H., Yoon, D., Park, H., Jung, K., Park, I., Tominaga, J., 2004, SPIE Proc. **5380**, 336.
Kim, J.H., Hwang, I., Yoon, D., Park, I., Shin, D., Kikukawa, T., Shima, T., Tominaga, J., 2003, Appl. Phys. Lett. **83**, 1701.
Kolobov, A., Fons, P., Frenkel, A., Ankudinov, A., Tominaga, J., Uruga, T., 2004, Nat. Mater. **3**, 703.
Kuwahara, M., Shima, T., Kolobov, A., Tominaga, J., 2004, Jpn. J. Appl. Phys. **43**, 8.
Lezec, H.J., Thio, T., 2004, Opt. Expr. **12**, 3629.
Lines, M.E., Glass, A.M., 1997, Principles and Applications of Ferroelectrics and Related Materials, Oxford Univ. Press, Oxford.
Nakano, T., Mashimo, E., Shima, T., Yamakawa, Y., Tominaga, J., 2005, Jpn. J. Appl. Phys. **44**, 3350.
Pendry, J.B., 2000, Phys. Rev. Lett. **85**, 3966.
Raether, H., 1988, Surface Plasmons, Springer, Hamburg.
Shima, T., Nakano, T., Kim, J.H., Tominaga, J., 2005, Jpn. J. Appl. Phys. **44**, 3631.

Smolyaninov, I.I., Elliott, J., Zayats, A.V., Davis, C.C., 2005, Phys. Rev. Lett. **94**, 57401.
Tominaga, J., 1985, MSc Thesis, Chiba University.
Tominaga, J., 2003, J. Phys. Condens. Matter **15**, R1101.
Tominaga, J., Fuji, H., Sato, A., Nakano, T., Atoda, N., 2000, Jpn. J. Appl. Phys. **39**, 957.
Tominaga, J., Nakano, T., 2005, Optical Near-Field Recording – Science and Technology, Springer, Heidelberg, Berlin.
Tominaga, J., Nakano, T., Atoda, N., 1998, Appl. Phys. Lett. **73**, 2078.
Tominaga, J., Nakano, T., Fukaya, T., Atoda, N., Fuji, H., Sato, A., 1999, Jpn. J. Appl. Phys. **38**, 4089.
Tominaga, J., Shima, T., Kuwahara, M., Fukaya, T., Kolobov, A., Nakano, T., 2004, Nanotchnology **15**, 411.
Tominaga, J., Tsai, D.P. (Ed.), 2003, Optical Nanotechnologies – The Manipulation of Surface and Local Plasmons (Topics in Applied Physics, vol. 88), Springer, Heidelberg, Berlin.
Tsai, D.P., Lin, W.C., 2000, Appl. Phys. Lett. **77**, 1413.
Tsai, D.P., Yang, C.W., Lin, W.C., Ho, F.H., Huang, H.J., Chen, M.Y., Teseng, T.F., Lee, C.H., Yeh, C.J., 2000, Jpn. J. Appl. Phys. **39**, 982.
Weinic, W., Pamungkas, A., Detemple, R., Steimer, C., Blugel, S., Wuttig, M., 2006, Nat. Mater **5**, 56.

Chapter 7

Metal stripe surface plasmon waveguides

by

Rashid Zia*, Mark Brongersma

Geballe Laboratory for Advanced Materials, Stanford University, Stanford, CA 94305
e-mail: Rashid_Zia@brown.edu

*Current Address: Brown University, Division of Engineering, Box D, Providence, RI 02912.

Nanophotonics with Surface Plasmons — *V.M. Shalaev & S. Kawata (Editors)*
Advances in Nano-Optics and Nano-Photonics — © *2007 Published by Elsevier B.V.*
ISSN: 1871-0018 — DOI: 10.1016/S1871-0018(06)02007-3

Contents

§ 1. Introduction

Metal nanostructures have received considerable attention for their ability to guide and manipulate electromagnetic energy in the form of surface plasmon-polaritons (SPPs) (Barnes et al., 2003; Takahara and Kobayashi, 2004). It has even been suggested that the unique properties of SPPs may enable an entirely new generation of chip-scale technologies, known as plasmonics. Such plasmonic devices could add functionality to the already well-established electronic and photonic device technologies. SPPs are surface electromagnetic ("light") waves supported by charge density oscillations along metal–dielectric interfaces (Raether, 1965). SPP excitations have been studied extensively on metal films. However, there has been renewed interest in the field, as researchers have begun to propose the use of patterned metal structures as tiny optical waveguides to transport electromagnetic energy between nanoscale components at optical frequencies (Takahara et al., 1997; Quinten et al., 1998; Weeber et al., 1999; Brongersma et al., 2000; Zia et al., 2005b).

The best-studied plasmonic waveguides to date have been finite width metal stripes on dielectric substrates (Krenn and Weeber, 2004), which have been the topic of numerous theoretical (Berini, 2000, 2001; Al-Bader, 2004; Zia et al., 2005a,b) and experimental studies (Charbonneau et al., 2000; Lamprecht et al., 2001; Weeber et al., 2001; Krenn et al., 2002; Weeber et al., 2003, 2004; Yin et al., 2005; Zia et al., accepted for publication). Such metal stripes have the desirable feature that they resemble traditional electronic interconnects, and thus they can enable simultaneous transport of photonic and electronic signals. Initial experimental results have suggested that subwavelength metal stripes may support highly confined surface plasmon modes (Krenn et al., 2002; Yin et al., 2005). If true, densely integrated systems could be realized in which the size of the information processing units (e.g., transistors) would be similar to the structures carrying information (the metal stripe waveguides).

Here we review a series of combined numerical and experimental studies of light propagation along metal stripe waveguides (Zia et al., 2005a,b; Zia et al., accepted for publication). The emphasis is placed on providing a physical understanding of the guiding mechanism of such waveguides and on highlighting their capabilities and ultimate limitations for use as nanoscale optical communication channels. In Section 2, we start with a discussion of the experimental methods used to launch SPPs onto metal stripes and discuss the near-field optical technique employed to image their propagation. Section 3 describes a full-vectorial, numerical method capable of solving for the SPP modes supported by metal stripe waveguides. Section 4 discusses the experimentally observed modal behavior and compares these experiments to simulations. Section 5 describes an intuitive picture for these modes that is analogous to the conservation of momentum picture used to describe light propagation in conventional dielectric waveguide structures in physical optics. In Section 6, we present a parametric study of SPP propagation as a function of waveguide width. This study demonstrates that the propagation of "light" along metal stripe waveguides is mediated by a discrete number of guided polariton modes as well as a continuum of radiation modes. We conclude in Section 7 by discussing the impact that these findings may have on the design and fabrication of future metal waveguide structures. A detailed discussion of the potential applications of these waveguides is deferred to a future publication.

§ 2. Experimental techniques

Theory has provided significant insight into the propagation of SPPs. The dispersion relations, which relate angular frequency (ω) and wave vector ($k = 2\pi/\lambda$), for the SPP modes supported by metal thin films have been studied in detail and are by now well established (Economou, 1969; Burke et al., 1986; Prade et al., 1991). Figure 1(a) shows the dispersion relations for light in homogenous regions of air (solid dark line) and glass (solid white line) as well as the SPP mode propagating along a metal–air interface (dotted line). These dispersion relations follow directly from Maxwell's equations with the appropriate boundary conditions. For light in a dielectric medium, the dispersion relation is given by $\omega = kc/n_r$, where c is the speed of light and n_r the optical refractive index. Due to the higher index of glass, the dispersion relation for light in glass has a slope that is about 1.5 times smaller than for the dispersion relation in air. The SPP modes propagating at the metal–air interface can be described by the dotted line dispersion relation (Raether, 1988). At low angular frequencies, this dispersion relation follows the light line in air. However, at angular frequencies close to the surface plasmon resonance frequency, ω_{sp}, the magnitude of the wave number, k_{sp}, diverges and the wavelength of the SPPs shortens. For many metals, the surface plasmon resonance

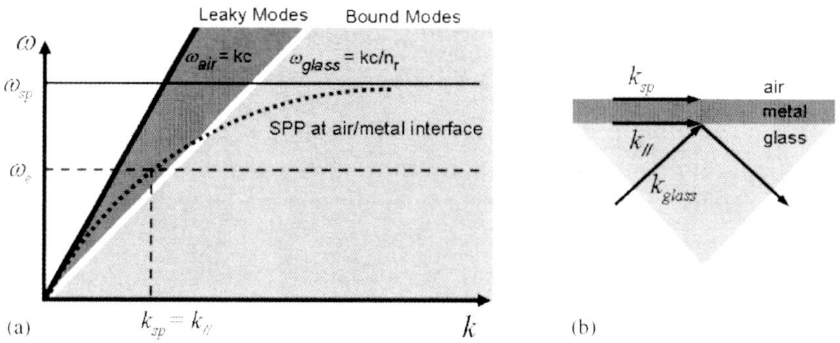

Fig. 1. (a) Dispersion relation for light in air (solid black line) and glass (solid white line) and the surface plasmon-polariton (SPP) propagating at the air–metal interface (dotted line). SPP modes in the dark gray region, to the left of the glass light line, are "leaky"and can radiate into the higher index glass. The modes in the light gray region are truly bound to the right of the glass light line. (b) A prism coupling (Kretschmann) setup is used to couple a free-space beam from a high-index glass to a SPP propagating along the metal–air interface. Optimum coupling is obtained when the phase matching condition ($k_{\text{sp}} = k$) is met for a specific angle of the incident light beam.

frequency at an air interface occurs in the ultraviolet (UV) regime or the blue region of the visible spectrum.

It is important to note that for every angular frequency the momentum of a SPP along an air–metal interface exceeds the momentum of a photon in air, i.e., $k_{\text{sp}} > k_{\text{air}}$. For this reason, free-space light in air cannot couple to SPPs while conserving ω (energy) and k (quasi-momentum). It is, however, possible to couple a free-space beam to SPPs along an air–metal interface using a high-index prism, and this method was pioneered by Kretschmann (1971). His experimental configuration is shown in fig. 1(b) and uses a far-field excitation beam that is angled through the glass prism such that the in-plane wave vector of the light in glass, $k_{//}$, corresponds to the associated SPP propagation constant, k_{sp}. This condition is known as the momentum or phase-matching condition.

The Kretchmann configuration only allows for the excitation of SPP modes that lie above the glass light line for which $k_{\text{sp}} < k_{\text{glass}}$ (Chen et al., 1976). As light from the high-index medium can couple to these SPP modes, the SPPs must also be able to leak back into that medium. This light is known as leakage radiation, and the modes located in the dark gray area in fig. 1(a) are known as leaky SPP modes. Energy loss due to leakage radiation and resistive heating causes the propagation constant of SPP modes to be complex. SPP modes below the glass light line (in the light gray region) cannot be addressed with prism coupling techniques because their propagation constant $k_{\text{sp}} > k_{\text{glass}}$. In turn, these modes do not give rise to leakage radiation and are truly bound, but nevertheless suffer considerable material losses.

To study finite width metal stripe waveguides such as shown in fig. 2(a), one can make use of the Kretschman configuration as well. To this end, samples are generated by electron beam lithography on glass substrates such that thin Au stripes protrude from larger thin film regions known as launchpads (Lamprecht et al., 2001; Krenn and Weeber, 2004). In order to excite SPPs along such metal stripes, SPPs are first excited on the surface of the launchpad. These SPPs are then directed though a tapered region to the stripe waveguides where they can excite a finite number of guided polariton modes as well as a continuum of radiation modes (as described later in Section 6). By scanning a near-field optical probe above the sample at a constant height, one can map the SPP propagation along the metal stripe. This method of imaging has been used in many experimental studies (Weeber et al., 2001, 2003; Krenn et al., 2002; Yin et al., 2005; Zia et al., accepted for publication), and the general technique is called photon scanning tunneling microscopy (PSTM) (Reddick et al., 1989).

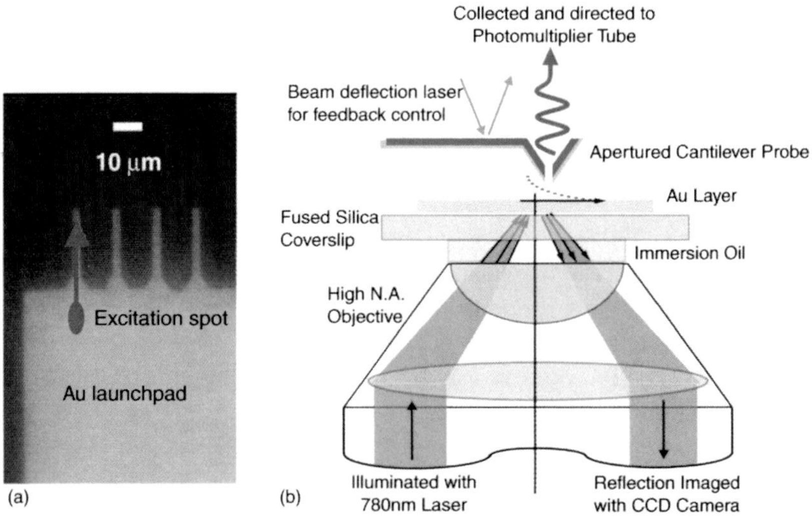

Fig. 2. (a) Optical microscopy image of a fused silica glass microscope coverslip with an array of 4.5 μm wide Au stripes attached to a launchpad generated by electron beam lithography. The white arrow pictorially illustrates the launching of SPPs from a small excitation spot. (b) Schematic of photon scanning tunneling microscope (PSTM). A partially illuminated, high numerical aperture objective is used to excite SPPs along the Au–Air interface via attenuated total reflection. Light is scattered from these surface waves by an aperture cantilever probe and detected in the far-field by a photomultiplier tube. The sample and illumination objective are rigidly mounted together on an x,y,z piezo-stage, which is scanned below the fixed cantilever.

A schematic of the PSTM used for this study is shown in fig. 2(b). This instrument has been constructed by modifying a commercially available scanning near-field optical microscope (α-SNOM; WITec GmbH; Ulm, Germany). The modified microscope is a variation on the conventional PSTM, which has been used extensively to characterize SPP propagation along extended films as well as metal stripe waveguides. In a conventional PSTM (Reddick et al., 1989), SPPs are excited via ATR using prism coupling, and the local optical fields are probed by scanning a tapered fiber tip above the sample. Our PSTM operates in a similar fashion, except for three modifications. First, in the place of a prism, a partially illuminated high numerical aperture total internal reflection fluorescence (TIRF) objective (Zeiss Alpha Plan-Fluar, $100\times$, NA $= 1.45$) is used to excite SPPs on the Au launchpad. Second, an apertured cantilever is used as an optical near-field probe as opposed to a tapered optical fiber (Mihalcea et al., 1996). Third, instead of scanning the cantilever above a stationary sample, the sample and illumination objective are scanned on an x,y,z piezo-stage beneath the apertured cantilever probe. The advantage of these modifications is that they can easily be incorporated into a conventional optical microscope to create a PSTM.

Since the PSTM makes use of a Kretschmann-like excitation scheme, only leaky modes can be excited. In fact, most experimental studies have focused on leaky modes for exactly this reason. In contrast, most theoretical investigations of metal stripe waveguides have been limited to bound modes (Berini, 2000, 2001; Al-Bader, 2004). While the bound modal solutions may be excited via end-fire excitation or scattering events (Charbonneau et al., 2000), their relevance in characterizing and understanding the behavior of leaky SPP modes has not been validated. Without the benefit of leaky modal solutions, it has been previously suggested that SPPs guided along metal stripes cannot be described by the conventional physical models for dielectric waveguides. In the next section, we will discuss a recently published numerical technique that can solve for both the leaky and bound modal solutions of plasmonic waveguides (Zia et al., 2005b).

§ 3. Numerical methods

Recent work has demonstrated that the bound modal solutions of plasmonic waveguides are hybrid transverse electric-transverse magnetic (TE-TM) modes, and therefore, proper analysis requires numerical solution of the full-vectorial wave equation (Al-Bader, 2004). In particular, Al-Bader demonstrated that the full-vectorial magnetic field-finite difference method (FVH-FDM) (Lusse et al., 1994) can successfully solve for the modes of metal stripe waveguides. However, with conventional Dirichlet, Neumann, or Robin boundary conditions, this technique can only solve for bound modes, because these boundaries cannot account for the non-zero radiating fields of leaky modes. Transparent or absorbing boundary conditions (TBCs or ABCs) are required to appropriately treat the semi-infinite extent of leaky modes into a high-index substrate. For scalar finite difference methods, it has been shown that perfectly matched layer (PML) ABCs allow for accurate solutions of leaky modes in planar waveguides (Huang et al., 1996). The technique we used for our study (Zia et al., 2005b) extended Al-Bader's technique for use with three-dimensional waveguides by implementing the generalized complex coordinate stretching (CCS) formulation of PML boundary conditions (Chew et al., 1997). To this end, we have made use of the work by Chew et al., who derived an elegant set of relationships describing electromagnetic waves in complex space that are isomorphic with respect to Maxwell's equations. This formulation has been previously implemented for a similar finite difference method in cylindrical coordinates (Feng et al., 2002). Borrowing this notation, we have implemented CCS-PML boundary conditions by modifying the Helmholtz equations to solve for the optical modes as follows:

$$\frac{\partial^2 H_x}{\partial \tilde{x}^2} + \frac{\partial^2 H_x}{\partial \tilde{y}^2} + (\varepsilon\beta_0^2 - (\beta + \mathrm{i}\alpha)^2)H_x = 0, \tag{3.1}$$

$$\frac{\partial^2 H_y}{\partial \tilde{x}^2} + \frac{\partial^2 H_y}{\partial \tilde{y}^2} + (\varepsilon\beta_0^2 - (\beta + \mathrm{i}\alpha)^2)H_y = 0, \tag{3.2}$$

where $H_{x,}$ and H_y are the magnetic fields in the x- and y-directions, respectively, $(\beta + \mathrm{i}\alpha)$ represents the complex propagation constant of the SPPs along the metal stripe, β_0 is the free-space wave number, and the complex coordinates $\tilde{x}$ and $\tilde{y}$are described by substituting x and y for ζ in the following change of variables:

$$\tilde{\zeta} = \int_0^{\zeta} s_\zeta(\zeta')\, d\zeta', \tag{3.3}$$

$$s_\zeta(\zeta) = \begin{cases} 1 + \mathrm{i}\frac{\sigma_\zeta(\zeta)}{\omega\mu_0}, & \text{within the PML,} \\ 1 & \text{elsewhere,} \end{cases} \tag{3.4}$$

and

$$\sigma_\zeta(\zeta) = \sigma_{\zeta,\,\max}\left(\frac{\zeta - \zeta_0}{d}\right)^m. \tag{3.5}$$

In the above equation, s_ζ is a complex stretching factor, and σ_ζ represents a conductivity-like loss term whose profile in the PML region of thickness d is increasing as an mth order polynomial from a minimum value of zero at the initial PML interface (ζ_0) to a maximum value ($\sigma_{\zeta,\max}$) at the simulation boundary. Rescaling the spatial coordinates of mesh points within the PML region by a complex number allows for attenuation of fields radiating from leaky waveguide modes without the introduction of an impedance mismatch and the associated reflective perturbation of the finite difference solution.

§ 4. Leaky modes supported by metal stripe waveguides

To investigate the propagation of SPPs along metal stripes, we performed a parametric numerical study of stripe modes. As a starting point we calculate the modal solutions for various width (W) of Au stripe waveguides on glass with a thickness (t) of 55 nm and for a free-space wavelength of 800 nm, as schematically shown in the inset of fig. 3(b) (Zia et al., 2005b). Figures 3(a) and (b) show the complex propagation constants ($\beta + i\alpha$) determined for the lowest order leaky quasi-TM modes. Several important trends can be discerned from these plots. First and the foremost there is a cutoff width below which no quasi-TM modes are allowed. This cutoff occurs when the propagation constant of the SPP becomes equal to the propagation constant in air, when the width approaches 1.3 μm. Second, for increasing stripe widths, higher order modes associated with additional maximums in the H_x field along the stripe width become accessible, while the propagation constants for all modes, β_n, asymptotically approach the propagation constant of a SPP localized on a metal film–air interface ($\beta_n/\beta_0 \rightarrow 1.02$). Third, for decreasing stripe widths, confinement of the quasi-TM modes within the metal stripe is reduced as the magnitude of β decreases. For the leaky SPP modes, this diminished confinement results in increased radiation losses into the high-index substrate (i.e., a larger α).

In addition to having a distinct complex propagation constant ($\beta_n + i\alpha_n$), each guided polariton mode is described by a unique transverse mode profile $\left(\overset{\omega}{\psi}_n(x,z)\right)$. Although the lowest order surface plasmon mode has

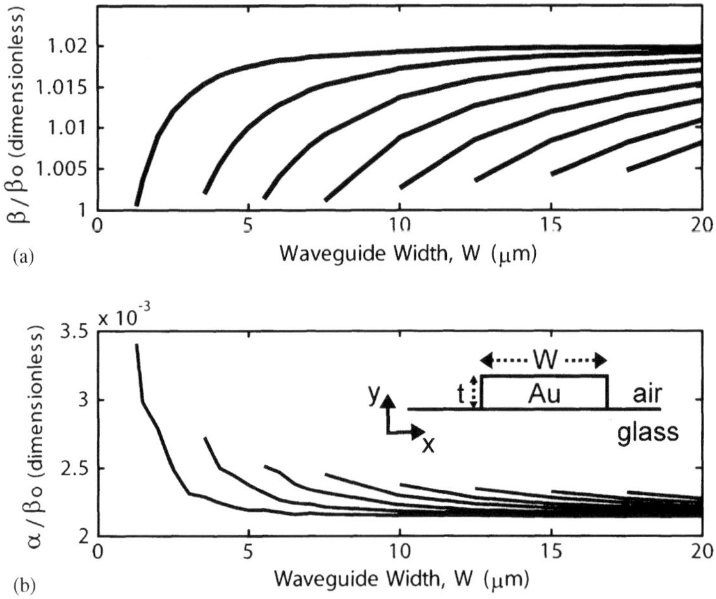

Fig. 3. (a) and (b) Calculated complex propagation constants (β+iα) for the 8 lowest order leaky, quasi-TM SPP modes of varying width Au stripe waveguides ($\varepsilon_{\mathrm{Au}} = -26.1437 + 1.8497i$ for $\lambda = 800$ nm).[35] For these calculations, the Au stripe thickness was $t = 55$ nm and the free-space excitation wavelength was chosen to be $\lambda = 800$ nm.The inset shows the simulated geometry.

only a single lateral maximum in the dominant H_x field, each higher order mode has an additional lateral peak. Figure 4 shows the simulated mode profiles for the leaky SPP modes supported by three Au stripe waveguides of different width. While the 2 μm wide stripe supports only a single mode with even lateral symmetry, the wider 4 and 6 μm stripes also support a second-order mode with an odd lateral symmetry.

By leveraging the lateral symmetries associated with the mode profiles, we have performed a parametric study of multimode interference for varying width stripes (Zia et al., accepted for publication). Multimode interference is commonly exploited in the design of couplers and dividers for applications in integrated optics (Soldano and Pennings, 1995). In the present context, multimode interference effects were used to characterize the modes supported by a complex waveguide structure (Campillo et al., 2003).

As shown in fig. 3(a), it is clear that the phase constant associated with a higher order mode is always smaller than that of the fundamental mode, and thus, if both modes were to be excited simultaneously, one

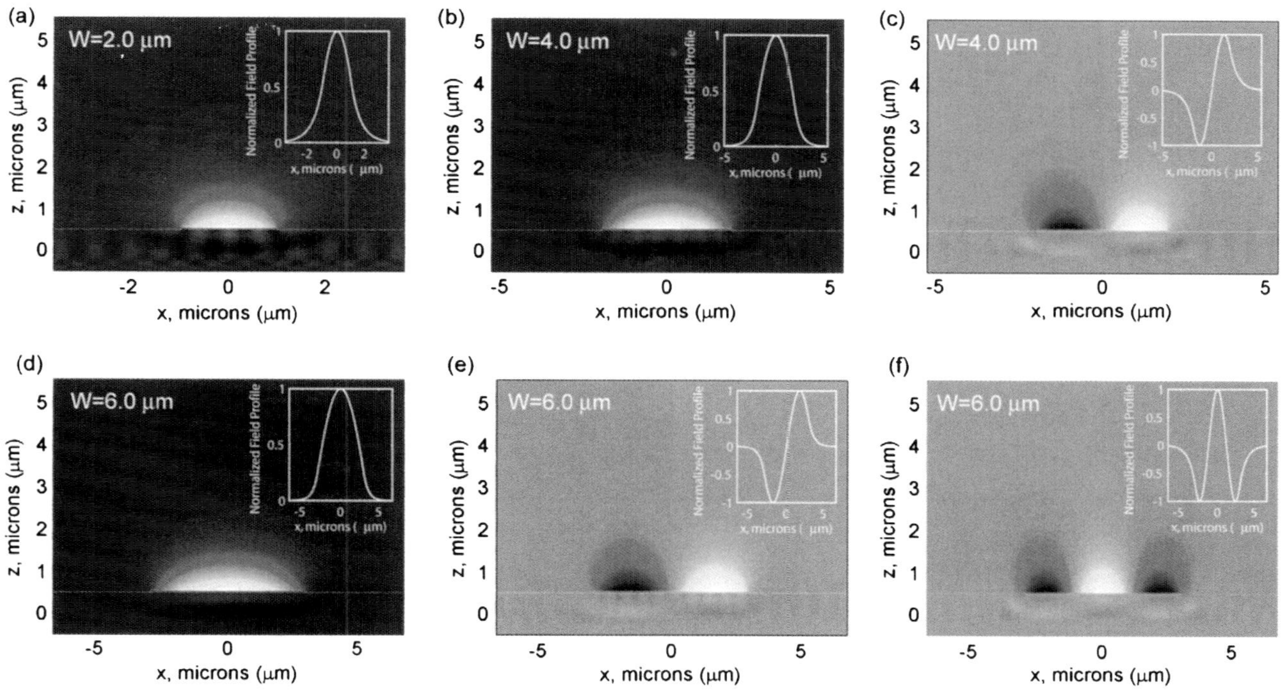

Fig. 4. Simulated field profiles for the leaky surface plasmon modes supported by 2, 4, and 6 µm wide Au stripe waveguides ($t = 48$ nm) using the full-vectorial finite-difference method described in Section 2 ($\varepsilon_{Au} = -24.13 + 1.725i$ for $\lambda = 780$ nm).[35] Insets depict the lateral mode profiles predicted by equivalent dielectric slab waveguides described later in this chapter. The core index of the dielectric slabs was determined by the effective index of the leaky SPP mode supported along an infinitely wide 48 nm thin Au film on a glass substrate ($n_{eff} = k_{sp}/k_0 = 1.022 + 0.003i$, calculated using the reflection pole method).[38]

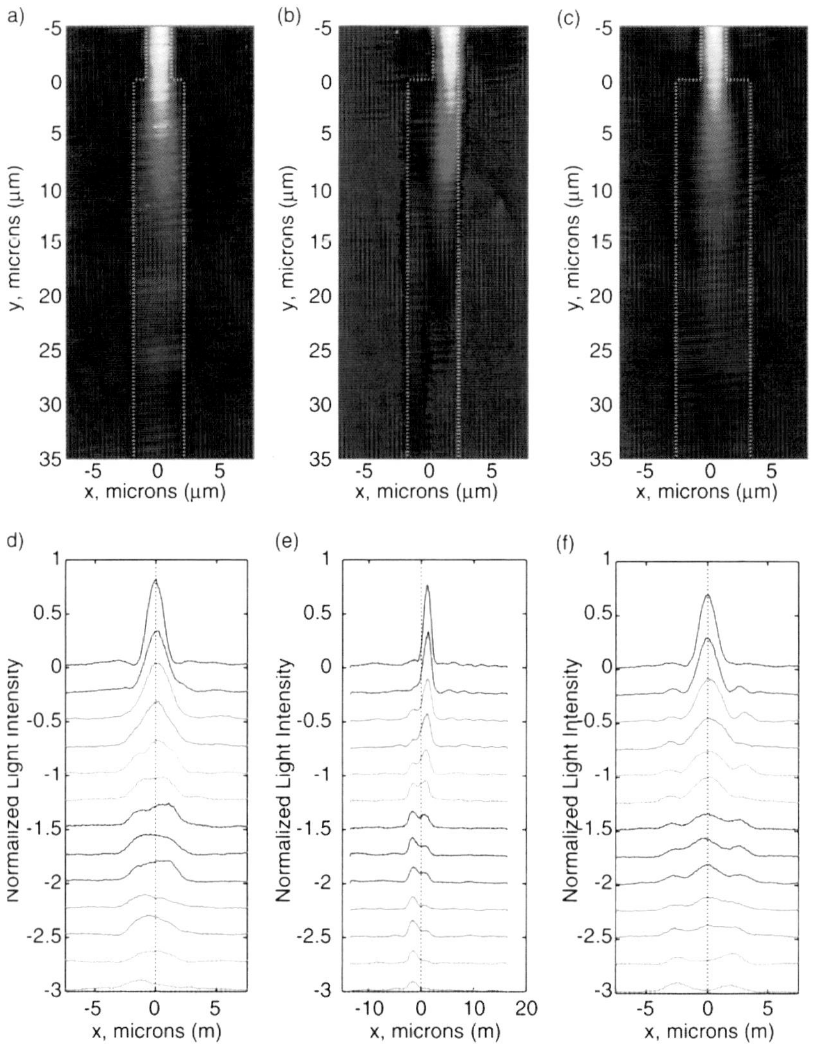

Fig. 5. Experimental demonstration of multimode interference between the guided modes supported by 4 and 6 μm wide Au stripes excited from a 2 μm wide input stripe ($t = 48$ nm, $\lambda = 780$ nm). The dashed white lines indicate the outline of the Au structures. Frames (a), (b), and (c) show near-field images of the multimode interference observed in the PSTM. Frames (d), (e), and (f) show lateral cross sections of the PSTM images shown in (a), (b), and (c), respectively. Initial cross sections show the intensity above the input stripe (acquired at $y = -1$ μm), and subsequent cross sections are taken at 2.5 μm intervals (beginning with $y = -2.5$ μm) and offset by -0.25 increments.

should observe a beating in the propagation direction. To demonstrate the interference related to the existence of these different modes, we can excite a 4 μm wide waveguide using a single mode 2 μm wide input stripe. When the input stripe is centered with respect to the wider waveguide as shown in fig. 5(a), we primarily excite the fundamental mode. This is seen in the PSTM image in fig. 5(a) and the corresponding set of lateral cross sections of the measured PSTM image in fig. 5(d). In these two figures, no major change in the mode profile is observed as a function of the propagated distance. However, when the input stripe is fabricated off-center with respect to the wider waveguide as shown in fig. 5(b), both supported modes are excited. As evidenced by this image and the corresponding lateral cross sections shown in fig. 5(e), there is a clear shift in the transverse intensity profile, as light propagates down the stripe. Near the input region, the lateral profile shows a single peak to the right of the dashed centerline; this profile is consistent with a superposition of the even first-order mode and the odd second-order mode. Further down the stripe though, the intensity of this initial peak diminishes, and an additional peak to the left of the centerline emerges. At the end of the 35 μm long stripe, it appears that the optical intensity has switched to the other side of the 4 μm waveguide. This lateral transition is consistent with a π-phase shift in the relative phases for the first-and second-order modes, and the length scale for this transition is in good agreement with the beat length that was predicted by our numerical simulations (i.e., $\pi/(\beta_1-\beta_2)\approx 38$ μm).

For the 6 μm wide stripe, a third-order leaky SPP mode is also supported. As shown in fig. 4(f), this mode has three lateral intensity peaks in the dominant H_x field profile and an even lateral symmetry. Again, we may exploit the parity difference to verify the existence of specific SPP modes by analyzing the observed multimode interference pattern. Similar to the previous case shown in fig. 5(a), we use a centered 2 μm wide stripe as the input for a wider waveguide to minimize excitation of the second-order odd mode. In figs. 5(c) and (f), we can observe multimode inference, as the excited 6 μm stripe supports two even surface plasmon modes, a fundamental mode with a single lateral peak as well as the third order with three peaks. Near the stripe input, the relative phase of the two modes is such that they interfere to form a single peak at the stripe center. With propagation though, a relative phase shift is incurred such that after propagating over 30 μm there is a local minima along the center of the stripe between two lateral peaks. Again, the observed beat length is in good agreement with the value predicted by full-vectorial simulations of the guided polariton modes (i.e., $\pi/(\beta_1-\beta_2)\approx 29$ μm).

§ 5. Analytical models for stripe modes

From the numerical solutions and experimental results of multimode interference, it is clear that the modes supported by a finite width stripe resemble the modes of dielectric waveguides. There are a finite number of guided modes, and each mode has a unique mode profile. However, to thoughtfully leverage potential analogies with conventional dielectric waveguides, it is important to develop a rigorous comparison.

We will start the discussion by deriving a simple model for the number of modes supported by a metal stripe waveguide (Zia et al., 2005b). From the numerical solutions shown in fig. 3, it is clear that the propagation constant of a SPP propagating along a metal stripe asymptotically approaches the value along an extended film for large stripe widths. It is therefore reasonable to assume that the in-plane momentum for a SPP supported by an infinitely wide metal–dielectric interface (k_{sp}) is conserved for a SPP on a finite width structure. However, along the finite width stripe, this in-plane wave vector can be separated into a component along the direction of propagation ($\beta = \Re\{k_z\}$) as well as a lateral component (k_x) such that

$$k_x^2 + \beta^2 = k_{sp}^2. \tag{5.1}$$

When considering sufficiently thick stripes, the in-plane wave vector can be approximated by the following expression for the SPP supported by a simple metal–dielectric interface (Lamprecht et al., 2001)

$$k_{sp} = \frac{\omega}{c}\sqrt{\frac{\varepsilon_d \varepsilon_m}{\varepsilon_d + \varepsilon_m}} \tag{5.2}$$

In the above, ε_d and ε_m are the relative permittivities of the dielectric and metal regions, respectively. For waveguiding to occur, there is a lower limit placed upon the propagating component of the momentum. The guided SPPs should not be able to couple with the radiation modes of the surrounding dielectric material; therefore, the propagating wave vector must exceed the effective wave number within the dielectric region (i.e., $\beta > (\omega/c)\sqrt{\varepsilon_d}$). At this guiding limit, we can use (eqs. (5.1) and (5.2)) to derive the maximum value for the lateral wave vector:

$$|k_x| < \left(\frac{\omega}{c}\right)\sqrt{\frac{\varepsilon_d \varepsilon_m}{\varepsilon_d + \varepsilon_m} - \varepsilon_d} \tag{5.3}$$

This maximum value for k_x limits quantization along the lateral direction, and thus, we can approximate the number of allowed surface plasmon modes (N) by relating the width (W) of the interface to the lateral wave vector. As the highest order supported mode would have N number of maximums in the lateral direction, the maximum lateral wave vector would be

$$k_{x,\max} = \frac{N\pi}{W} \tag{5.4}$$

Using the above, eq. (5.3) can be simplified to derive an approximate analytical expression for the number of supported modes

$$N < \frac{2W}{\lambda}\sqrt{\frac{-\varepsilon_{\rm d}^2}{\varepsilon_{\rm d}+\varepsilon_{\rm m}}} \tag{5.5}$$

where λ is the free-space wavelength (i.e., $2\pi c/\omega$). Substituting into this expression the relative dielectric constants of air and glass for the leaky and bound modes, respectively, we find our expression in good agreement with the FVH-FDM results as shown in fig. 6. In this figure, the dashed lines indicate the number of supported modes predicted by our simple analytical approximation and the solid squares give the FVH-FDM results.

Note that this description is directly analogous to the conservation of momentum used to describe the propagation of light along conventional dielectric waveguides in physical optics. To build upon this analytical approximation, we need only to recognize how the view of momentum conservation described in eq. (5.1) is intimately related to the ray-optics used in conventional dielectric optics (Zia et al., 2005a).

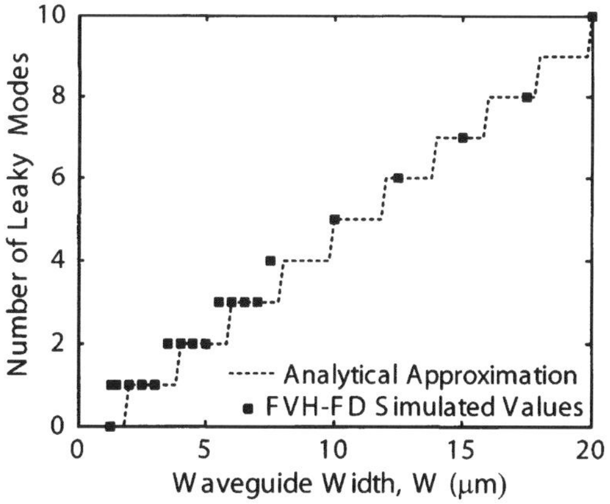

Fig. 6. Comparison with FVH-FDM simulation results for the analytical approximation of the number of leaky quasi-TM SPP modes supported by a metal stripe waveguide as a function of waveguide width. The analysis was performed for a 48 nm thick stripe and an excitation wavelength, $\lambda = 800$ nm.

In the remaining part of this section, we review our demonstration that an equivalent dielectric slab waveguide can be used to approximate the solutions of guided polariton modes (Zia et al., 2005a). By definition, a guided mode is an eigen-state representation of an electromagnetic field profile that propagates in a specified direction (e.g., z-direction) with a unique propagation constant (i.e., $k_z \equiv \beta + i\alpha$). For dielectric slab waveguides, an exact analytical formulation for the modal solutions is possible. Nevertheless, to acquire a physical intuition for these waveguides, an interpretation based upon ray-optics is often utilized. Here, the guided mode is defined by the superposition of plane-wave solutions within the high-index core. To constitute a mode, these waves with ray-like paths must satisfy two conditions: (1) total internal reflection (TIR) at the core–cladding interface and (2) constructive interference following the reflections at both such interfaces. Note that these plane-wave rays, whose use is consistent with application of the isotropic wave equation in parts, have a wave vector magnitude determined by the core's refractive index (i.e., $|k| = (\omega/c)\sqrt{\varepsilon_{\text{core}}} = k_0\, n_{\text{core}}$). Thus, the calculation of the phase and amplitude of reflections at each interface can be obtained from the Fresnel relations.

For surface polariton reflection at the edge adjoining two distinct metal surfaces, it has been shown that Fresnel-like relations provide good approximations when the in-plane wave vectors for infinitely wide polariton modes (k_{sp}) are considered (Stegeman et al., 1983). For example, polariton TIR occurs when the projection of the incident wave vector along such an edge exceeds the maximum magnitude allowed for a transmitted polariton. Conservation of momentum, therefore, stipulates that an associated critical angle ($\theta_c = \sin^{-1}(k_{\text{sp},t}\ /k_{\text{sp},i.})$) can be anticipated. This analysis is equivalent to treatment of each surface polariton region with an effective refractive index, defined as follows

$$n_{\text{eff}} = \frac{k_{\text{sp}}}{k_0} \tag{5.6}$$

where k_{sp} is the in-plane wave vector of a surface polariton supported by an infinitely wide structure.

Extension of the above analysis to reflections at the termination of a polariton supporting surface is straightforward. Consider, for example, the TIR of a surface polariton at the metal film edge depicted in fig. 7(a). The effective refractive index of the polariton (n_{eff}) represents the magnitude of the incident in-plane wave vector, while the refractive index of the dielectric region ($n_{\text{d}} = \sqrt{\varepsilon_{\text{d}}}$) represents the wave vector of a transmitted homogenous wave. Thus, a critical angle ($\theta_c = \sin^{-1}(n_{\text{d}}/n_{\text{eff}})$) is

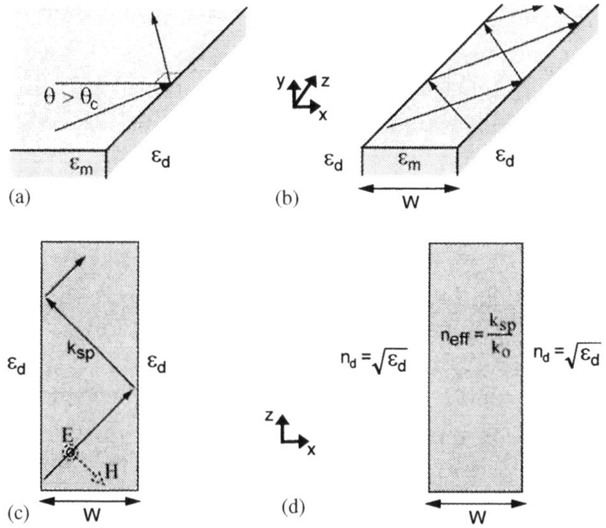

Fig. 7. Dielectric waveguide treatment of surface polaritons along finite width interfaces: (a) TIR of surface polariton wave; (b) ray-optics interpretation of surface polariton mode; (c) top-view of ray-optics interpretation; (d) equivalent two-dimensional dielectric slab waveguide.

expected. If such an effective index treatment accurately represents the internally reflected surface polariton, an approximate ray-optics model for guided polariton modes can be derived. For the continuous and constructive TIR of polaritons along a finite width interface (shown in figs. 7(b) and (c)), the guided mode resembles, and can be modeled by, the TE mode of an equivalent dielectric slab waveguide (shown in fig. 7(d)). We call this model approximate because the TM nature of a surface polariton requires electric field components both normal to the supporting surface and along the direction of propagation. However, for surface polaritons that propagate any significant distance, the dominant electric field component is normal to the film as drawn in fig. 7(c), and thus, the analogous dielectric waveguide modes in our two-dimensional model should be TE.

To investigate the applicability of this model, we compare its approximate solutions with solutions obtained by the full-vectorial magnetic-field finite-difference method discussed in Section (3). Specifically, we consider the SPP modes supported by coupled top and bottom interfaces of a metal stripe embedded within a dielectric matrix. The width of the model waveguide is identical to the finite width (W) of the stripe, and the effective refractive index of the core is determined by eq. (5.6). However, it is important to note that the wave vector along an infinitely wide structure (k_{sp}) depends upon the spatial separation (t) of the coupled interfaces. So, for this case the value of k_{sp} that should be used in eq. (5.6) to determine the effective index must be determined by solving for the modes of a two-dimensional metal slab waveguide (Zia et al., 2005b). Here, the solutions for both the field-symmetric modes were determined by use of the reflection pole method (Anemogiannis et al., 1999).

We have used the FVH-FDM to solve for the so-called M_{00} and M_{10} modes as a function of varying stripe width and thickness. The notation for these modes was proposed by Al-Bader et al. (2004), and the two indices indicate the number of nodes in the TM field in the x- and y-directions. The corresponding dielectric slab modes are TE_0 and TE_1, respectively. To facilitate a straightforward comparison with the dielectric waveguide model, we present these solutions in the form of normalized, dispersion curves (Haus, 1984). Such a curve is obtained by plotting the normalized frequency (V) against the normalized guide index (b), which are defined as follows

$$V \equiv k_0 W \sqrt{n_{\mathrm{eff}}^2 - n_{\mathrm{d}}^2} \tag{5.7}$$

$$b \equiv \frac{\left(\frac{(\beta+\mathrm{i}\alpha)^2}{k_0^2} - n_{\mathrm{d}}^2\right)}{(n_{\mathrm{eff}}^2 - n_{\mathrm{d}}^2)} \tag{5.8}$$

n_{eff} is effective refractive index for our two-dimensional dielectric model as defined by eq. (5.6).

Figure 8 shows that when normalized, the lowest order field-symmetric modes of a metal stripe waveguide (solid and open symbols) are in good agreement with the universal solutions for the equivalent TE dielectric slab waveguide (solid and dashed curves). This agreement extends to the prediction of the cutoff frequency for the higher order (M_{10}) mode.

Moreover, the lateral confinement of the three-dimensional guided polariton is well predicted by the two-dimensional model. As a representative example, we have plotted the lateral power density (i.e., $\int \Re\{S_z\} \mathrm{d}y$) for a metal stripe waveguide and the equivalent dielectric waveguide in the inset of fig. 8. Despite the discontinuity of the Poynting vector for the polariton mode, the dielectric approximation anticipates the power density profile. That such physical behavior for a three-dimensional surface wave can be predicted by a volume electromagnetic waveguide has implications on many debated topics in guided polariton optics. For example, the minimum optical mode size of a polariton stripe waveguide has been a subject of much interest and, like that of a dielectric waveguide, must be determined by an uncertainty principle. Without

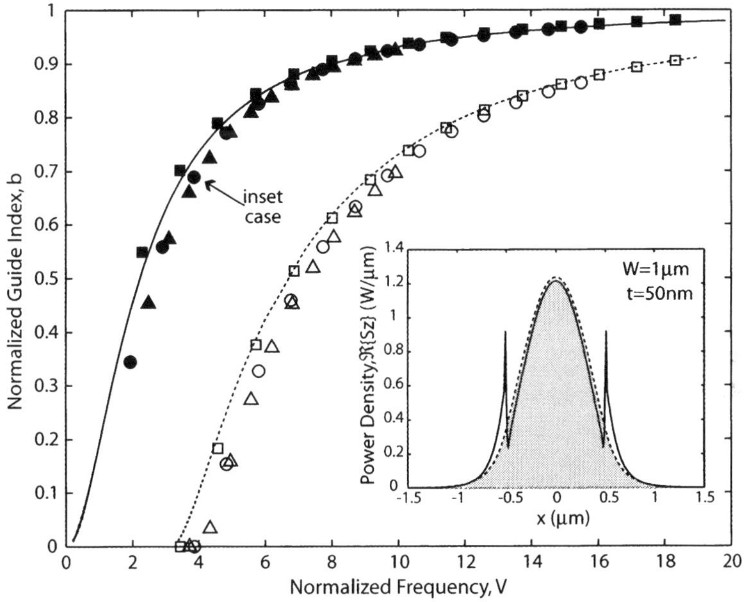

Fig. 8. Normalized dispersion curves for SPP modes supported by an Ag stripe waveguide. Filled and hollow markers denote solutions obtained by FVH-FDM for the M_{00} *and* M_{10} modes, respectively. For three stripe thicknesses ($t = 25$ nm (Δ), 50 nm (○), 100 nm (□)), the width (W) was varied between 0.5 and 4 μm. Solid and dashed lines represent the TE_0 and TE_1 modes of a dielectric slab waveguide, respectively.[14] Inset compares the lateral power densities for surface plasmon waveguide (solid line, $W = 1$ μm, $t = 50$ nm) and approximate dielectric waveguide (dashed line with gray shading, $n_{\text{eff}} = 3.628 + 0.002267\text{i}$) normalized to unit power.

validation of an appropriate wave vector basis set, such a diffraction limit is difficult to formulate (Zia et al., 2005b). However, on the basis of index guiding, a diffraction limited mode size (Δx) in the lateral dimension can be derived for modes accurately approximated by our model, as

$$\Delta x \geq \frac{\lambda_0}{2n_{\text{eff}}} \tag{5.9}$$

§ 6. Propagation along metal stripe waveguides

If a diffraction limit for metal stripe waveguides indeed exists, then how is one left to interpret the initial experimental results, which suggested that subwavelength metal stripes may support highly confined surface plasmon modes (Lamprecht et al., 2001; Yin et al., 2005)? Our numerical studies have shown that there is a cutoff condition for metal stripe waveguides and that no guided surface polariton modes are supported along subwavelength metal stripes (Zia et al., 2005b). Thus, a comparison with previous experimental results suggests that guided polariton modes alone are insufficient to describe the observed behavior. In the following section, we review a recent study demonstrating that the propagation of light along surface plasmon waveguides is mediated by a continuum of radiation modes as well as a discrete number of guided polariton modes (Zia et al., accepted for publication).

It is well known that a complete description of light propagation in dielectric waveguides requires a continuum of radiation modes in addition to a discrete number of guided solutions (Marcuse, 1982, 1991). Here, we demonstrate that a similar description may be used for the propagation of "light" along metal waveguides. To introduce this description, we leverage the dielectric waveguide model for guided surface polaritons in fig. 7 (Zia et al., 2005a). In fig. 7(a), we showed how the TIR of SPPs at the edge of a metal film may lead to the existence of guided surface polariton modes in a metal stripe of finite width, as shown in fig. 7(b). Similar to the modes of a dielectric slab waveguide, SPPs along the finite width stripe must constructively interfere upon TIR to form a guided mode. This interference condition establishes an eigen-value problem with a discrete set of modal solutions. In contrast, SPPs incident below the critical angle for TIR may be transmitted into the external dielectric region. This transmission forms the basis for radiation modes. Given the continuous range of possible angles below the critical condition forTIR, these solutions form a continuum of radiation modes. Although the fields associated with the radiation continuum extend well beyond the finite width stripe, these solutions may contribute to the local optical field in the vicinity of the waveguide. To distinguish the guided modes from the radiation continuum, we perform a parametric study of propagation length as a function of stripe width.

Using the aforementioned PSTM, we have mapped the propagation of SPPs along varying width of Au stripes on glass substrates. Fifteen different stripe widths were investigated ranging from 500 nm to 6 μm. Figure 9 shows characteristic near-field images for the eight narrowest stripe widths.

Similar to previous far-field measurements along Ag stripes (Lamprecht et al., 2001), it appears that the observed propagation length decreases as a function of decreasing stripe width. This general behavior is also in good agreement with previous numerical solutions for the leaky SPP modes supported by the top air–metal interface of Au stripe waveguides (Zia et al., 2005b). In analyzing the propagation distances for different waveguide widths, we can anticipate additional features related to the discrete nature of guided polariton transport in such structures. In

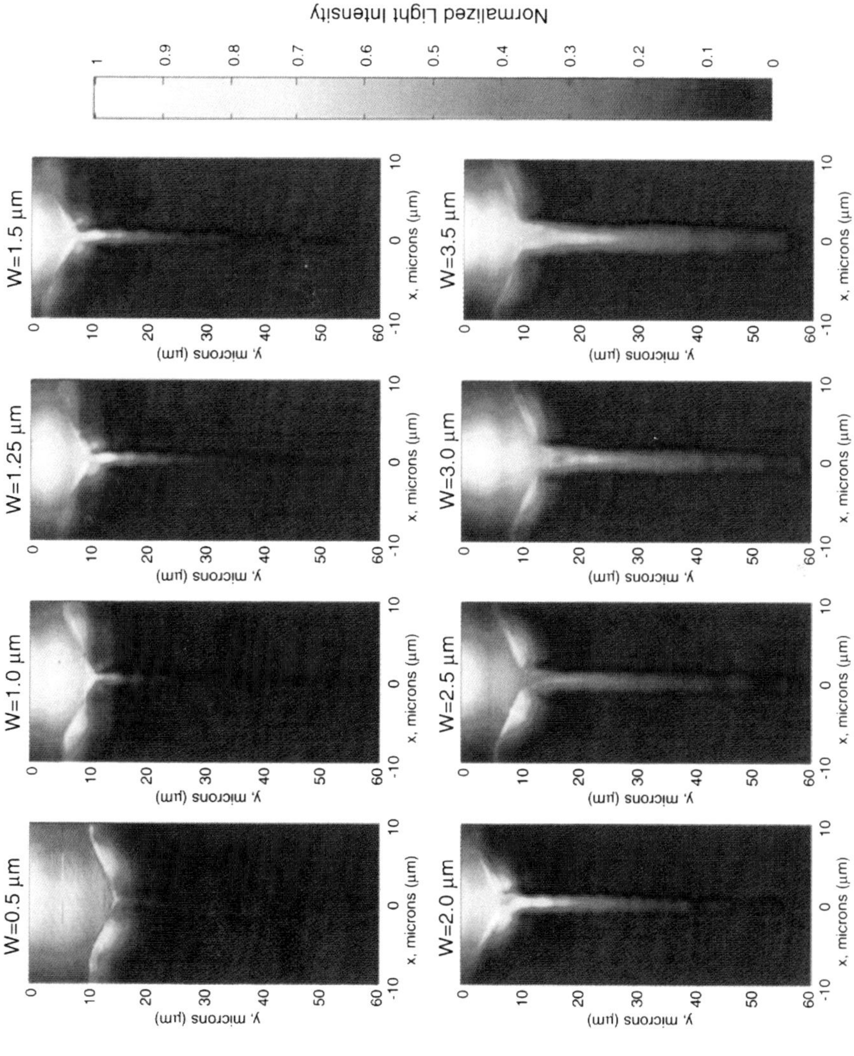

Fig. 9. Experimental near-field images of SPP propagation along metal stripe waveguides with widths ranging from 0.5 to 3.5 μm.

particular, a metal stripe waveguide can support a finite number of guided modes. While wide stripes may support multiple guided modes, narrower stripes may support none. In the following, we will show how these finite variations in mode number may be used to distinguish the discrete guided modes from the radiation continuum.

To quantify the propagation length, previous studies have commonly fit observed intensity profiles into a single decaying exponential function (i.e., $|E|^2 \approx A\mathrm{e}^{-y/L} + c$, where y denotes position along the direction of propagation, L is the 1/e decay length for intensity, and c is an offset constant generally associated with background noise). Such analysis provides a qualitatively useful measure of propagation length, but the single exponential decay is an imprecise description when multiple modes are excited. As we vary the stripe width, we anticipate this inaccuracy will be most noticeable in regions where there is a transition in the number of allowed modes. In the context of waveguide theory, the electric field in a region supporting N number of guided modes may be described by the following expression

$$\overset{\varpi}{E}(x,y,z) = \sum_{n=1}^{N} a_n \overset{\varpi}{\psi}_n(x,z)\mathrm{e}^{\mathrm{i}(\beta_n+\mathrm{i}\alpha_n)y} + \int_0^{+\infty}\int_0^{k_0} b_{k',k''}\overset{\varpi}{\psi}_{k',k''}(x,z)\mathrm{e}^{\mathrm{i}(k'+\mathrm{i}k'')y}\,\mathrm{d}k'\,\mathrm{d}k'' \tag{6.1}$$

where the summation and integral terms denote the contribution of the guided modes and the radiation continuum, respectively (Snyder and Love, 1983). Each guided (radiation) mode is described by an amplitude coefficient $a_n[b_{k',k''}]$, transverse mode profile $\overset{\omega}{\psi}_n(x,z)\left[\overset{\omega}{\psi}_{k',k''}(x,z)\right]$, and complex propagation constant $(\beta_n + \mathrm{i}\alpha_n)[(k' + \mathrm{i}k'')]$. Note that the summation term in eq. (6.1) reflects the discrete nature of the guided polariton modes. Ignoring for the time being the contribution of the radiation continuum, we may recognize how this discrete nature influences the relationship between the physical decay constants (α_n) and the phenomenological propagation length. For the case of a single guided mode (e.g., $N = 1$), the propagation length is directly related to the mode's decay constant (e.g., $L = 1/2\alpha_1$). For the case of multiple modes, however, the fit will depend upon the relative intensities of the supported modes, and the propagation length more closely approximates a weighted average of the decay constants. Therefore, for varying stripe widths, the propagation length will reflect not only changes in the decay constants, but also changes in the number of supported modes. As cutoff of a guided mode may dramatically alter a weighted average of the decay constants, we may anticipate discontinuities in the observed propagation length as a function of stripe width.

While discontinuities were not reported in previous far-field measurements (Lamprecht et al., 2001), the enhanced spatial resolution offered by near-field techniques has enabled us to observe such behavior. To calculate propagation lengths from the near-field images, we have integrated the light intensity along the width of each stripe and fit the resulting curve with a simple exponential decay, as follows

$$\int_{-W/2}^{W/2} \left|E(x, y, z = h_0)\right|^2 \mathrm{d}x \approx A\mathrm{e}^{-(y)/L} + c. \tag{6.2}$$

The integration here serves to average the light intensity along the stripe width. To avoid artifacts associated with the finite stripe length, the fit region for each stripe did not include the 5 μm region closest to the end of each stripe. Figure 10 shows the fit propagation lengths as a function of stripe width.

Using the full-vectorial finite difference method described previously in Section 3 (Zia et al., 2005b), we have also solved for leaky SPP modes supported by these stripes. Alongside the experimental data, we plot the calculated decay behavior for the three lowest order surface plasmon modes. Vertical lines have been used to denote the calculated cutoff widths

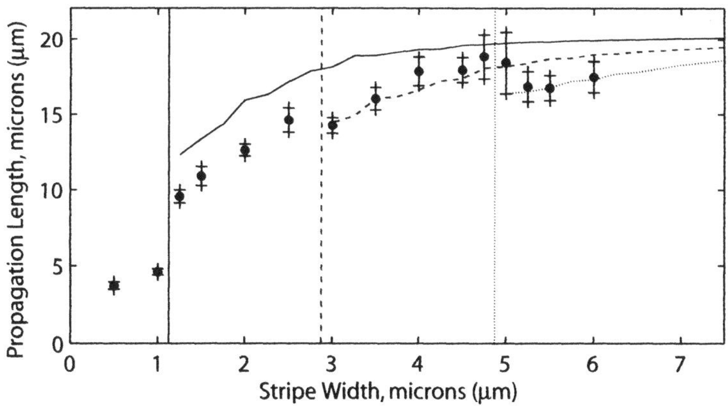

Fig. 10. Surface plasmon propagation length as a function of stripe width at 780 nm for Au stripes on glass substrates. Circular markers denote experimental data with error bars determined by 95% tolerance intervals. The solid, dashed, and dotted curves show the simulated decay behavior for the three lowest order leaky surface plasmon modes supported by these stripes, and the associated vertical lines denote the predicted cutoff widths for these modes. Note that numerical solutions were obtained using the full-vectorial finite-difference method described in Zia, R. 2005b for 48 nm thick Au stripes ($\varepsilon_{\mathrm{Au}} = -24.13 + 1.725\mathrm{i}$) on a glass substrate ($\varepsilon_{\mathrm{glass}} = 2.25$).

for the first-, second-, and third-order modes near 1.25, 3, and 5 μm, respectively. Note that for wider stripes (i.e., $W \geqslant 3\,\mu m$), the observed propagation length falls within the range of values predicted by these numerical simulations. For these stripe widths, we also observe the expected discontinuities near cutoff for the higher order guided modes. While the propagation length tends to decrease with decreasing stripe width, there are two increases that oppose this trend. The propagation lengths for stripe widths between 4 μm and 5 μm are higher than those for wider stripes, and the propagation length increases slightly as stripe widths are reduced from 3 μm to 2.5 μm. These increases are consistent with cutoff for the lossier third- and second-order SPP modes, respectively. As higher order modes are cutoff, the observed propagation length increasingly reflects the lower loss fundamental mode. Below the predicted cutoff for the fundamental guided mode, we observe a severe discontinuity. As stripe widths are reduced from 1.25 to 1.0 μm, the propagation length drops significantly from 9.6 to 4.8 μm. Unlike cutoff for the higher order modes where there still exist lower order modes with reduced losses, it is not surprising that cutoff for the fundamental guided mode results in a severe decrease in propagation length. This third discontinuity again reflects the discrete nature of the guided solutions, but also represents a transition to a new regime in which there are no guided modes at all.

Below the predicted cutoff width for the fundamental surface plasmon mode, it is not surprising that we continue to observe finite propagation lengths. The lack of guided modes along narrow stripes does not imply that light cannot propagate nor that propagation cannot be observed. Even in the absence of a metal stripe (i.e. the limiting case of infinitesimal stripe width), the termination of the tapered launchpad presents a discontinuity which should scatter SPPs into propagating radiation, and this scattered light may be detected by our PSTM at short distances from the launchpad edge. In the context of modal theory, such non-guided pathways are described by the integral term for the continuum of radiation modes in eq. (6.1). Although such radiation modes are beyond the scope of the guided modes reviewed here, we have demonstrated elsewhere that the finite propagation lengths observed for narrow stripes are consistent with both experimental and numerical models for the contribution of the radiation continuum. (Zia et al., submitted)

§ 7. Summary

In this chapter, we presented work from recent studies on SPP propagation along metal stripe waveguides at optical frequencies. In the past, experimental results had been interpreted to suggest that the SPP modes supported by finite width metal stripe waveguides were fundamentally different from those of conventional dielectric waveguides and required a radical new way of thinking. It was thought that the propagation along metal stripes was dominated by a single optical mode and that this mode was inconsistent with the standard ray-optics interpretation of guided waves. Some researchers had suggested that the propagation of light along metal stripes was not limited by diffraction. Given the complex boundary conditions that exist at the edges of a metal stripe, no appropriate basis set for such a diffraction limit seemed to exist.

To address the open questions posed above, we presented a combined near-field optical and numerical study of the transport properties of metal stripe waveguides. Multimode interference studies on such stripes were employed to provide direct evidence for the existence of multiple guided modes and demonstrate modal cutoff in narrow stripes. This cutoff behavior was clearly observed in the parametric study of the dependence of SPP transport as a function of waveguide width. In this study it was also found that propagation length decreases with decreasing waveguide width until finally cutoff is observed in the narrowest stripes. Experimental and numerical investigations confirm that the finite propagation lengths observed along stripes below cutoff are in good agreement with an intuitive model for the radiation continuum. It is important to note that these studies do not indicate that the propagation of SPP waves with deep subwavelength mode diameters is impossible. While weakly guided stripe waveguides may not achieve this goal, there are alternative geometries, which can provide strong confinement. For example, Takahara's original paper on SPP modes of a metal cylinder shows that subwavelength mode diameters are possible and propagation over short distances can be realized (Raether, 1965). Moreover, waveguides consisting of two closely spaced metals also support deep subwavelength modes that can propagate over micron-sized distances (Veronis and Fan, 2005; Zia et al., 2005b). For all types of waveguides though, there is a clear tradeoff between confinement and propagation distance (loss). The use of one type of waveguide over another will thus depend on application specific constraints.

To compliment the presented full-vectorial model, we also considered a more intuitive physical picture of SPP guiding along plasmon stripe waveguides. This picture allowed for a simple estimate of the number of modes supported by a metal stripe. More generally, our study suggests that existing dielectric waveguide theory and many of the existing dielectric simulation tools can be employed to model the behavior of more complex plasmonic structures. For example, beam propagation tools could be leveraged to model coupled waveguides, splitters, multimode interference couplers, etc. The success of this dielectric waveguide model is based on the important notion that guided polariton modes supported by finite width interfaces can be related to the solutions for infinitely wide structures. This description is directly analogous to the conservation of momentum picture used to describe light propagation in conventional dielectric waveguide structures in physical optics.

It was our aim to develop and present a unified theory of the guided polariton optics for surface plasmon waveguides that is consistent with conventional guided wave optics for dielectric waveguides. Although this formulation belies some of the previous claims for novel optical physics in this regime, we hope that this work will provide a basis from which to leverage decades of research on dielectric integrated optics for the development of surface plasmon optics and to expand into new and exciting directions.

References

Al-Bader, S.J., 2004, Optical transmission on metallic wires-fundamental modes, IEEE J. Quantum Electron. **40**, 325–329.

Anemogiannis, E., Glytsis, E.N., Gaylord, T.K., 1999, Determination of guided and leaky modes in lossless and lossy planar multilayer optical waveguides: reflection pole method and wavevector density method, J. Lightwave Technol. **17**, 929–941.

Barnes, W.L., Dereux, A., Ebbesen, T.W., 2003, Surface plasmon subwavelength optics, Nature **424**, 824–830.

Berini, P., 2000, Plasmon-polariton waves guided by thin lossy metal films of finite width: bound modes of symmetric structures, Phys. Rev. B **61**, 10484–10503.

Berini, P., 2001, Plasmon-polariton waves guided by thin lossy metal films of finite width: bound modes of asymmetric structures, Phys. Rev. B Condens. Matter Mater. Phys. **63**, 125415–125417.

Brongersma, M.L., Hartman, J.W., Atwater, H.A., 2000, Electromagnetic energy transfer and switching in nanoparticle chain arrays below the diffraction limit, Phys. Rev. B Condens. Matter **62**, R16356–R16359.

Burke, J.J., Stegeman, G.I., Tamir, T., 1986, Surface-polariton-like waves guided by thin, lossy metal films, Phys. Rev. B Condens. Matter **33**, 5186–5201.

Campillo, A.L., Hsu, J.W.P., Parameswaran, K.R., Fejer, M.M., 2003, Direct imaging of multimode interference in a channel waveguide, Opt. Lett. **28**, 399–401.

Charbonneau, R., Berini, P., Berolo, E., Lisicka-Shrzek, E., 2000, Experimental observation of plasmon-polariton waves supported by a thin metal film of finite width, Opt. Lett. **25**, 844–846.

Chew, W.C., Jin, J.M., Michielssen, E., 1997, Complex coordinate stretching as a generalized absorbing boundary condition, Microw. Opt. Technol. Lett. **15**, 363–369.

Chen, W.P., Ritchie, G., Burstein, E., 1976, Excitation of surface electromagnetic waves in attenuated total-reflection prism configurations, Phys. Rev. Lett. **37**, 993–997.

Economou, E.N., 1969, Surface plasmons in thin films, Phys. Rev. **182**, 539–554.

Feng, N.N., Zhou, G.R., Xu, C.L., Huang, W.P., 2002, Computation of full-vector modes for bending waveguide using cylindrical perfectly matched layers, J. Lightwave Technol. **20**, 1976–1980.

Haus, H.A., 1984, Waves and Fields in Optoelectronics, Prentice-Hall, Englewood Cliffs, NJ, USA.

Huang, W.P., Xu, C.L., Lui, W., Yokoyama, K., 1996, The perfectly matched layer boundary condition for modal analysis of optical waveguides: leaky mode calculations, IEEE Photonics Technol. Lett. **8**, 652–654.

Krenn, J.R., Lamprecht, B., Ditlbacher, H., Schider, G., Salerno, M., Leitner, A., Aussenegg, F.R., 2002, Non-diffraction-limited light transport by gold nanowires, Europhys. Lett. **60**, 663–669.

Krenn, J.R., Weeber, J.C., 2004, Surface plasmon polaritons in metal stripes and wires, Philos. Trans. R. Soc. Lond. A –Math. Phys. Eng. Sci. **362**, 739–756.

Kretschmann, E., 1971, The determination of the optical constants of metals by excitation of surface plasmons, Zeitschrift fur Physik A (Atoms and Nuclei) **241**, 313–324.

Lamprecht, B., Krenn, J.R., Schider, G., Ditlbacher, H., Salerno, M., Felidj, N., Leitner, A., Aussenegg, F.R., Weeber, J.C., 2001, Surface plasmon propagation in microscale metal stripes, Appl. Phys. Lett. **79**, 51–53.

Lusse, P., Stuwe, P., Schule, J., Unger, H.G., 1994, Analysis of vectorial mode fields in optical wave-guides by a new finite-difference method, J. Lightwave Technol. **12**, 487–494.

Marcuse, D., 1982, Light Transmission Optics, 2nd ed., Van Nostrand Reinhold, New York, NY, USA.

Marcuse, D., 1991, Theory of Dielectric Optical Waveguides, 2nd ed., Academic Press, San Diego, CA, USA.

Mihalcea, C., Scholz, W., Werner, S., Munster, S., Oesterschulze, E., Kassing, R., 1996, Multipurpose sensor tips for scanning near-field microscopy, Appl. Phys. Lett. **68**, 3531–3533.

Prade, B., Vinet, J.Y., Mysyrowicz, A., 1991, Guided optical waves in planar heterostructures with negative dielectric constant, Phys. Rev. B Condens. Matter **44**, 13556–13572.

Quinten, M., Leitner, A., Krenn, J.R., Aussenegg, F.R., 1998, Electromagnetic energy transport via linear chains of silver nanoparticles, Opt. Lett. **23**, 1331–1333.

Raether, H., 1965, Tracts of Modern Physics, Springer-Verlag, Berlin.

Raether, H., 1988, Surface Plasmons on Smooth and Rough Surfaces and on Gratings, Springer-Verlag, New York.

Reddick, R.C., Warmack, R.J., Ferrell, T.L., 1989, New form of scanning optical microscopy, Phys. Rev. B Condens. Matter **39**, 767–770.

Snyder, A.W., Love, J.D., 1983, Optical Waveguide Theory, Chapman & Hall, London, UK.

Soldano, L.B., Pennings, E.C.M., 1995, Optical multi-mode interference devices based on self-imaging: principles and applications, J. Lightwave Technol **13**, 615–627.

Stegeman, G.I., Glass, N.E., Maradudin, A.A., Shen, T.P., Wallis, R.F., 1983, Fresnel relations for surface polaritons at interfaces, Opt. Lett. **8**, 626–628.

Takahara, J., Kobayashi, T., 2004, Low-dimensional optical waves and nano-optical circuits, Opt. Photonics News **15**, 54–59.

Takahara, J., Yamagishi, S., Taki, H., Morimoto, A., Kobayashi, T., 1997, Guiding of a one-dimensional optical beam with nanometer diameter, Opt. Lett. **22**, 475–477.

Veronis, G., Fan, S.H., 2005, Guided subwavelength plasmonic mode supported by a slot in a thin metal film, Opt. Lett. **30**, 3359–3361.

Weeber, J.C., Dereux, A., Girard, C., Krenn, J.R., Goudonnet, J.P., 1999, Plasmon polaritons of metallic nanowires for controlling submicron propagation of light, Phys. Rev. B Condens. Matter **60**, 9061–9068.

Weeber, J.C., Krenn, J.R., Dereux, A., Lamprecht, B., Lacroute, Y., Goudonnet, J.P., 2001, Near-field observation of surface plasmon polariton propagation on thin metal stripes, Phys. Rev. B Condens. Matter Mater. Phys **64**, 045411–045419.

Weeber, J.C., Lacroute, Y., Dereux, A., 2003, Optical near-field distributions of surface plasmon waveguide modes, Physical Review B (Condensed Matter and Materials Physics) **68**, 115401.

Weeber, J.C., Lacroute, Y., Dereux, A., Devaux, E., Ebbesen, T., Girard, C., Gonzalez, M.U., Baudrion, A.L., 2004, Near-field characterization of Bragg mirrors engraved in surface plasmon waveguides, Phys. Rev. B Condens. Matter Mater. Phys, **70**, 2354061.

Yin, L.L., Vlasko-Vlasov, V.K., Pearson, J., Hiller, J.M., Hua, J., Welp, U., Brown, D.E., Kimball, C.W., 2005, Subwavelength focusing and guiding of surface plasmons, Nano Lett. **5**, 1399–1402.

Zia, R., Chandran, A., Brongersma, M.L., 2005a, Dielectric waveguide model for guided surface polaritons, Opt. Lett. **30**, 1473–1475.

Zia, R., Schuller, J.A., Brongersma, M.L., accepted for publication, Near-field characterization of guided polariton propagation and cutoff in surface plasmon waveguides, Phys. Rev. B.

Zia, R., Selker, M.D., Brongersma, M.L., 2005, Leaky and bound modes of surface plasmon waveguides, Phys. Rev. B Condens. Matter Mater. Phys, **71**, 165431.

Chapter 8

Biosensing with plasmonic nanoparticles

by

Thomas Arno Klar

School of Electrical and Computer Engineering and Birck Nanotechnology Center, Purdue University, West Lafayette, IN

On leave from Photonics and Optoelectronics Group, Physics Department and CeNS, Ludwig-Maximilians-Universität München, Germany

Nanophotonics with Surface Plasmons
Advances in Nano-Optics and Nano-Photonics
ISSN: 1871-0018

V.M. Shalaev & S. Kawata (Editors)

DOI: 10.1016/S1871-0018(06)02008-5

Contents

§ 1. The current need for new types of biosensors

Knowledge in molecular biology and medical diagnostics has experienced a tremendous increase over the past few decades. Laboratory techniques have made a huge step forward, especially in automation and parallelisation. The decoding of the human genome has set a milestone in genomics (International Human Genome Sequencing Consortium, 2001; Venter et al., 2001), and proteomics has also made significant progress. Both areas have matured to appreciable industrial branches, especially in molecular diagnostics, which is the art of detecting diseases on the molecular level. In most cases, the primary step of detection is performed by biomolecules. Single-stranded DNA can be naturally 'detected' by its complementary strand and proteins are recognized by antibodies. As the primary steps of molecular recognition take place on the molecular level, which is on the nanometre scale, one might ask how to efficiently record these events. We need a 'reporter' that tells us that the primary step of molecular recognition has taken place. Owing to the nanoscopic size of proteins and DNA, plasmonic nanoparticles naturally present themselves as potential reporters. In order to perform the desired task, they have to translate the molecular-recognition events into our macroscopic world. This means, they have to send out a signal that can be easily detected by us (macroscopic) humans.

Apart from detecting proteins or DNA, plasmonic nanoparticles may also be used for the detection of haptens (small molecules) or in environmental sensing. The number of substances that can be detected by sensors is rapidly growing each year. As mentioned, breathtaking progress in proteomics and genomics sets pressure on engineers and physicists to come up with new and innovative concepts to detect the molecular-recognition events, especially because the concentrations of analytes of interest become smaller and even undetectable with the conventional methods. There is also a general tendency to miniaturize the sensor formats in order to facilitate the simultaneous detection of a large number of different antigens on one single biochip. In this chapter I will show that plasmonic nanoparticles can tackle this challenge.

Section 2 will introduce the nanoparticle plasmon resonance and its physical consequences of field enhancement, absorption, and scattering, with special attention to biosensing. Section 3 will show how plasmonic nanoparticles are used to detect antigens in heterogeneous immunoassays. In these detection schemes, the sheer presence of the plasmonic nanoparticles tells about the antigens. Differently, in Sections 4 and 5 a spectral shift of the plasmon resonance is used to detect molecular-binding events. In Section 4, the molecular-binding event leads to a coupling of two or more plasmonic resonators and therefore to a spectral shift. In difference, the spectral shifts in Section 5 are caused by a change in the dielectric surrounding of the plasmonic nanoparticles, originating by molecular-recognition events. Finally, Section 6 reviews approaches to make use of the extremely high fields close to plasmonic nanostructures. This allows for surface-enhanced Raman scattering (SERS). Returning back to our initially introduced metaphor of a plasmonic nanostructure as a reporter of biomolecular-recognition events, the nanostructure 'talks' in its own words in Sections 3, 4, and 5, but rather acts as a 'loudspeaker' in Section 6.

§ 2. Nanoparticle plasmons

Plasmonic materials have been used for at least 1700 years, although in those early days craftsman did certainly not understand the physics behind them. One of the oldest plasmonic glass materials is the 'Lycurgus' cup from the fourth century AD on display in the British Museum. It appears red when transilluminated, but shines green when imaged in reflection. Also world famous are the medieval glass windows in French and German gothic cathedrals, which also contain plasmonic metal nanoparticles.

Early understanding of the physics of nanoparticle plasmons dates back to Faraday (1857) and Mie (1908), and excellent reviews may be found in the books by Kreibig and Vollmer (1995) or Bohren and Huffman (1983) or in the articles by Mulvaney (1996) or by Link and El-Sayed (2000) just to name a few. It is not the aim of this chapter to rewrite their work. I just want to give enough insight into the theory of nanoparticle plasmons and their relatives, the volume plasmons and the surface plasmons, which is necessary to understand the working principles of plasmonic biosensors.

The conduction band electrons in metals can undergo a coherent oscillation, the so-called plasma oscillation. The conduction band electrons can be considered as essentially free electrons where the presence of the periodic distribution of positively charged core atoms is subsumed in their effective mass m_{eff} (Ashcroft and Mermin, 1976). The electromagnetic field of an incoming light wave can induce polarisation of the conduction electrons; this means that the electrons are displaced with respect to the much heavier positive core ions. Depending on the dimensionality of the metal body, one has to distinguish between different modes of plasmonic oscillations: There exist volume plasmons in a large three-dimensional metallic body. The so-called surface plasmon resonances (SPRs) are of importance at a metal–dielectric interface, and the nanoparticle plasmon resonances (NPPRs) are of decisive influence on the optical spectra of nanoscopic metallic particles. Sometimes the NPPRs are also called localized surface plasmonic resonances (LSPR), but I would like to stick to the term NPPR because, as we will see below, the NPPRs show significant differences to surface plasmons.

At this point I would like to make a comment on nomenclature: I will use the terms 'plasmonic resonance' and 'plasmon' as synonyms throughout this chapter. The term 'plasmonic resonance' stems from a more electrodynamically motivated picture and the term 'plasmon' clearly points to the particle nature from a more quantum mechanical point of view. Strictly speaking, one should also term these quanta 'plasmon-polaritons' as they are mixed entities made from photons and plasmons; however, I will drop the 'polariton' for sake of brevity, as it does not add to the understanding of nanoparticle *biosensors*.

2.1. *Volume plasmons*

The dielectric response of a metal to electrodynamic radiation is given by the complex dielectric constant

$$\varepsilon_{met} = \varepsilon'_{met} + i\varepsilon''_{met}, \tag{2.1}$$

where the real part, ε'_{met}, determines the degree to which the metal polarizes in response to an applied external electric field and the imaginary part, ε''_{met}, quantifies the relative phase shift of this induced polarization with respect to the external field and it includes losses. The dielectric constants of metals are frequently approximated by the Drude–Sommerfeld model. However for noble metals, which are the most important metals used in biosensing applications, the Drude–Sommerfeld model must be corrected for the influence of d-band electrons. Especially in the case of gold, the electronic excitations of d-band electrons into the sp-band at the X-point (1.8 eV) and the L-point (2.4 eV) give rise to substantial deviations from the Drude–Sommerfeld model. Therefore, it is a common practice to use experimentally measured values for ε_{met} rather

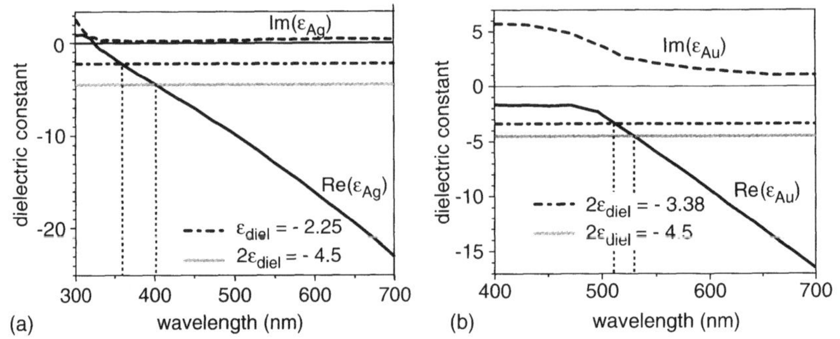

Fig. 1. (a) Dielectric constant of silver (solid and dashed lines) and the negative of the dielectric constant of a dielectricum ($n = 1.5$; $\varepsilon = 2.25$) fulfilling the SPR condition $\varepsilon'_{\mathrm{Ag}} = -\varepsilon_{\mathrm{diel}}$ (dash-dotted line) and the NPPR condition $\varepsilon'_{\mathrm{Ag}} = -2\varepsilon_{\mathrm{diel}}$ (grey line). (b) Dielectric constant of gold (solid and dashed lines) and the negative of the dielectric constants of two dielectrica ($n = 1.3$; $\varepsilon = 1.69$, dash-dotted line) and ($n = 1.5$; $\varepsilon = 2.25$, grey line), both fulfilling the NPPR condition $\varepsilon'_{\mathrm{Ag}} = -2\varepsilon_{\mathrm{diel}}$.

than calculated ones. Measured values for $\varepsilon_{\mathrm{met}}$ can be found for example in Johnson and Christy (1972) or in Palik (1985). The real part of the dielectric constant of the noble metals silver and gold is negative in the visible and near infrared region as can be seen in fig. 1 (Johnson and Christy, 1972).

An important quantity in a metal's dielectric response is the plasmon frequency. A metal does not transmit light with frequencies below the volume plasmon frequency, but becomes transparent for higher energetic radiation. The volume plasmon frequency ω_{P} is given by the equation

$$\omega_{\mathrm{P}}^2 = \frac{n_{\mathrm{e}} e^2}{m_{\mathrm{eff}} \varepsilon_0}, \tag{2.2}$$

where n_{e} is the density of electrons, e is the charge of an electron, and ε_0 is the vacuum dielectric constant.

Because the volume plasmon 'lives' inside a, strictly speaking, infinitely large metallic crystal, it is not good for sensor applications, but the discussion was necessary to provide the basis of the following two subsections that deal with surface plasmons and nanoparticle plasmons. Both can be used for sensing, as we shall see.

2.2. *Surface plasmons*

Maxwell's equations allow for a special surface-bound mode of plasmons at the interface between a metallic and a dielectric medium. These surface

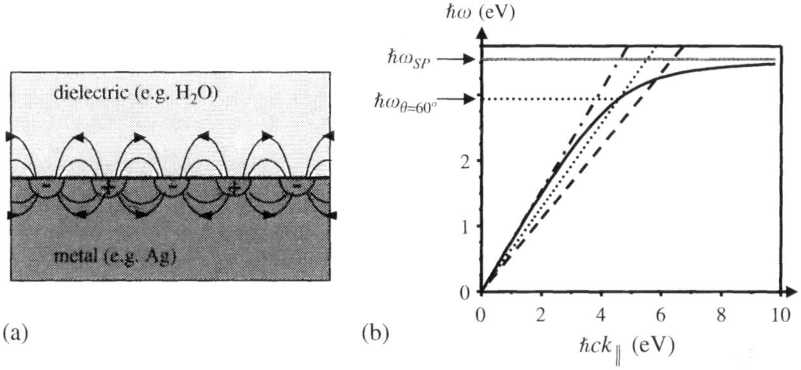

Fig. 2. (a) Scheme of a surface plasmon resonance. (b) Dispersion relation of a surface plasmon resonance (black solid line) that converges to $\hbar\omega_{SP}$ for large k (grey line). Water ($n = 1.3$) has been taken as the dielectric and silver as the metal. The dispersion relation for photons (the so-called 'light line') in water (dash-dotted line) does not cross the SPR dispersion relation. The light line for photons in a high refractive index glass ($n = 1.8$, dashed line) crosses the SPR dispersion relation. The dashed line is also the light line for photons coming from the prism side ($n = 1.8$) in the Kretschmann configuration (see Fig. 3) for $\theta = 90°$. The light line for the Kretschmann configuration, $n = 1.8$ and $\theta = 60°$, is shown by the dotted line.

plasmons travel along the surface and they decay exponentially into both media. Further more, they are longitudinal waves and consist of charge density fluctuations and the associated fields (fig. 2a). Early literature on these was wirtten in the early 20th century by Zenneck (1907) and Sommerfeld (1909) in connection with radio broadcasting. While they used the interface between the soil (as 'conducting' medium for radio waves) and air, we will consider the interface between a metal and water for biosensing. The mathematics, however, is the same.

A simplified derivation of the plasmonic resonance condition can be given as follows: We assume a coordinate system with the z-direction perpendicular to the surface and the x-direction shall be the direction of the travelling surface wave. According to the boundary conditions for the electric and magnetic fields at an interface, the following equations must hold for the wave vectors k^z_{met} (wave vector on the metal side in z-direction), k^z_{diel} (wave vector on the dielectric side [e.g. water] in z-direction), k^x_{met} (wave vector on the metal side parallel to the interface), and k^x_{diel} (wave vector on the dielectric side parallel to the surface):

$$k^x_{\text{diel}} = k^x_{\text{met}} \quad ; \quad k^z_{\text{diel}} \cdot \varepsilon_{\text{met}} = k^z_{\text{met}} \cdot \varepsilon_{\text{diel}}, \tag{2.3}$$

where $\varepsilon_{\text{diel}}$ is the dielectric constant of the dielectric. Together with the mandatory relations of the wave propagation (c: vacuum speed of light),

$$(k^x_{\text{diel}})^2 + (k^z_{\text{diel}})^2 = \varepsilon_{\text{diel}}\left(\frac{\omega}{c}\right)^2 \;;\quad (k^x_{\text{met}})^2 + (k^z_{\text{met}})^2 = \varepsilon_{\text{met}}\left(\frac{\omega}{c}\right)^2, \tag{2.4}$$

one readily obtains the dispersion relation for the in-plane wave vector $k_{\|}$ of the surface plasmon:

$$k_{\|} = \frac{\omega}{c}\sqrt{\frac{\varepsilon_{\text{met}}\varepsilon_{\text{diel}}}{\varepsilon_{\text{met}} + \varepsilon_{\text{diel}}}}. \tag{2.5}$$

This dispersion relation is shown in fig. 2b (solid line) for an interface between silver and water. It is important that ε_{met} is frequency dependent (fig. 1) and therefore the dispersion relation is not a straight line. Furthermore, the real part $\varepsilon'_{\text{met}}$ is negative for visible wavelengths and hence the denominator in eq. (2.5) could become very small (actually it does not become zero due to the finite imaginary part, $\varepsilon''_{\text{met}}$). Nevertheless, the condition $-\varepsilon'_{\text{met}} = \varepsilon_{\text{diel}}$ marks a resonance that occurs at the frequency of the horizontal asymptote (grey line in fig. 2b) to which the dispersion relation converges at large $k_{\|}$.

The dash-dotted line in fig. 2 marks the light line, which is the dispersion relation of light in the dielectric medium:

$$\omega_{\text{light}} = \frac{c}{\sqrt{\varepsilon_{\text{diel}}}} \cdot k_{\text{light}} = \frac{c}{\sqrt{\varepsilon_{\text{diel}}} \cdot \sin\theta} \cdot k_{\|\text{light}}, \tag{2.6}$$

where $k_{\|\text{light}}$ is the component of the wave vector of light parallel to the interface and θ is the angle of the incoming light path with the normal to the surface. The light line shown in fig. 2b (dash-dotted line) is for $\sin\theta = 1$. This corresponds to the gracing incidence for which the light line in fig. 2b has its smallest possible steepness. For all other angles, the light line is even steeper than the dash-dotted line shown in fig. 2b. It is seen that the both dispersion relations, eqs. (2.5) and (2.6), do not intersect for any chosen angle θ. The physical consequence is that a SPR between a metal and a dielectric cannot be excited with light impinging from the side of the dielectric.[1]

[1] Coincidence of dispersion relations plays a similar role in solid state physics as energy and momentum conservation in classical mechanics. Only if both, the quasi-momentum $\hbar k$ and the energy $\hbar\omega$, are conserved a photon can excite a surface plasmon.

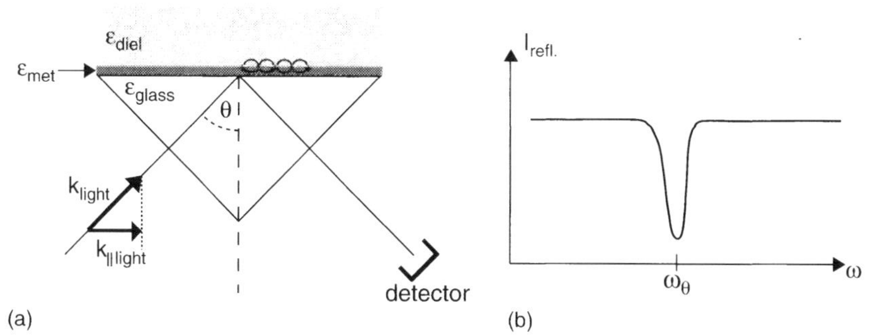

Fig. 3. (a) Scheme of the Kretschmann configuration for exciting SPRs. A thin metal film is evaporated onto glass and illuminated through the glass. Evanescent modes penetrate through the thin metal film and excite SPRs on the upper metal–dielectric interface, provided the frequency of the light and the illumination angle θ match the dispersion relation. The dielectric above the metal film must have a smaller dielectric constant than the glass substrate; it may be, e.g. water. (b) Sketch of the detected intensity. If white light is used for illumination at a fixed angle, the reflection is usually high. At the specific frequency, where surface plasmons can be excited, the reflection is low because energy is consumed by SPRs.

A way out of this dilemma has been suggested by Kretschmann and is depicted in fig. 3 (Kretschmann, 1972). Light is not shone onto the interface from the side of the dielectric, but from the rear side, through the metal. This requires that the thickness of the metal film be in the range of the penetration depth of light into metal, which is some tens of nanometres, depending on the metal. The thin film is evaporated on top of a glass prism with a refractive index larger than the refractive index of the dielectric on top ($n_{\text{glass}} = \sqrt{\varepsilon_{\text{glass}}} > \sqrt{\varepsilon_{\text{diel}}} = n_{\text{diel}}$). In this case, the light line (dashed line in fig. 2b) is less steep than the previous light line where the light impinged the interface from above (dash-dotted line in fig. 2b). Therefore, an intersection with the SPR dispersion relation exists. The dashed line in fig. 2b represents the case of grazing incidence ($\theta = 90°$). The light line and the SPR dispersion relation intersect at a well-defined frequency $\omega(\theta = 90°)$. If θ is reduced, the intersection will shift to lower frequencies because, according to eq. (2.6) (exchange suffix 'diel' with 'glass'), the light line becomes steeper. The case of $\theta = 60°$ is shown by the dotted line in fig. 2b.

The essence of this consideration is simple: Assume light impinges from below at a certain angle θ (fig. 3a) and the light shall have the corresponding frequency ω_θ where the light line and the SPR dispersion relation (eq. (2.5)) intersect. Because we choose a thin metallic film, the light will penetrate the metallic film and will be able to excite surface plasmon modes on the upper metal–water interface. The reflected light of frequency ω_θ will have a reduced intensity because some of the incident power is used to excite the SPR at ω_θ. Light of different frequencies (but same θ) is not able to excite SPRs because the light line and the SPR dispersion relation do not coincide for $\omega \neq \omega_\theta$ and therefore light of frequencies $\omega \neq \omega_\theta$ will be reflected completely. Hence, shining white light onto the sample at angle θ leads to a spectrum of the reflected light as depicted in fig. 3b.

2.3. Nanoparticle plasmons

We now turn to plasmonic excitations in metal nanostructures. The dimensions of metallic nanoparticles are so small that light can easily penetrate the whole nanoparticle and grasp at all conduction band electrons. The result is that the sea of conduction band electrons is displaced with respect to the positively charged ions that form the metallic lattice (fig. 4). The resulting electric dipole on the particle represents a restoring force and hence the nanoparticle can be considered (in a first approximation) as a harmonic oscillator, driven by a light wave and damped by some losses such as ohmic losses (essentially the production of heat) and radiative (scattering) losses. The latter are equivalent to the re-emission of a photon on the expense of nanoparticle plasmon (NPP) excitation. Only light with a wavelength in resonance with an eigenmode of NPP oscillation is able to excite the NPPRs. Therefore, the NPPR manifests itself in two different ways in the optical spectra of solutions or glasses containing metallic nanoparticles: first, there is a pronounced extinction band in the extinction spectrum and second, there is a substantial amount of scattered light. As we will see, both effects can be used for biosensing.

Let me highlight one more distinct difference between SPRs and NPPRs: we have seen that it is not trivial to excite SPRs. Experimental tricks like the Kretschmann configuration must be applied in order to be able to excite SPRs. The bottom line in the excitation of SPRs is that the

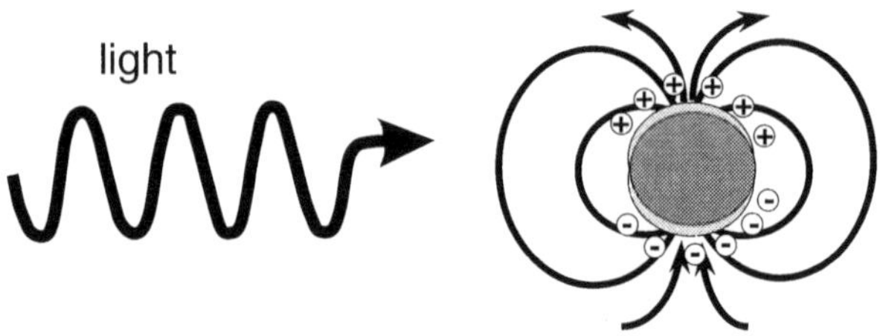

Fig. 4. Plasmon resonance in a metallic nanoparticle, excited by a light wave.

wave vectors of the travelling SPRs and the photons must match. In contrast, NPPRs do not travel. On the scale of the wavelength of light, they are perfectly localized, and therefore NPPRs are also often called localized plasmon resonances (LPR). In essence, we do not have to care about wave vectors in the excitation of NPPRs. We can always excite an NPPR of a spherical metal nanoparticle, regardless of the direction we shine the light onto it. The only condition that must be met is to choose the right wavelength (and polarisation in case of non-spherical nanoparticles) in resonance with the NPP oscillation. This is a huge advantage of NPPR biosensors compared to SPR biosensors.

Gustav Mie solved Maxwell's equations for the case of an incoming plane wave interacting with a spherical particle (Mie, 1908). In essence, the electromagnetic fields are expanded in multipole contributions and the expansion coefficients are found by applying the correct boundary conditions for electromagnetic fields at the interface between the metallic nanoparticle and its surrounding. For very small spherical particles (diameter $<30\,\mathrm{nm}$) it is sufficient to consider only the first term of the expansion, which is the dipolar term. This solution is also called the quasi-static or Rayleigh limit. Let us use this approximate solution, because it is much more lucid than the full Mie expansion and it is sufficient for an understanding of the working principles of biosensors using metal nanoparticles. Readers who are interested in the full Mie theory may turn to the books of Kreibig and Vollmer (1995) or Bohren and Huffmann (1983).

The scattering, extinction, and absorption cross sections of a spherical nanoparticle are given in the Rayleigh limit by the following expressions:

$$\sigma_{\mathrm{sca}} = \frac{3}{2\pi}\left(\frac{\omega}{c}\right)^4 \varepsilon_{\mathrm{diel}}^2 \; V^2 \; \frac{(\varepsilon'_{\mathrm{met}} - \varepsilon_{\mathrm{diel}})^2 + (\varepsilon''_{\mathrm{met}})^2}{(\varepsilon'_{\mathrm{met}} + 2\varepsilon_{\mathrm{diel}})^2 + (\varepsilon''_{\mathrm{met}})^2}, \tag{2.7}$$

$$\sigma_{\mathrm{ext}} = 9\frac{\omega}{c}(\varepsilon_{\mathrm{diel}})^{3/2} \; V \frac{\varepsilon''_{\mathrm{met}}}{(\varepsilon'_{\mathrm{met}} + 2\varepsilon_{\mathrm{diel}})^2 + (\varepsilon''_{\mathrm{met}})^2}, \tag{2.8}$$

$$\sigma_{\mathrm{abs}} = \sigma_{\mathrm{ext}} - \sigma_{\mathrm{sca}}, \tag{2.9}$$

where ω is the frequency of light, c is the speed of light in vacuum, V is the volume of the particle, $\varepsilon_{\mathrm{diel}}$ is the (purely real) dielectric constant of the surrounding medium, and $\varepsilon_{\mathrm{met}} = \varepsilon'_{\mathrm{met}} + i\varepsilon''_{\mathrm{met}}$ is the complex dielectric constant of the metallic nanoparticle. We note that the use of the bulk values for $\varepsilon_{\mathrm{met}}$ gives the correct result at least down to nanoparticle diameters of 20 nm.

We recognize that the denominator in the fractions of eqs. (2.7) and (2.8) can have a minimum in the case $\varepsilon'_{\mathrm{met}} + 2\varepsilon_{\mathrm{diel}} = 0$ because the real

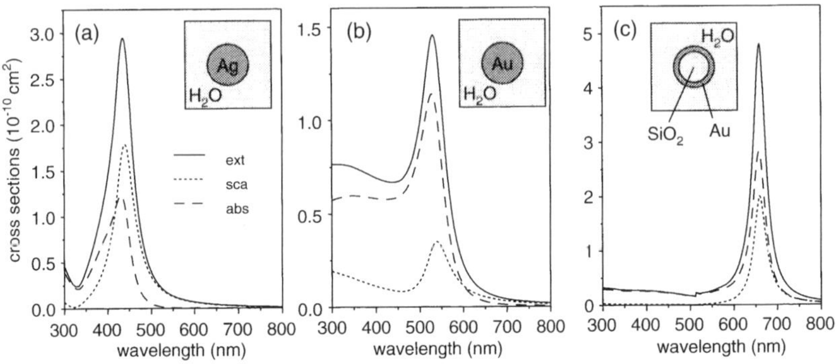

Fig. 5. Extinction, scattering, and absorption cross sections of (a) a silver sphere of 60 nm diameter, (b) a gold sphere of 60 nm diameter, and (c) a 5 nm gold shell on a SiO_2 core of 50 nm diameter. All three nanoparticles are immersed in water.

part of the refractive index of noble metals is negative in the visible range (fig. 1). We note that the resonance condition is similar, but not equal, to the condition we found for SPRs (eq. (3.5)). The two different conditions are sketched in the example of silver in fig. 1a: for a dielectric of refractive index $n = 1.5$, the resonance condition for the SPR occurs at a wavelength of 359 nm, while the resonance condition for NPPRs occurs at 401 nm (see dash-dotted and grey lines in fig. 1a, respectively).

Figures 5a and b show the extinction, scattering, and absorption cross sections for a silver and a gold sphere, respectively. The spectra are calculated using the Mie theory (calculations were carried out with the MQMie software package). The diameters of the spheres are 60 nm each and the physiologically relevant case of immersion of the spheres in water as the surrounding dielectric is shown.

In the limit of very small nanoparticles (eqs. (2.7) and (2.8)), the spectral position of the resonance is independent of the shape of the nanoparticles because no geometric factor occurs in the denominator. The only geometrical factor that enters is the volume, but it only determines the magnitude of the scattering and absorption cross sections, but not their spectral position. Note that the scattering cross section depends quadratically on the volume, while the volume enters only linearly in the extinction cross section, which is the sum of the absorption and the scattering cross sections. Therefore, for very small nanoparticles (less than 20 nm in diameter) it is very hard to see any scattered light because most of the light is absorbed. However, if the diameter becomes too large, the scattered spectrum eventually becomes very broad and broad spectra are usually not desirable for biosensing applications as we shall see below.

A different geometry than the solid sphere is that of noble metal shells (Zhou et al., 1994; Averitt et al., 1997). The scattering, absorption, and emission spectra of a core-shell nanoparticle comprising a dielectric core with a diameter of 50 nm and a gold shell of 5 nm thickness are shown in fig. 5c. The spectra were calculated using an extended Mie theory for concentric spheres. It is seen that the spectra shift to the red, eventually approaching the 'biological spectral window' from 700 to 1000 nm where absorption by both the heme and the water are low. Interestingly, the full widths at half maximum of the spectra of the gold shells are narrower as the spectra of the solid gold sphere of comparable total diameter. This is because the spectral redshift eliminates NPP damping due to interband absorption and because the reduced metal content of a shell compared to a sphere results in reduced radiative damping (Raschke et al., 2004).

The scattering and extinction cross sections of non-spherical nanoparticles are much more complicated to calculate and analytic solutions have been found for a few select cases only. For example, the scattering cross sections for rod-like spheroidal particles can be given in the quasistatic limit (Gans, 1925; Bohren and Huffmann, 1983). For even more complicated structures, numerical simulations must be used.

§ 3. Metal nanoparticles replacing fluorophores in assays

3.1. Greyscale-assays

I will use the term 'greyscale-assays' for biomolecular detection comprising metal nanoparticles where the readout signal is of the simple black/white type, i.e. metal nanoparticles are present or not present. The readout may occur with different methods like transmission or reflection. The common feature of the assays in this chapter is that any information in excess to the shear 'nanoparticle present/not present' information is not used. However, we allow for greyscale readout, i.e. the amount of detected gold nanoparticles relates to the amount of analytes. More elaborate information which greyscale sensors do not make use of may be, for instance, the spectrum of the reflected or transmitted light, the angular dependence of the scattered light or its polarisation. Assays that make use of this information (maybe, we should call them 'coloured' assays) will be described in the Section 3.2 and in the following paragraphs 4 and 5.

Gold nanoparticles have first been labelled with antibodies in order to specifically mark proteins in cells or on cell membranes so that they can be used as contrast agents in electron microscopy (Faulk, 1971;

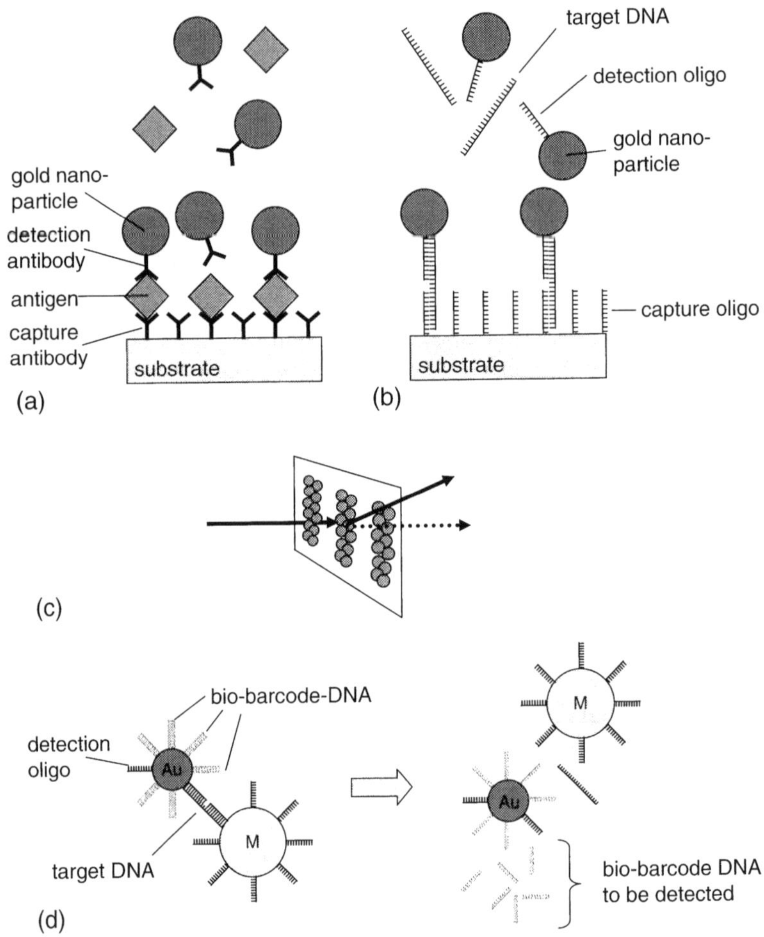

Fig. 6. (a) Sandwich immunoassay: detection antibodies are labelled with gold nanoparticles, capture antibodies are immobilized on a substrate. If antigens are present, they are sandwiched between the two types of antibodies and gold nanoparticles accumulate on the substrate. (b) Sandwich DNA assay. (c) Readout of a sandwich assay in form of a diffraction grating by the diffraction of a laser beam. (d) Bio barcode assay. For details see text.

Bauer et al., 1972). The use of noble metal nanoparticles in immunoassays dates back to the early 1980s (Leuvering et al., 1980b; Moeremans et al., 1984). A standard scheme of a gold nanoparticle immunoassay is as follows: antigens that are present in human body liquids are detected in a sandwich-type immunoassay. Two different types of antibodies directed against the same sort of antigens are used (fig. 6a). Capture antibodies are fixed on a substrate and detection antibodies are labelled with gold nanoparticles. When the matching antigens are present in the body liquid, the antigens are sandwiched between the immobilized capture antibodies and the detection antibodies carrying gold nanoparticles. As a consequence, the detection area on the substrate is covered with many gold nanoparticles that can easily be detected because of the huge number of nanoparticles, and therefore the detection spot appears brownish in colour even for the naked eye. Commercialized test systems using this scheme include pregnancy tests and tests for heart attacks (Frey et al., 1998).

As already discussed in Section 1 there is a need to parallelise assays. The driving force is to detect many different species in parallel on so-called biochips. In the year 2000, there were two publications that showed that gold nanoparticles are an excellent label for DNA chips (Reichert et al., 2000; Taton et al., 2000). The idea is to multiplex the detection of several different oligonucleotides on a chip by using an array of specific detection fields and miniaturize the whole sensor. The detection scheme is again of the sandwich-type; this means the target DNA is complementary to a capture strand immobilized on the substrate and it is also complimentary to a detection DNA that carries a marker (fig. 6b). Fluorescent labels had been established as markers, but they were prone to some disadvantages such as bleaching during illumination. The replacement of the fluorescent markers with gold nanoparticles circumvented the disadvantages. It has further been shown that the assays using gold nanoparticles instead of fluorophores as markers show a much sharper DNA melting curve (Taton et al., 2000). It has also been pointed out that transmission or reflection measurements used to read out a gold nanoparticle DNA chip is generally less technically demanding compared to fluorescence measurements and yields a better signal-to-noise ratio (Taton et al., 2000; Köhler et al., 2001). This can be further improved using silver enhancement after the molecular recognition, i.e. additional silver is reduced on the gold nanoparticles (Taton et al., 2000). An interesting variant of read out utilizes a diffraction grating where the diffractive strips of the grating are covered with capture antibodies (fig. 6c). Without an analyte, light is not scattered at the array of fields that carry capture antibodies or capture DNA only, while the array acts as a strong diffraction grating after the analyte sandwiches have been formed (Bailey et al., 2003). Another interesting readout scheme is the electrical readout. When enough gold nanoparticles are accumulated in the sandwich area (and eventually silver enhancement is applied), a current can flow between two contacts close to the sandwich area (Park et al., 2002).

The sensitivity of gold nanoparticle DNA assays was dramatically enhanced up to the level of PCR-type assays (PCR: polymerase chain reaction) using the 'bio-barcode' technique (Nam et al., 2004) (fig. 6d). Gold nanoparticles are bi-functionalized with a small number of DNA complimentary to the target and a large number ($100 \times$ more) of 'barcode' DNA. Magnetic nanoparticles are functionalized with target-complimentary DNA as well. If the target is present in the analyte solution, the gold and magnetic nanoparticles cluster. These agglomerates can be separated from the solution simply by a static magnetic field. After such a separation step, the barcode DNA is de-hybridised and probed by a usual gold nanoparticle sandwich assay. Because the gold nanoparticles in the first step carry 100 times more barcode DNA than target DNA, the method has an intrinsic amplifying step that renders PCR obsolete (Nam et al., 2004). This same technique was also applied to an immunoassay (Georganopoulou et al., 2005). The gold nanoparticles in the first step carried barcode DNA and a few antibodies against Alzheimer's disease. The antigens are taken from cerebrospinal fluid of Alzheimer's disease patients. Because the antigens are present in a very low concentration (<1 pM) it is generally hard to detect them using other assay formats. However, they are detectable using the bio-barcode technique.

3.2. Single metal nanoparticles as labels

Techniques were described in the previous section where the presence of a multitude of metal nanoparticles is used as a detector signal. This 'multitude' can easily amount to billions of particles. In sharp contrast, in this chapter we describe sensor techniques that use the presence or absence of a single or only a few metal nanoparticles as the sensor signal. The central question, of course, is how to detect a single metal nanoparticle. Near-field microscopy has been used to measure the scattering spectrum of a single gold nanoparticle and to determine its homogeneous linewidth (Klar et al., 1998). In practice, it turned out to be more convenient to use dark filed microscopy (Yguerabide and Yguerabide, 1998a, b; Schultz et al., 2000; Sönnichsen et al., 2002b) or evanescent illumination microscopy (Sönnichsen et al., 2000; Taton et al., 2001).

Figure 7a shows a transmission dark field microscope. In this setup, the sample is illuminated with white light from below. The light is directed towards the sample under very steep angles, actually steep enough that the light propagating straight cannot be collected by the objective lens. Light is scattered out of the straight path and into the objective lens only if a scattering object such as a metallic nanoparticle is present. Importantly, not all wavelength components of the white light used for illumination are scattered with the same cross section. Actually, the spectral

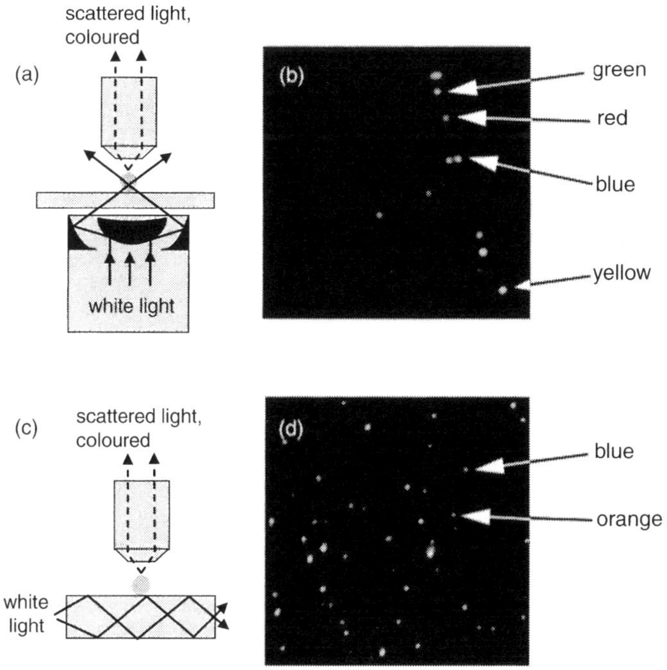

Fig. 7. Methods of single gold nanoparticle spectroscopy: Dark field microscope: setup (a) and image (b). A coloured version of the image would show bright blue, green, yellow, and red nanoparticles. Evanescent field illumination (c) setup and (d) image. The imaged samples in (b) and (d) contain silver and gold nanoparticles of different sizes and shapes. Blue light is scattered from small, spherical silver nanoparticles; green light is scattered from small, spherical gold nanoparticles. Orange and red light are scattered either from pairs of particles or rod-shaped particles.

composition of the scattered light is given by the scattering cross section as shown in fig. 5. As it becomes clear from fig. 5, nanoparticles of different metals, shapes, or sizes scatter different components of the visible spectrum. Figure 7b shows an image of a test sample where all kinds of different nanoparticles are dispersed on the substrate. A coloured version of fig. 7b would show bright blue, green, yellow, and red colours. Judging from the calculated spectra shown in fig. 5 we can assign the blue spots to small silver nanoparticles. The green light is scattered by small gold nanoparticles and the orange and red spots indicate either large gold or silver nanoparticles or pairs of nanoparticles, rods of different length (Sönnichsen et al., 2002b, Sönnichsen et al., 2002a), or other geometrical shapes (Mock et al., 2002). This opens a new dimension in multiplexing assays (Taton et al., 2001). The assays reported in the previous subsection simply used the presence or absence of a multitude of nanoparticles as the

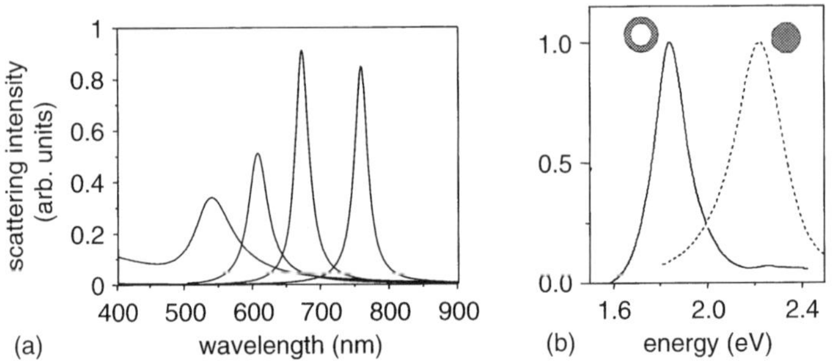

Fig. 8. (a) Calculated scattering spectra of (left to right): a solid sphere of 30 nm diameter, core shell structures with a 20 nm dielectric core ($n = 1.34$) and a gold shell of 5, 3, and 2 nm thickness. (b) Measured spectra of a single nanoshell (left, core 30 nm diameter, 5 nm gold shell) and a single solid gold sphere of 40 nm diameter.

detector signal. The assays are read out using some ‘greyscale’ method. Multiplexing must be accomplished by the spatial distinction of sensor regions, e.g. in a grid like fashion. In the dark field microscope, however, it is possible to distinguish between different sorts of nanoparticles by colour. If, for example, silver nanoparticles report on the presence of antigen A and gold nanoparticles report on antigen B, multiplexing can take place on one and the same spot on a biochip.

It is promising to colour-multiplex gold nanoparticle biosensors by using gold shells with a dielectric core (Zhou et al., 1994; Averitt et al., 1997). The particle plasmon resonance shifts to the red with a decreasing aspect ratio of the shell thickness to the diameter of the dielectric core. Figure 8a shows the calculated scattering spectra for some examples of core-shell gold nanoparticles. Figure 8b shows the measured scattering spectra of a single nanoshell with an Au_2S dielectric core of approximately 30 nm diameter and a 5 nm thick gold shell (solid line) (Raschke et al., 2004). In comparison, the NPP of a solid gold sphere of 40 nm diameter resonates at higher energies and shows a broader spectrum (dotted line).

Dark field microscopy using epi-illumination is a technique similar to transmission dark field microscopy. In this case, the dark field illumination is facilitated from the same side as the light collection takes place. Epi-illumination dark field microscopy must be used in the case of opaque substrates. A technique related to dark field microscopy is shown in fig. 7c: in total internal reflection microscopy, the sample is either placed on a prism or on a waveguide (or the sample substrate itself acts as a waveguide). The illumination light cannot exit the glass in the direction of the collecting objective lens due to total internal reflection. However, evanescent modes are present above the glass–air interface. If a metal nanoparticle rests on top of this interface, it can scatter light out of the evanescent field modes into propagating modes that can be collected by the objective lens. Again, the scattered light is coloured because the efficiency with which the nanoparticles convert evanescent modes into propagating modes is given by the scattering cross section as shown in fig. 5. Figure 7d shows a sample containing several different types of metal nanoparticles illuminated with a waveguide as shown in fig. 7c.

It becomes clear from figs. 7b and 7d that metal nanoparticles can be used for coloured labelling similar to fluorophores. However, the scattering cross sections of gold nanoparticles are generally much larger than the absorption cross sections of fluorophores. Furthermore, gold nanoparticles are practically infinitely photostable and do not blink. Therefore, they show clear advantages over fluorophores, for example in delicate sensor formats that rely on single fluorophore or nanoparticle detection.

The detection of tiny amounts of molecules by counting single metal nanoparticles has been reported by Schultz and co-workers (Schultz et al., 2000). They realized an immunocytology assay and a sandwich immunoassay where the sensor signal is given by the counted number of metal nanoparticles in a dark field microscopic image. An extended study including histology, immunoassays, and DNA assays has been reported by Yguerabide and Yguerabide (2001).

The detection of metal nanoparticles below 20 nm in diameter is challenging because of their small scattering cross section. However, it is sometimes required that an antibody or a specific oligonucleotide sequence is attached to a very small gold nanoparticle because larger nanoparticles reduce their ability to recognize their target. Still such small labels can be used in gold nanoparticle counting assays. In order to count very small gold nanoparticles in assays, several strategies have been developed. For example, the small gold nanoparticles may be silver enhanced (Schultz et al., 2000; Storhoff et al., 2004b). In this case, silver is reduced on the surface of the gold nanoparticles and larger noble metal nanoparticles form that can be detected easily in a dark field microscope. A different strategy is to use capture oligonucleotides that are linked to a biotin molecule. Large gold nanoparticles functionalized with antibiotin antibodies are applied after the biomolecular recognition sandwich has formed on a substrate (Oldenburg et al., 2002).

Some interesting techniques have been suggested recently to increase the sensitivity of single nanoparticle detection. Gold nanoparticles with diameters below 10 nm have been detected using an interferometric detection technique in confocal reflection microscopy (Lindfors et al., 2004).

The absorption of single gold nanoparticles down to 5 nm in diameter was detected using spatial modulation microscopy (Arbouet et al., 2004). An elegant technique to directly probe sub 5 nm gold nanoparticles is the photothermal imaging technique (Boyer et al., 2002; Bericaud et al., 2004). It has very recently been successfully applied for the optical read-out of nanoparticle-based DNA microarrays comprising small gold nanoparticles (Blab et al., 2006).

§ 4. Coupled NPP resonances as sensor signal

4.1. The basic idea

When gold nanoparticles approach towards each other, nanoparticle resonances start to couple and consequently, the eigenfrequencies shift (Kreibig et al., 1981; Quinten et al., 1985; Rechberger et al., 2003; Su et al., 2003). This is illustrated in fig. 9 where the calculated extinction spectra of two spheres are shown for different interparticle distances. It is seen that the longitudinal eigenmode (i.e. the electric field is polarized parallel to the

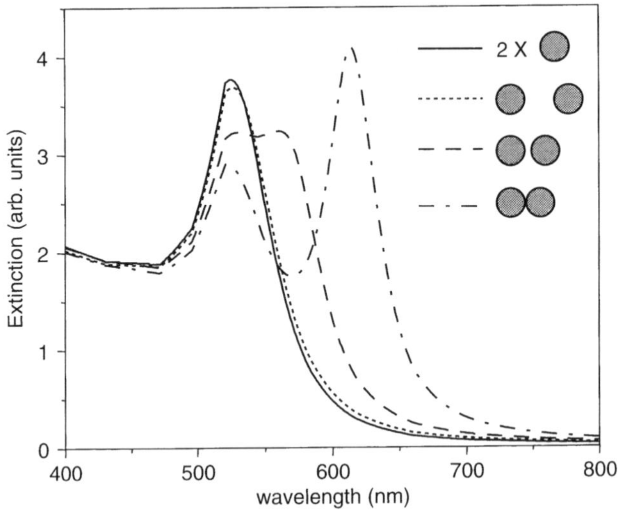

Fig. 9. Extinction spectra of two infinitely spaced gold nanoparticles of 40 nm diameter (solid line), two gold nanoparticles, separated by 40 nm surface to surface (dotted line), 4 nm (dashed line), and almost touching (0.4 nm separation, dash-dotted line). In all cases the spectra were averaged over the twofold degenerate polarisation perpendicular to the connecting axis and the polarisation parallel to the connecting axis. The short wavelength resonance corresponds to the perpendicular optically allowed eigenmode, the long wavelength resonance corresponds to the longitudinal optically allowed eigenmode.

axis of the pair of particles) shifts substantially to longer wavelengths. The perpendicular eigenmode concomitantly shifts to shorter wavelengths. However, the blueshift of the perpendicular modes is by far less pronounced than the redshift of the parallel mode. Let us now assume that a multitude of nanoparticles, all with different but close spacings to each other, is present in a solution. In this case, the maximum of the extinction spectrum of the solution weakens and the overall extinction spectrum redshifts when the average distance between the nanoparticles is reduced.

4.2. Using the extinction spectrum

4.2.1. Immunoassays

The effect described above can be used as a sensor that uses coupled NPPRs. For instance, coupled NPPRs have been used in immunoassays already in the early 1980s. Those days the prefix 'nano' was not yet that commonly used and the sensors were called 'sol particle agglutination assays' (Leuvering et al., 1980a, 1981; Goverde et al., 1982; Leuvering et al., 1983). Gold nanoparticles were functionalized with two different monoclonal antibodies against human chorionic gonadotrophin (HCG, present in the urine of pregnant women). When the gold nanoparticles are dissolved in urine samples, the solutions have a very distinct colour depending on the presence or absence of HCG. This colour change is obvious when a spectrometer is used (Leuvering et al., 1981); however, the colour change is so pronounced that a colour readout by the naked eye gave correct results with more than 99% confidentiality (Goverde et al., 1982; Leuvering et al., 1983). A more recent report on an immunoassay using nanoparticle aggregation in order to detect anti-protein A was reported in Thanh and Rosenzweig (2002). A related work has been published by Kim and co-workers who showed that heavy metal ions can be detected with gold nanoparticles that are functionalized with 11-mercaptoundecanoic acid (Kim et al., 2001).

Note that the colour changes in this section are different in nature from the colour changes reported in Section 3. In Section 3 a colour (or a greyscale) change was initiated because of the assembly of a specific number of gold nanoparticles that have not been present in the readout area before. In contrast, the number of gold nanoparticles does not change in the aqueous solutions reported in this section. Only their average distance changes which leads to a change of the spectral properties of the solution.

4.2.2. Oligonucleotide sensors

In 1996, the groups of Chad Mirkin and Paul Alivisatos simultaneously published two seminal papers where they showed that gold nanoparticles can be assembled in three dimensions using oligonucleotides (Alivisatos et al., 1996; Mirkin et al., 1996). Gold nanoparticles are functionalized with complimentary oligonucleotides that are terminated with a thiol group at the 3' end. When the solution is cooled down below the hybridisation temperature of the two complimentary strands, the gold nanoparticles cluster and the average distance is reduced. It was reported by Mirkin et al. that this DNA mediated clustering leads to a colour change of the gold nanoparticle solution (Mirkin et al., 1996).

A DNA sensor was developed soon after the principle investigations on assemblies of gold nanoparticles and oligomers (Elghanian et al., 1997). Instead of functionalizing the gold nanoparticles with two complimentary strands, the nanoparticles were functionalized with two different oligonucleotide sequences that are not complimentary to each other; however, they are both complimentary to some part of the target oligonucleotide. Only when the target oligonucleotides are present, the gold nanoparticles cluster and the extinction spectrum is changed. This process is visualized in fig. 10a. It shows the extinction spectra of a solution containing functionalized gold nanoparticles and the target oligonucleotide. Above melting temperature, the gold nanoparticles are far apart and the extinction spectrum resembles the sum of all single particle extinction spectra in solution. Below melting temperature, the

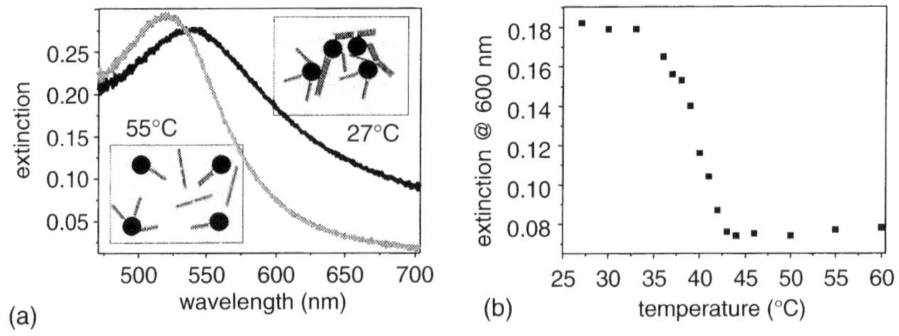

Fig. 10. (a) Extinction spectra of gold nanoparticles, functionalized with oligonucleotides in the presence of the target DNA. Above melting temperature (37°C) the gold nanoparticles in solution are far apart and an extinction spectrum similar to single gold nanoparticles is observed. Below melting temperature, the gold nanoparticles are bound together via the hybridised oligonucleotides and the extinction spectrum redshifts due to coupling of nanoparticle plasmon renonances. (b) Extinction at 600 nm as a function of temperature. The melting profile is steep (< 5°C).

nanoparticles cluster and the extinction spectrum is slightly weaker in its maximum and is redshifted. Of course, one does not have to take the full extinction spectrum in a practical sensor device. A single wavelength (e.g. from a cheap laser diode) is sufficient to monitor the spectral shift. This is shown in fig. 10b, where the extinction at 600 nm is related to the solution temperature. One clearly sees the melting transition of the DNA. This melting transition is very sharp (Elghanian et al., 1997; Jin et al., 2003) such that even single base imperfections can be monitored (Storhoff et al., 1998).

4.3. Using light scattering

So far we have discussed sensors that exploit a change in the *extinction* spectrum when gold nanoparticles cluster in the presence of the analyte. Apart from the extinction spectrum, one could also use the *scattered* light of the gold nanoparticles. Upon agglomeration, both the *spectrum* of the scattered light as well as the *angular dependence* of the scattered light changes. Both features can be used as a sensor signal as shown in the following two subsections.

4.3.1. Scattering spectrum

One of the largest advantages of measuring the scattered rather than the transmitted spectrum is the following: a large number of gold nanoparticles is usually required in solution in order to obtain a robust extinction signal. This is because an intensity decrease must be detected against a large background of transmitted light. On the contrary, a scattering sensor detects scattered photons against a (ideally) dark background. Therefore, as already explained in the Section 3.2, even single gold nanoparticles can be detected in light scattering experiments. Of course, pairs of nanoparticles can also be detected in dark field or evanescent field microscopes. Despite the fact that a pair of nanoparticles might be separated by a few nanometres only, and hence the nanoparticles cannot be spatially resolved in a far-field microscope, their distance can be measured because their scattering spectrum changes with distance. Storhoff and coworkers have shown that this can be used for a DNA sensor (Storhoff et al., 2004a). Again, the distance of DNA functionalized gold nanoparticles is smaller if the complimentary DNA target is present as compared to the distance without the target. Because in principle, only a single closely spaced pair of gold nanoparticles is necessary to report on the presence of the target DNA, the amount of DNA which is needed to obtain a sensor signal is fairly low, and unamplified genomic DNA detection has become possible (Storhoff et al., 2004a).

Other reports that use the scattered spectrum of clustered gold nanoparticles rather than the extinction spectra include a glucose sensor (Aslan et al., 2005c) and an immunoassay where gold nanoparticles functionalized with anti-EGFR antibodies are used for oral cancer diagnosis (El-Sayed et al., 2005).

4.3.2. Angular distribution of scattered light

Light which is Mie-scattered from a nanoparticle agglomerate does not only change its spectral content when the geometry of the agglomerate changes, but also the angular distribution of the scattered light strongly depends on the size and the composition of the agglomerate. In general, a single small particle scatters light similar to the emission profile of a point-like emitter, which is a $\cos^2(\theta)$ dependence on the scattering angle measured from the direction of straight light propagation. In contrast, the angular dependence of the scattered light differs from that $\cos^2(\theta)$ law in the case of a large particle or an agglomerate of small particles. This fact was used in DNA assays (Souza and Miller, 2001) and in biotin-streptavidin recognition (Aslan et al., 2005a). It was also shown that the depolarisation of the scattered light can be used as a sensor signal in addition to the angular dependence of the scattered light (Aslan et al., 2005b).

4.4. The nanoruler

After the report on several sensor formats that employ the coupling of nanoparticle plasmons, let me end this section with another interesting application of coupled nanoparticle resonances. On the nanometre domain it is inherently difficult to directly measure the distance between two nanoscopic objects, because the resolution power of far-field microscopy is limited by the diffraction limit (Abbe, 1873), and also in near-field microscopy (Pohl et al., 1984) the practical resolution limit is approximately 50 nm. In principle, one can label two points of interest with two different fluorophores which allows for the measurement of distances in the nanometre range by applying wavelength-selective localisation

(Bornfleth et al., 1998). It is also possible to use the same sort of fluorescence molecules and to apply stimulated emission depletion (STED) far-field microscopy (Klar et al., 2000) which is capable of resolution down to 15 nm (Westphal and Hell, 2005). Förster resonant energy transfer (FRET) is an alternative to measure distances in the 5 nm range because it shows a strong R^{-6} dependence of the FRET signal on the molecular distance R (Förster, 1948). This renders the FRET sensor very sensitive around 5 nm but useless above approximately 8 nm.

An interesting alternative was suggested by Sönnichsen and co-workers (Sönnichsen et al., 2005). As described above, the scattering spectrum of a pair of gold nanoparticles depends on their mutual distance and hence, one can use the spectral response as a nanoruler. Interestingly, it covers the range from 5 nm to 75 nm (Reinhard et al., 2005) and therefore bridges the gap between FRET sensors and the typical resolution range of near-field or STED microscopy. The change of the persistence length of ssDNA upon hybridisation and dehybridisation was monitored, and it was shown that gold nanoparticle rulers can be used for almost 1 hr, which is hard to achieve with FRET sensors that are prone to photobleaching (Sönnichsen et al., 2005).

§ 5. Dielectric environment plasmonic biosensors

In Section 4, the coupling of two or more NPPRs, initiated by some target biomolecules, was used as a sensor signal. It was shown that the coupling of nanoparticle resonances shifts their spectral responses (extinction as well as scattering) and therefore the presence of the target molecule is detected.

In this section, a change of the extinction or scattering spectra will be used as well to report on the presence of analytes. However, this time the spectral response is not caused by a coupling of two or more (metallic) resonators, but it is caused by the change of the dielectric environment around the nanoparticles. Such a change arises when some biomolecules assemble on the surface of the nanoparticle and therefore cause an increase of the refractive index in the nanoenvironment of the nanoparticle.

5.1. Surface plasmon resonance sensors

Although this book chapter is devoted to biosensing with plasmonic *nanoparticles*, I would like to give a very brief survey of SPR sensors

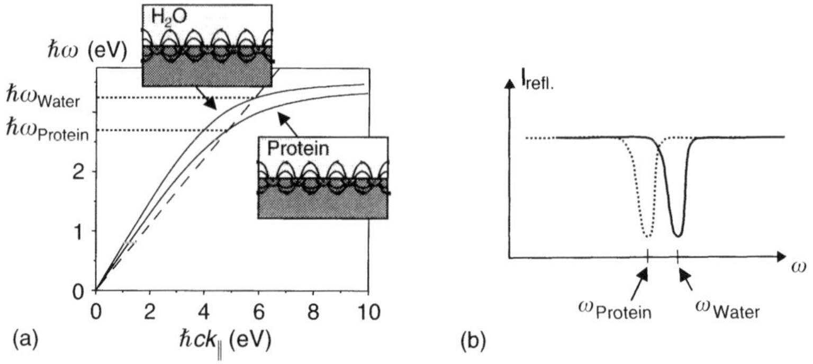

Fig. 11. Working principle of a surface plasmon resonance biosensor. (a) Dispersion relation of a SPR of a silver film covered by water (upper solid line) and by proteins (lower solid line). The change in refractive index ($n = 1.33$ for water, $n = 1.5$ for proteins) causes the change of the dispersion relation. The crossing of the dispersion relation with the light line (dashed line) defines the frequency of light that can excite surface plasmons in a Kretschmann configuration. (b) A change of the refractive index from 1.33 to 1.5 leads to a redshift (shift to lower frequency) of the extinction band in the spectrum of the reflected light (see also Figs. 2 and 3).

because they are conceptually similar and have been invented 15 years earlier. Because of this, they have already been successfully commercialized, while the nanoparticle sensors are mostly in the research and development stage. However, nanoparticle-dielectric-environment sensors have some decisive advantages as will be outlined at the end of Section 5.

It has been noticed by Pockrand and co-workers in 1978 that the SPR of silver films can be tuned when the silver films are covered by a protein layer (Pockrand et al., 1978). However it was not until 1983 when Liedberg and co-workers revealed the potential of this effect for biosensing (Liedberg et al., 1983). The working principle of a surface plasmon biosensor is outlined in fig. 11. The reader is advised to turn back to Section 2.2, where we found that the dispersion relation of the SPR depends critically on the dielectric constant of the dielectric above the metal film (eq. (2.5)). If this dielectric constant is changed (e.g. because proteins attach to the silver surface) then the dispersion relation changes (fig. 11a). It actually becomes flatter. In Section 2.2 we saw that SPRs cannot easily be excited because the light line has to meet the SPR dispersion relation in order to excite SPRs. One possible method to excite SPRs has been outlined in Section 2.2: the so-called Kretschmann configuration (fig. 3) (Kretschmann, 1972). A thin silver film on a high refractive glass substrate is illuminated from below. In this geometry, the light line has an intersection with the SPR dispersion relation (fig. 2b). As a consequence, the light that is reflected from the silver film shows a minimum in intensity at a frequency that corresponds to the crossing point of the light line and the SPR dispersion relation (fig. 3). If the dispersion relation is changed because of a change of the refractive index of the dielectric, the intersection with the light line will also change (fig. 11a). Therefore the minimum in reflectance will occur at a different light frequency. In a real biosensor, the silver film will be covered with the analyte solution (i.e. essentially with water of refractive index $n = 1.33$). The silver layer is prefunctionalized with capture-antibodies or oligomers similar to what was shown in figs. 6a, b. This will slightly increase the average refractive index $n = \sqrt{\varepsilon_{\text{diel}}}$ on top of the silver layer. In fact, the relevant dielectric constant $\varepsilon_{\text{diel}}$ that enters into eq. (2.5) is the dielectric constant averaged from 0 nm to approximately 500 nm above the silver film surface. If now, in the sensing step, the analytes attach to the capture antibodies, the average dielectric constant will further increase and the spectral position of the minimum of the reflected light will further redshift (fig. 11b).

Soon after the first reports, SPR sensors have been applied to many different biosensing schemes such as antigen–antibody sensors (Karlsson et al., 1991) and real-time sensors (Jonsson et al., 1991) and the system has been commercialized (Löfas et al., 1991). Protein–DNA interactions have also been studied with SPR sensors (Brockman et al., 1999). During the last two decades, the field has expanded greatly and it is not even remotely possible to give an overview of all the recent work within the framework of this chapter. The interested reader is referred to reviews specialized on SPR sensors (e.g. Homola et al., 1999; Rich and Myszka, 2000; Homola, 2003).

5.2. Nanoparticle plasmon resonance sensors

5.2.1. Working principle

We noticed in Section 2 that (two-dimensional) SPRs and (zero-dimensional) NPPRs have very similar but not equal resonance conditions. Referring to eqs. (2.5), (2.7), and (2.8), we see that both resonances depend on the refractive index of the surrounding dielectric. In case of the SPR the resonance condition is $\varepsilon'_{\text{met}} + \varepsilon_{\text{diel}} = 0$ while for the NPP resonance the condition is $\varepsilon'_{\text{met}} + 2\varepsilon_{\text{diel}} = 0$. The tiny difference is marked in fig. 1a where it is shown that the NPPR occurs at longer wavelengths compared to the SPR.

Figure 1b shows the resonance conditions for the two cases where a gold nanoparticle is surrounded by water ($n = 1.33$, dash-dotted line) or a

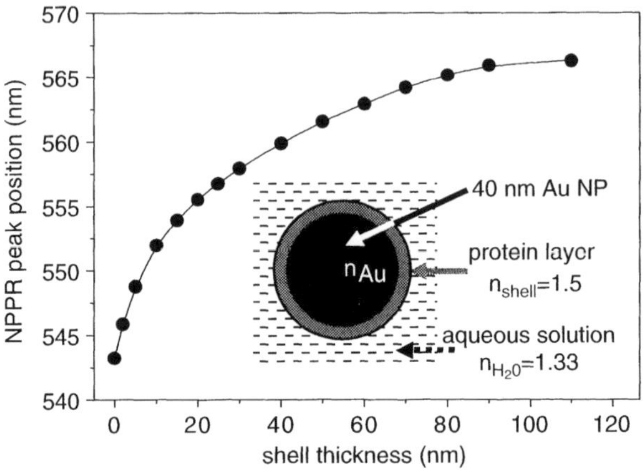

Fig. 12. Mie calculations of the peak wavelength of the nanoparticle scattering cross section. A gold nanoparticle of 40 nm diameter is assumed. The nanoparticle is submerged in water and covered with a variable layer of proteins. The maximum of the scattering cross-section redshifts when the protein shell grows and finally saturates.

thick protein shell ($n = 1.5$, grey line). It becomes clear that the assembly of proteins around a gold nanoparticle leads to a spectral shift of the NPPR and consequently to a spectral shift of the extinction and scattering spectra. Figure 1b shows the extreme case of a gold nanoparticle in aqueous solution without any proteins on its surface, and of a gold nanoparticle that has been wrapped with an infinitely thick layer of proteins. In a sensor, of course, only a thin layer of proteins would assemble around the nanoparticle. This situation is shown in fig. 12 (Raschke et al., 2003). Using an extension of the Mie theory (Bohren and Huffmann, 1983), we calculated the peak position of the scattering spectrum of a gold nanoparticle with respect to the thickness of the protein layer. The gold nanoparticle was assumed to have a diameter of 40 nm and is submerged in water ($n = 1.33$). A gradual increase of the protein shell redshifts the scattering spectrum.

We now assume that the gold nanoparticle is pre-functionalized with a thin layer of antibodies (approximately 3 to 7 nm thickness). The protein layer will swell to 15 to 30 nm after the recognition of the antigens, which should result in a shift of the scattering spectrum by a few nanometres. The extinction spectrum will show a similar shift. Having a closer look at fig. 12 we notice that the shift of the NPPR saturates at layer thicknesses that are, roughly speaking, twice as large as the nanoparticle's diameter. This is an important finding, because in a real sensor one wants to make sure that the nanoparticle's spectra only react to analytes that are recognized by the antibodies on the nanoparticle surface, and that they do not respond to any other non-target proteins that are present in the analyte solution. From fig. 12 one can estimate that other proteins swimming around at a distance of at least 100 nm from the nanoparticle surface do not cause any (false) signal. A more detailed study on the relationship between the NPPR position and the protein layer thickness may be found in Malinsky et al. (2001) and Haes et al. (2004b, c).

So far, we have only discussed a spectral change of the NPPR that is caused by a change of the dielectric constant of the surrounding medium. There are other influences that may contribute to a spectral shift of the NPPR as well. One of them is a change of the electron density in the gold nanoparticles (Kreibig et al., 1997; Henglein and Meisel, 1998; Hilger et al., 2000). As can be seen in eq. (2.2), the volume plasmon resonance frequency depends on the square root of the electron density n_e and the NPPR depends on the electron density in a similar way. If the attachment of organic molecules increases or decreases the electron density in the metal nanoparticle, the scattering and the extinction spectra will shift. However, such an effect is only expected when the analyte molecules will directly dock to the gold nanoparticles. In a real biosensor, there will be an intermediate layer of antibodies or oligonucleotides that guarantee the specificity of the sensor and it is less likely that charges from the analyte will be transferred across that intermediate layer onto the gold nanoparticles. Therefore, the charging effect may play a minor role in real biosensors. However, it may be used to sense the ionic strength of a solution (Linnert et al., 1993).

5.2.2. Ensemble sensors

Patrick Englebienne first used the above described NPP sensor format as an immuno-sensor (Englebienne, 1998). He observed that the extinction spectrum of an ensemble of gold nanoparticles, functionalized with antibodies, redshifts upon the addition of the appropriate antigens. Later he and his co-workers used the technique to infer the affinity constants from the interactions between protein antigens and antibodies (Englebienne, 1998) and to use it as a rapid homogeneous immunoassay for human ferritin (Englebienne et al., 2000).

While the Englebienne group used solutions of spherical gold nanoparticles, the Van Dyne group used triangular-shaped nanoparticles in a hexagonal pattern on a solid substrate as the nanoparticular plasmonic resonators for biosensing (Malinsky et al., 2001). Such triangles can be produced by nanosphere lithography as originally established by Fisher and Zingsheim (1981). In their first reports, the group of Van Duyne measured the spectral shift of the NPPR relative to a nitrogen environment. Because of the large difference in refractive index of nitrogen ($n = 1$) to proteins ($n = 1.5$) the reported shifts of the plasmonic resonances were comparatively large (some tens of nanometres) (Haes and Van Duyne, 2002). Later, the same group showed applications in physiological buffers, after they had improved the nanoparticle adhesion on the substrate (Riboh et al., 2003). In this case the plasmonic shift was less pronounced because of the smaller refractive index contrast between water ($n = 1.33$) and proteins ($n = 1.5$).

Nath and Chilkothi used spherical gold nanoparticles which where rigidly bound to a glass surface via silane anchors. They followed specific biomolecular recognition in real time (Nath and Chilkoti, 2002). They also investigated the optimal size of gold nanoparticles for this type of sensor and found that gold nanoparticles of approximately 40 nm show the best results in terms of 'analytical volume' and maximal sensitivity to a change in the refractive index (Nath and Chilkoti, 2004).

5.2.3. Single nanoparticle sensors

It has been suggested by Klar et al. (1998) and experimentally verified by Mock et al. (2003) that a single gold nanoparticle can sense its dielectric nano-environment. In the early study by Klar et al., a near-field microscope has been used to image single gold nanoparticles. However, dark field microscopy and related techniques have proven to be much more convenient to image the scattered light of a single noble metal nanoparticle (Yguerabide and Yguerabide, 1998a,b; Schultz et al., 2000; Sönnichsen et al., 2000, 2002b; Taton et al., 2001). Please refer to Section 3.2 for a detailed discussion of these techniques.

The first *single* gold nanoparticle biosensor has been realized by Raschke et al. (2003). A single gold nanoparticle was used which was functionalized with biotinylated albumin in order to specifically detect streptavidin. The main result is shown in fig. 13. The scattering spectrum of a single functionalized gold nanoparticle is monitored over more than 1 hr using a transmission dark field microscope, similar to the setup outlined in fig. 7a but with a water immersion lens as the collecting lens. For the first 15 min ('negative' time scale in fig. 13) the spectral position of the NPP resonance is measured without the addition of the analyte. It is

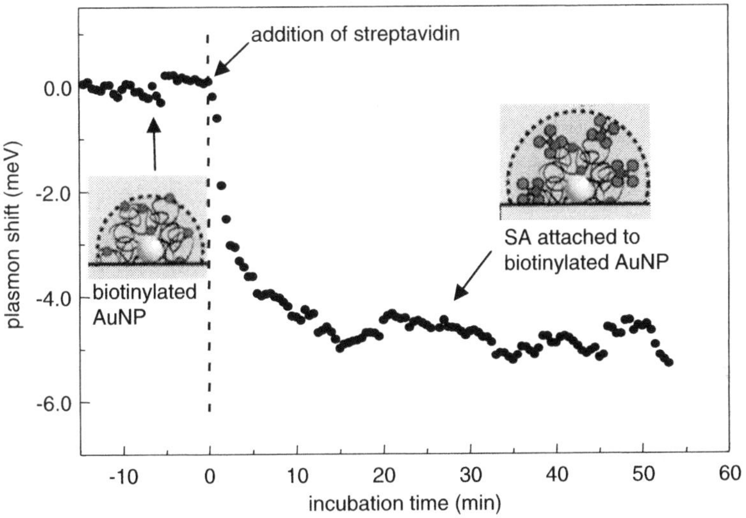

Fig. 13. Single gold nanoparticle biosensor. A gold nanoparticle has been functionalized with biotinylated albumin. A single gold nanoparticle is first imaged for 15 min in a dark field microscope. The peak energy of the scattering spectrum does not change. At $t = 0$ min streptavidin is added. The NPP resonance immediately starts to redshift and saturates at a 5 meV redshift. This redshift corresponds to a total streptavidin coverage of approximately 200 molecules.

seen that the spectrum stays the same within ± 0.5 meV. At $t = 0$, streptavidin is added to the water droplet under the immersion lens. Immediately after the addition of streptavidin, the spectral position of the NPPR redshifts (to lower energies). We note, that the saturated spectral shift of 5 meV stems from the binding of about only 200 streptavidin molecules as determined by steric arguments comparing the nanoparticle surface and the size of streptavidin molecules (Raschke et al., 2003). The addition of an analyte solution containing other proteins but no streptavidin showed no spectral response and therefore proves the selectivity of the single gold nanoparticle sensor.

The unspecific detection of 60,000 hexadecanthiol molecules using single gold nanoparticles has been published by McFarland and Van Duyne (2003). Very recently, Liu, Doll, and Lee have multiplexed the single metal nanoparticle sensor (Liu et al., 2005). Because of the ultimate small size of a single gold nanoparticle, the area that is needed for the detection of one specific sort of protein on a biochip is essentially limited by the resolution power of the microscope. The information density of single metal nanoparticle sensors is therefore on the order of 1 μm^{-2}. This holds a great potential for future miniaturized biochip technology.

5.2.4. Nanohole sensors

Nanoscopic holes in noble metal films can be considered as the 'inverse' of metal nanoparticles. Indeed they behave very similar in many respects (Prikulis et al., 2004). Nanoholes also show a zero-dimensional plasmonic resonance, which in this case is usually called a LPR rather than an NPPR because of the absence of a real particle. However, the physical behaviour of hole-like LPRs and NPPRs is very similar. Most importantly for hole-like plasmonic rensonances, no dispersion argument has to be matched, as is also the case for NPPRs, and they similarly redshift when the refractive index of the dielectric inside the hole is increased (Prikulis et al., 2004). As a consequence, nanoholes are also able to serve as biosensors (Dahlin et al., 2005). Because of their geometry, the noble metal films with holes are ideal for use in membrane-bound biosensor applications. For example, a double lipid layer is spread out on a silver film containing holes. It is then possible to study lipid-membrane mediated reaction kinetics in the area of the holes (Dahlin et al., 2005). Shortly after this first application of an assembly of nanoholes for biosensing, the team around Sutherland and Käll extended their research to *single* nanohole biosensing (Rindzevicius et al., 2005).

5.2.5. Analytical applications

Though most of the feasibility studies of gold nanoparticle dielectric environment bio-sensors have been carried out only recently, some applications have already been reported. For example, the Van Duyne group used NPPR immunoassays to detect Alzheimer's disease (Haes et al., 2004a, 2005). Kreuzer et al. report on a sensor for doping substances (Kreuzer et al., 2006).

5.2.6. Nanoparticles for spectroscopy in the biophysical window

The biological window for spectroscopy spans the wavelength range from 700 to 1100 nm where the absorption of the heme and the water are simultaneously low. However, the 'natural' NPPRs of spherical silver and gold nanoparticles are at around 400 and 530 nm in aqueous solution and therefore far off the biological window. However, it may still be accessed by tuning the shape of the nanoparticles. First, one can elongate the nanoparticles in order to retrieve rod-like nanoparticles. These nanorods show redshifted plasmonic resonances (Bohren and Huffmann, 1983) and can be used for biosensing at 700 nm (Alekseeva et al., 2005).

Nanoshells that comprise a dielectric core and a thin noble metal shell (Zhou et al., 1994; Averitt et al., 1997) can also be used as red NPP resonators as it is shown in figs. 5c and 8. They have been used in ensemble-based sensors (actually using coupled NPPRs) (Hirsch et al., 2003) and in single gold nanoshell biosensors (Raschke et al., 2004). A more advanced geometry of nanoshells for biosensor applications was suggested recently: multilayered nanospheres that consist of several layers of dielectrics and metals may show ultrasharp resonances up to the near IR (Chen et al., 2005). Noble metal nanorods (Sönnichsen et al., 2002b) and nanoshells (Raschke et al., 2004) show narrower NPPRs compared to the NPPRs of spherical nanoparticles. In this case the nanoparticle plasmons are not damped by the interband absorption because their resonance frequencies are too far in the red. Furthermore, the radiation damping, which scales with the metal volume, is reduced compared to solid nanospheres. Clearly, the sharper resonances of these structures are an advantage in NPP sensors.

5.3. A short comparison of SPR and NPPR sensors

In the two preceding subsections we have discussed SPR sensors and NPPR sensors. A direct comparison of the two sensor formats was carried out by the group of Van Duyne (Yonzon et al., 2004). It was found that during the association phase (when the antigen binds to the antibody) the SPR and the NPPR sensors exhibit qualitatively similar kinetics. However, in the dissociation phase, the NPPR sensor shows only a weak loss of signal. Backed by simulations, it was concluded that the antigens first dissociate from sites on the nanoparticles that have less influence on the actual spectral position of the NPPR. Alternatively, a partial removal of some antigens could lead to a more dense packing of the remaining antigens. This would lead to a thinner but higher index layer; however, both tendencies would cancel each other and the NPPR shifts only a little during the early dissociation phase.

A more technical and comprehensive comparison of the two sensor formats was given in (Haes and Van Duyne, 2004). The main advantage of the SPR sensor is its far more mature state of development. It has been commercialized and is used in many analytical and diagnostic laboratories. However, a major advantage of the NPPR biosensor is the simplicity of the excitation of NPPRs in comparison to the excitation of SPRs. The latter needs lavish optical designs such as a Kretschmann prism, because of restrictions in the dispersion relation (see Section 2.2). NPPR ensemble measurements may be carried out using a simple and cheap UV-Vis spectrometer.

Single metal nanoparticle sensors may become important in biochip applications because of their small size, and high density multiplexing may be achieved (Liu et al., 2005). The required area per detection spot is basically given by the resolution power of the imaging apparatus and may be as small as $1\,\mu m^2$. In contrast, SPR sensors typically need a minimal area of $100\,\mu m^2$. Owing to their extreme sensitivity to even smallest changes in the refractive index, the SPR sensors need to be temperature stabilized or otherwise they would sense the temperature gradient of the refractive index of water. In the case of NPPR sensors an elaborate temperature stabilisation is not necessary. Overall, this leads to possibly much cheaper setups for NPPR sensors as compared to SPR-based sensors (Haes and Van Duyne, 2004; Kreuzer et al., 2006).

Research has been performed on biosensors where the analyte molecules are sandwiched between a metal surface and metal nanoparticles. What may look like a combination of both worlds at first sight turns out to be a SPR sensor where the metal plasmonic nanoparticles are used in order to dramatically enhance the SPR shift upon analyte binding. The technique has been used for ultra sensitive immunoassays (Lyon et al., 1999) and DNA detection (He et al., 2000).

§ 6. Biosensing with surface-enhanced Raman scattering

We discussed in Section 3 how metal nanoparticles can be used as markers and in Sections 4 and 5, how a change of their spectral properties reports about a molecular recognition event that takes place on nanoscopic dimensions. All these applications required that the nanoparticles were pre-functionalized with molecules such as oligonucleotides or antibodies, and the selectivity was predominantly ensured by the quality of these functionalizing molecules. In this section, we would like to discuss a different technique that provides additional specific information inherent to the detected molecules (the analyte) such that, in principle, sensing is possible even without prior functionalization. The technique is SERS, which provides information about the vibronic 'fingerprint' of a molecule located close to the plasmonic nanostructures and therefore allows for a unique classification of the sort of analyte detected.

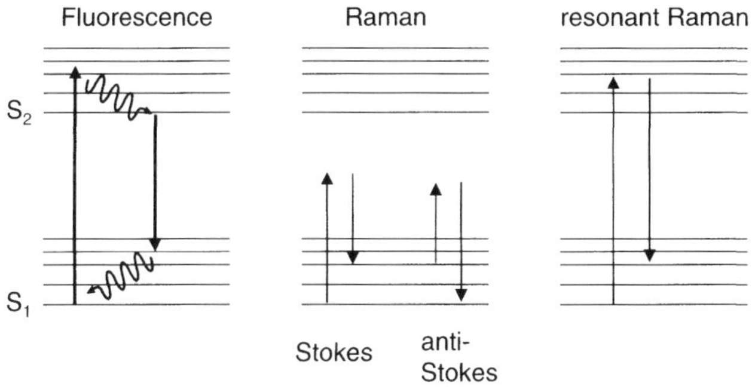

Fig. 14. Scheme of molecular excitations leading to fluorescence, Raman scattering, and resonant Raman scattering.

6.1. *SERS mechanism*

6.1.1. *Raman scattering*

When light is scattered by a molecule, almost all of the photons are scattered elastically, i.e. without changing their frequency. Only a very small fraction of the photons may change their frequency because they pick up or loose one quantum of vibronic excitation $\hbar\Omega$ of the molecule in a process that is called Raman scattering. This inelastic scattering of light was first published by Raman and Krishnan (1928) and independently observed and correctly explained as a deposition of an infrared energy quantum by Landsberg and Mandelstam (1928). Details can be found in many textbooks like the one of Ferraro (1994) or Schrader (1995). The process of Raman scattering is depicted in fig. 14 (centre sketch) whereby the creation of a vibronic excitation is called a 'Stokes' type event and the uptake of a vibronic quantum is called an 'anti-Stokes' type event. Because the spectrum of vibronic eigenmodes is characteristic to each type of molecule, one also speaks of the 'vibronic fingerprint' of the molecule. It can be evaluated by the determination of the difference in frequency of the Raman scattered light ω_{Raman} to the frequency of the illumination light ω_{ill}: $\hbar\Omega = \hbar(\omega_{\mathrm{ill}} - \omega_{\mathrm{Raman}})$.

Unfortunately, the Raman scattering cross section is very low, typically in the range of $10^{-31}\,\mathrm{cm}^2$ and therefore a huge number of molecules is needed in order to get a detectable signal. In comparison, the absorption cross section of a fluorescent molecule (fig. 14, left scheme) is typically in the range of $10^{-16}\,\mathrm{cm}^2$. The Raman cross section can be slightly increased up to $10^{-29}\,\mathrm{cm}^2$ if the Raman process involves real molecular levels. In this case, one speaks of 'resonant Raman' scattering. However, the detection of resonant Raman signals is often complicated due to the strong signal from the competing fluorescence. Of course, one can compensate for the small Raman cross sections by looking at a large number of molecules at the same time. However, in biosensing there are quite often only a few molecules available. In fluorescence spectroscopy, single molecule detection is meanwhile a state of the art technique. In order to bring Raman scattering to the same single molecule sensitivity, one has to increase the cross section dramatically. Indeed it was shown that Raman cross sections can be enhanced up to 10^{14}-fold using surface enhancement (Kneipp et al., 1996) and even single molecule detection became feasible (Kneipp et al., 1997; Nie and Emory, 1997; Kneipp et al., 1998a; Michaels et al., 1999; Xu et al., 1999).

6.1.2. Surface enhancement

Surface-enhanced Raman scattering (SERS) uses the strongly enhanced electromagnetic fields in the vicinity of metal nanostructures in order to increase the Raman signal substantially. SERS was first discovered in the 1970s (Fleischmann et al., 1974; Albrecht and Creighton, 1977; Jeanmarie and Van Duyne, 1977). The basis of the SERS effect is the substantial field enhancement in the vicinity of metal nanoparticles (Moskovits, 1978; Chen and Burstein, 1980; McCall et al., 1980). A SERS effect which is even better than the one provided by single nanoparticles can be obtained from closely spaced nanoparticles or rough surfaces, provided the average radii of curvature of the surface roughness is on the nanometre scale (Bergmann et al., 1981; Weitz et al., 1982). Perfectly flat metallic films, however, do not show a substantial electromagnetic enhancement effect. In the following, we will subsume any of the assemblies of nanoparticles, surfaces with nanoscopic roughness, and so on as a *SERS substrate*. Apart from the electromagnetic enhancement which we will discuss in this subsection in more detail, there is also another effect called chemical enhancement (Pockrand et al., 1983; Otto, 1984, 2001; Kambhampati et al., 1998), which will not be discussed here. It provides typically an enhancement factor of 10 to 10^3 in addition to the electromagnetic enhancement, which contributes a factor of 10^3 to 10^{14}. This subsection will briefly introduce the SERS effect to an extent that is necessary to understand its applications in biosensing. For a more detailed treatment, I refer to books and review articles such as Chang and Furtak (1982), Moskovits (1985), and Garrell and Pemberton (1993).

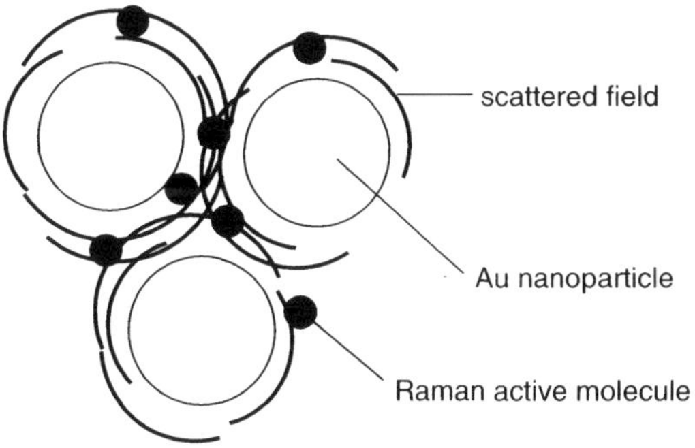

Fig. 15. Scheme of surface enhancement in Raman scattering.

The detailed model of the electromagnetic enhancement effect is as follows: Light of frequency ω_{ill} and intensity I_{ill} is shone on a SERS substrate with organic molecules in its nanoscopic vicinity (fig. 15). Because of nanoplasmonic resonances, the field of the incoming light is enhanced in the close vicinity of the SERS substrate. This will cause an enhancement of the Raman effect because the Raman signal depends linearly on the intensity of the original light, which is, in the case of nearby metallic nanostructures, proportional to the square of the enhanced near field. Then, the molecule will eventually alter the frequency of a photon by adding or subtracting a vibronic quantum $\hbar\Omega$ and create a photon with frequency ω_{Raman}. Now, in reversal of the above argument of a locally increased illumination field, the SERS substrate will act as an antenna to increase the Raman field and efficiently radiate the Raman photon. The efficiency of this radiative effect is again proportional to the local intensity, but this time at the frequency ω_{Raman}. Therefore, the total enhancement of the Raman signal of a molecule close to a SERS substrate in comparison to a molecule without a SERS substrate is given by (Shalaev, 2000)

$$G_{\mathrm{SERS}} \propto \left|\frac{E_{\mathrm{ill}}}{E^0_{\mathrm{ill}}}\right|^2 \left|\frac{E_{\mathrm{Raman}}}{E^0_{\mathrm{Raman}}}\right|^2, \tag{6.1}$$

where E^0_{ill} and E^0_{Raman} are the illumination field and the Raman scattered field without a SERS substrate and E_{ill} and E_{Raman} are the corresponding fields in the vicinity of a SERS substrate. In specific cases, the local enhancements of both the illumination and the Raman fields are so strong that G_{SERS} reaches 10^{10} or even 10^{12}, and single molecule Raman spectroscopy is possible. However, in general, the optimum field enhancement for the illumination field and the Raman field may not spatially coincide on SERS substrates because both field enhancements independently show strong spatial fluctuations. Therefore, one has to take the spatial average of eq. (6.1) over the SERS substrate's nanometric volume V in order to determine the effective enhancement factor of an extended SERS substrate (Shalaev, 2000):

$$G_{\mathrm{SERS}}^{\mathrm{eff}} \propto \frac{1}{V}\int_V \left|\frac{E_{\mathrm{ill}}}{E_{\mathrm{ill}}^0}\right|^2 \left|\frac{E_{\mathrm{Raman}}}{E_{\mathrm{Raman}}^0}\right|^2. \tag{6.2}$$

In the case of metal films of statistical roughness, and provided that the Raman shift $\omega_{\mathrm{ill}}-\omega_{\mathrm{Raman}}$ is large compared to the homogeneous linewidth of nanoparticle plasmons, we can assume that the positions of the 'hot spots' of large field enhancement are not related for the two frequencies. Then, the spatial averaging decouples and equation (6.2) becomes

$$G_{\mathrm{SERS}}^{\mathrm{eff}} \propto \frac{1}{V^2}\cdot \int_V \left|\frac{E_{\mathrm{ill}}}{E_{\mathrm{ill}}^0}\right|^2 \cdot \int_V \left|\frac{E_{\mathrm{Raman}}}{E_{\mathrm{Raman}}^0}\right|^2. \tag{6.3}$$

Equation (6.3) also implies that in this limit the SERS efficiency is proportional to the product of the extinction spectra at the illumination and at the Raman frequency (Chen et al., 1979; Chen and Burstein, 1980; Bergmann et al., 1981; Weitz et al., 1982):

$$G_{\mathrm{SERS}}^{\mathrm{eff}} \propto \mathrm{Ext}(\omega_{\mathrm{ill}})\cdot \mathrm{Ext}(\omega_{\mathrm{Raman}}). \tag{6.4}$$

It is important to note that, while single hot spots on a SERS substrate may show enhancements of 10^{12}, the spatially averaged SERS enhancement factor is typically in the order of 10^5 to 10^6 only, because averaging also includes 'cold' areas on the SERS substrate.

6.1.3. SERS substrates

There has been a long search for the 'ideal' SERS substrate, i.e. for a substrate where the averaged enhancement factor $G_{\mathrm{SERS}}^{\mathrm{eff}}$ is large. As mentioned, a perfectly flat metallic film only shows chemical SERS enhancement but no field enhancement. To the other extreme, a single, small noble metal nanoparticle is far from being optimal. This is because it enhances only one of the two fields (if the illumination and SERS frequencies are not very close together) and furthermore, the maximal field enhancement of a single spherical gold nanoparticle is at most a factor of 10 (Wokaun et al., 1982; Sönnichsen et al., 2002b). Therefore a SERS enhancement of 100 is expected from single spherical nanoparticles. In contrast, a large electric field enhancement can be achieved between two metallic spheres (Inoue and Ohtaka, 1983; Hillenbrand and Keilmann, 2001), or with a self-similar chain of spheres of decreasing radius (Li et al., 2003), or between the tips of two triangles forming a bowtie structure (Fromm et al., 2004; Hao and Schatz, 2004). In this case the fields in the gap between the metallic nanostructures can be enhanced by many orders of magnitude.

Rough metallic films may show averaged SERS enhancements of up to 10^9. For example, roughened silver electrodes were used in the first observations of SERS (Fleischmann et al., 1974; Albrecht and Creighton, 1977; Jeanmarie and Van Duyne, 1977). Also, aggregated silver island films show excellent Raman enhancement (Bergmann et al., 1981; Weitz et al., 1982) and so do semicontinous films near the percolation limit (Shalaev et al., 1996; Gadenne et al., 1997; Shalaev, 2000; Sarychev and Shalaev, 2000). Metal films with high surface roughness have also been fabricated by the over-coating of a dense layer of dielectric spheres with a thin metallic layer (Van Duyne et al., 1993). Technically related is the method of nanosphere lithography first applied by Fischer and Zingsheim where dielectric nanospheres are arranged in a hexagonal pattern and serve as an evaporation mask for the metal (Fischer and Zingsheim, 1981). After the evaporation of the metal, the dielectric spheres are removed and a hexagonal structure of metallic triangles is left behind on the glass surface. It has been shown that these structures can serve as SERS substrates (Haynes and Van Duyne, 2003). High field enhancements can also be obtained from metallic nanoshells on dielectric cores (Jackson et al., 2003).

It has been of some dispute whether to use gold or silver as SERS metal. Gold has the advantage of being more inert than silver, especially when it comes to oxidation, but silver shows sharper resonances in the limit of small particles (see fig. 5) and therefore a larger field enhancement is expected. The latter is certainly true in the Rayleigh limit, when the gold NPP is damped by interband transitions, while the silver NPP is not. However, in the case of coupled NPPRs, as they are used in SERS substrates, the resonances are shifted to longer wavelengths, away from the gold interband absorption, and consequently gold has been shown to be similarly suitable for SERS substrates such as colloidal gold clusters (Kneipp et al., 1998b) or gold nanoparticle shells (Jackson et al., 2003).

A very recent, and for biophysical applications appealing, method of growing SERS substrates is sketched in fig. 16 (Drachev et al., 2005b): A thin layer of SiO_2 is evaporated onto a glass substrate, followed by a submonolayer of silver (fig. 16a). The silver forms granules that stick only weakly to the SiO_2 layer. Then, a solution that contains proteins (e.g.

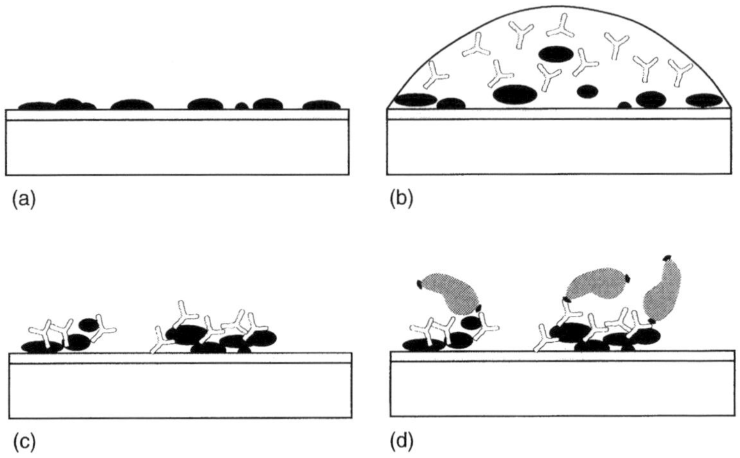

Fig. 16. Self-adaptive silver films: (a) A percolation film is evaporated onto a thin SiO_2 layer. (b) A solution with antibodies is drop casted onto the percolation film. The silver grains rearrange. (c) Islands of silver grains and antibodies form. (d) Antigens bind to the antibodies. The Raman signal from the pure antibodies and the antibodies bound to the antigens is distinct enough to report about the antigen–antibody binding.

antibodies) is drop casted onto the granular silver film. Some of the silver granules are detached from the surface and coalesce with the proteins forming clusters of silver nanoparticles and proteins. It turns out that these clusters show a good SERS enhancement and simultaneously retain the functionality of at least some of the proteins. Therefore, the proteins (antibodies in the case shown in fig. 16) arrange their own SERS substrate in a self-assembly style process. The product shown in fig. 16c is called an 'adaptive silver film'.

6.2. Biosensing with SERS

SERS has been applied in many fields of biological and medical sensing. It is certainly not possible to give a comprehensive list of all publications in the framework of this chapter as there are far too many. Rather, I would like to give some distinctive examples in order to show how powerfully the SERS technique can be used in biological and medical applications.

6.2.1. Applications in cell and molecular biology

SERS has been applied in cell biology. For example, it was shown that silver colloids can be grown inside and on the cell wall of *Escherichia coli* bacteria (Efrima and Bronk, 1998). It has been shown that peptides and polysaccharides inside the cell wall and its membrane can be identified by their Raman signal that is amplified by the silver colloids. Another study (Wood et al., 1997) reports that SERS can be used to monitor transport processes through membranes. The 'original' SERS substrate, namely an anodized silver electrode, turned out to be a good tool to perform simultaneous voltammetric and SERS spectroscopic measurements on cytochrome c and to study its redox behaviour (Niki et al., 1987). It is also possible to study the dynamics of the electron transfer reactions of the heme proteins using time-resolved SERS spectroscopy (Murgida and Hildebrandt, 2001).

As mentioned above, resonant Raman spectroscopy often has the drawback that the competing fluorescence overwhelms the resonant Raman signal. Therefore, it is difficult to take advantage of the increased Raman cross section in the case of resonant Raman compared to non-resonant Raman. However, it has been shown that metallic nanostructures quench fluorescence very efficiently if the fluorophores are very close to the metallic surface (Dulkeith et al., 2002). This, however, is the regime where SERS works best. Thus, resonant Raman spectroscopy at rough metal surfaces takes the full advantage of surface enhancement, while the competing fluorescence is quenched. Therefore, surface-enhanced resonant Raman spectroscopy (SERRS) is an ideal tool to study fluorescent biomolecules such as photosynthetic complexes (Cotton and Van Duyne, 1982; Thomas et al., 1990; Picorel et al., 1991; Seibert et al., 1992; Lutz, 1995).

6.2.2. Diagnostics with SERS labels

SERS has also proven to be a very powerful tool for diagnostics. Despite the fact that SERS has the potential of being a label-free sensor format, an additional molecule was utilized as a SERS active reporter of the antigen–antibody recognition event in the first SERS immunoassay (Rohr et al., 1989) and most of the SERS immunoassays that are reported up to now do so as well. The principal idea is sketched in fig. 17a. Capture antibodies are attached to a rough metal film and detection antibodies are functionalized with a Raman active label without loosing their ability to detect the antigen. When the antigen is sandwiched between the two types of antibodies, the Raman active label is brought into the field enhancement zone of the rough metal film and a Raman signal is detected. An improved version is displayed in fig. 17b (Ni et al., 1999): gold or silver

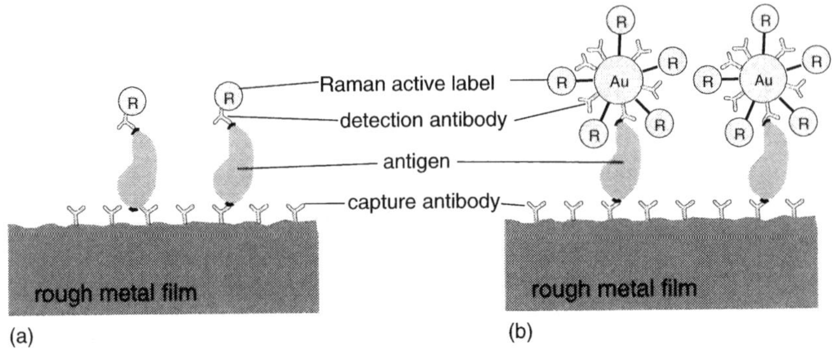

Fig. 17. Scheme of an immunoassay using Raman labels. (a) Capture antibodies are attached to a rough metal film; detection antibodies are functionalized with a Raman active label. When the antigen is sandwiched between the two types of antibodies, the Raman active label is in the field enhancement zone of the rough metal film. (b) Improved assay: gold nanoparticles are co-functionalized with detection antibodies and Raman active labels. The SERS enhancement is improved in comparison to the method described in case (a) because the field enhancement in case (b) stems from the rough metal film and the clustered gold nanoparticles.

nanoparticles are co-functionalized with antibodies and Raman active labels. As long as the noble metal nanoparticles are isolated in solution, the field enhancement provided by the nanoparticles is only weak and the Raman signal from the Raman active label is weak. When antigens are present, they form a sandwich layer and the fields between the rough silver film and the nanoparticles become substantially more enhanced compared to the case without metal nanoparticles. This leads to an increase in the Raman signal from the Raman active molecule (fig. 17b). The technique was further refined and eventually reached femtomolar sensitivity in the detection of prostate-specific antigen (Grubisha et al., 2003) and also allows for the detection of viral pathogens (Driskell et al., 2005).

Enzymatic immunoassays comprising SERS have been developed as well (Dou et al., 1997). They consists of a sandwich-type immunoassay similar to the one shown in fig. 17a, but the detection antibody is labelled with an enzyme rather than with a Raman active molecule. After the sandwich layer is formed and superfluous antibodies are washed away, the assay is incubated in some solution. This solution contains clusters of silver nanoparticles and it allows the enzyme to produce its product. The product of the enzymatic reaction subsequently attaches to the silver clusters and gives rise to a SERS signal. A similar protocol has been used to detect not only antibodies but whole cells such as human hepatocellular carcinoma cells (Hawi et al., 1998).

One might question what the advantages of SERS immunoassays using Raman active labels are compared to other assays using labels such as fluorophores of metal nanoparticles. The expenditure seems, and in deed is, similar. However, SERS immunoassays using Raman labels have a big advantage when multiplexing is considered. It is the goal to detect not only one antigen, but as many antigens as possible in parallel. The number of fluorophore or metal nanoparticle labels that can be clearly distinguished from each other within the range of visible frequencies is limited because fluorescence spectra or nanoparticle scattering spectra are usually fairly broad, say in the 100 nm range. Therefore, the number of analytes to be detected in parallel is limited using the visible spectrum. In contrast, Raman labels show very sharp lines that are characteristic to each kind of Raman active label, and therefore SERS immunoassays using Raman labels hold great promise for immunoassay multiplexing (Ni et al., 1999; Cao et al., 2003).

Similar to immunoassays, nucleic acids and short strands of DNA may be specifically detected (Vo-Dinh et al., 1994, 1999; Culha et al., 2003). An interesting variant has been reported by Cao et al. (2002). Gold nanoparticles are co-functionalized with ssDNA and a Raman active molecule. After these probes have been fixed at a substrate due to hybridisation, a silver enhancement step forms larger noble metal nanoparticles near the Raman active molecules and the SERS effect is enhanced. A very recent variant of a Raman-labelled DNA-SERS sensor was reported by Wabuyele and Vo-Dinh (2005). They used a molecular-beacon-like assay where one end of a ssDNA is attached to a silver nanoparticle and the other end is attached to a Raman active label. They showed that the HIV virus can be detected with this assay.

In many cases, the PCR is used in DNA assays. It has been shown that also in this case Raman-labelled primers can be used in combination with SERS to detect the Gag gene of the HIV virus (Isola et al., 1998). As explained above, different Raman active primers are prone to multiplexing due to their characteristic fingerprint spectra. With this technique, the three different genotypes of the cystic fibrosis transmembrane regulator gene can be distinguished (Graham et al., 2002). The amount of double stranded DNA in a PCR reaction can also be controlled by the addition of the dye molecule DAPI, which is also Raman active, and silver colloids. DAPI weakly attaches to the silver cluster surfaces giving rise to a Raman signal. However, it is much more likely to intercalate in double stranded (ds) DNA, so the more dsDNA present, the weaker the SERS signal becomes (Dou et al., 1998a).

6.2.3. Label-free SERS diagnostics

It has been mentioned that SERS holds the potential for immunoassays without any Raman active labels. This is because of the following reasons: The antigens and the antibodies have their own characteristic Raman spectra. So it should, in principle, be possible to either detect the Raman spectrum of the antigen or to detect a change of the Raman spectrum of the antibody upon antigen recognition, or some convolution of both. However, antigens and antibodies cannot be specifically optimized to deliver very strong Raman signals as it was possible with the choice of the Raman active molecules. Nevertheless, SERS turned out to be powerful enough to boost the Raman signal of proteins to such an extent that they become detectable, and additionally, small changes of the Raman signal upon antigen–antibody binding are detectable unambiguously (Dou et al., 1998b). A label-free immunoassay using an adaptive silver film has been reported recently (Drachev et al., 2005a). The adaptive silver films were prepared as described above (figs. 16a–c). The mixed islands of antibodies and silver clusters as shown in fig. 16c show a characteristic Raman signal. The Raman spectrum changes significantly upon addition of the matching antigen (fig. 16d) (Drachev et al., 2005a).

6.2.4. Other selected biomedical applications

We will concentrate on small physiologically and medically relevant molecules in this subsection, whereas the previous two subsections were devoted to the 'classical' immunoassays and DNA assays which respond to fairly large molecules. Small molecules are also called haptens. Actually, the literature on this topic is large and only a selection can be given. In general, SERS applications on small biological molecules rely on the Raman fingerprints of the analytes and do not use labels; so this chapter is conceptually related to the previous one, with the difference being analyte size.

Examples of SERS detection of small molecules include the detection of drugs in the urine (Ruperez et al., 1991) and the label-free detection of bilirubin and salicylate in whole blood samples (Sulk et al., 1999). SERS has also been used for investigations of neurotransmitters (Kneipp et al., 1995; Lee et al., 1988; Volkan et al., 2000). In the latter report, a fibre SERS probe was introduced that holds great promise for an in vivo detection of dopamine. The tip of the fibre probe was coated with silver nanoparticles to initiate a SERS signal that is then guided by the probe.

An interesting application of SERS in cancer research is reported by Nabiev and co-workers. They use SERS in order to study the interaction of an antitumor drug with DNA in vitro as well as in cancer cells (Morjani et al., 1993, Nabiev et al., 1994). Enzyme activity was monitored using a SERRS active molecule initially linked to some quencher that suppresses SERRS. Active enzymes may cleave the linkage and the SERRS active molecules produce a Raman signal (Moore et al., 2004). It has also been shown that SERS has potential to be used in glucose sensing, which is an important issue because of the large number of patients suffering from diabetes (Shafer-Peltier et al., 2003). In this context, another report on adaptive SERS substrates is of relevance (Drachev et al., 2004): it has been shown that self-adaptive silver films that were prepared with two different isomers of insulin can be distinguished by their SERS spectrum. This is remarkable because the two isomers used in the study, human insulin and insulin lispro, differ only in the interchange of two neighbouring amino acids. However, this leads to a conformational change of the protein, which obviously is unaffected by the formation of the adaptive silver clusters as shown in fig. 16c.

§ 7. Concluding remarks

We have discussed several formats of biosensors comprising plasmonic nanoparticles, including Raman biosensors, and also briefly touched surface plasmon sensors. We have seen that plasmonics can offer quite a bit to tackle the challenges set by proteomics and genomics with respect to novel, sensitive, and biochip-compatible sensor formats. Some of the presented sensor formats, such as test-strip sensors for pregnancy or the surface plasmon immunosensor, have already proven successful by entering the market. As other formats are still in the research and development stage, continued research is vital in this area. Over the last seven years the annual number of papers on plasmonic nanoparticle biosensors has been growing exponentially (actually doubling every 13 months), indicating that there is still great research potential.

In this chapter, several sensor formats have been described: heterogeneuos sensors that use sheer presence or absence of plasmonic nanoparticles as the sensor signal and sensors that rely on a change of the scattering or extinction spectra of plasmonic nanoparticles upon analyte binding. The potential of nanoparticle clusters for SERS sensors has also been described. However, the given list of formats is not complete. Other physical phenomena related with plasmonic nanoparticles are capable to be exploited in biosensors. For example, gold nanoparticles can substitute organic acceptor molecules in molecular-beacon-like DNA assays (Dubertret et al., 2001, Maxwell et al., 2002). It has been shown that the large quenching efficiency of gold nanoparticles is not only due to their strong absorption band associated with the plasmonic resonance, but it is also because the gold nanoparticles affect the radiative lifetime of adjacent dye molecules (Dulkeith et al., 2002, 2005).

Beyond biosensing, plasmonic nanoparticles have further applications in biology and medicine. For example, they simultaneously give contrast in electron and optical microscopy such that a labelled cell can be imaged in optical and subsequently in electron microscopy (Geoghegan et al., 1978). A modified version uses fluorophore-gold nanoparticle conjugates for simultaneous contrast in electron and fluorescence microscopy (Powell et al., 1997, 1998). Gold nanoparticles are also promising in hypothermal cancer therapy (Hüttmann and Birngruber, 1999; O'Neal et al., 2004). In these experiments, nanoparticles that are attached to cancer cells are heated by the absorption of light until the cancer cell overheats and dies.

The variety of biosensing applications explained in detail in this chapter and the additional applications outlined briefly in this last section together with the exponentially growing number of publications clearly point out that we can expect even more biologically and medically relevant applications in the near future.

Acknowledgements

It would have been impossible for me to work on biosensing with noble metal nanoparticles without the help of, and fruitful discussion with, many colleagues and friends. I would like to thank all of them, in particular Jochen Feldmann, Konrad Kürzinger, Vladimir Shalaev, Gunnar Raschke, Alfons Nichtl, Vladimir Drachev, Moritz Ringler, Joachim Stehr, Carsten Sönnichsen, Sandra Brogl, Thomas Franzl, and Stefan Kowarik. Reuben Bakker and Josh Borneman helped in proofreading.

References

Abbe, E., 1873, Arch. Mikroskop. Anatom. **9**, 413–420.

Albrecht, M.G., Creighton, J.A., 1977, J. Am. Chem. Soc. **99**, 5215–5217.

Alekseeva, A.V., Bogatyrev, V.A., Dykman, L.A., Khlebtsov, B.N., Trachuk, L.A., Melnikov, A.G., Khlebtsov, N.G., 2005, Appl. Op. **44**, 6285–6294.

Alivisatos, A.P., Johnsson, K.P., Peng, X., Wilson, T.E., Loweth, C.J., Bruchez, M.P., Schultz, P.G., 1996, Nature **382**, 609–611.
Arbouet, A., Christofilos, D., Del Fatti, N., Vallee, F., Huntzinger, J.R., Arnaud, L., Billaud, P., Broyer, M., 2004, Phys. Rev. Lett. **93**, 127401.
Ashcroft, N.W., Mermin, N.D., 1976, Solid State Physics, Saunders College, Philadelphia.
Aslan, K., Holley, P., Davies, L., Lakowicz, J.R., Geddes, C.D., 2005a, J. Am. Chem. Soc. **127**, 12115–12121.
Aslan, K., Lakowicz, J.R., Geddes, C.D., 2005b, Appl. Phys. Lett. **87**, 234108.
Aslan, K., Lakowicz, J.R., Geddes, C.D., 2005c, Anal. Chem. **77**, 2007–2014.
Averitt, R.D., Sarkar, D., Halas, N.J., 1997, Phys. Rev. Lett. **78**, 4217–4220.
Bailey, R.C., Nam, J.-M., Mirkin, C.A., Hupp, J.T., 2003, J. Am. Chem. Soc. **125**, 13541–13547.
Bauer, H., Horisberger, M., Bush, D.A., Sigarlakie, E., 1972, Arch. Mirkobiol. **85**, 202–208.
Bergmann, J.G., Chemla, D.S., Liao, P.F., Glass, A.M., Pinczuk, A., Hart, R.M., Olson, D.H., 1981, Opt. Lett. **6**, 33–35.
Bericaud, S., Cognet, L., Blab, G.A., Lounis, B., 2004, Phys. Rev. Lett. **93**, 257402.
Blab, G.A., Cognet, L., Berciaud, S., Alexandre, I., Husar, D., Remacle, J., Lounis, B., 2006, Biophys. J. Biophys. Lett., **90**, L13–L15.
Bohren, C., Huffmann, D., 1983, Absorption and Scattering of Light by Small Particles, John Wiley & Sons, New York.
Bornfleth, H., Satzler, K., Eils, R., Cremer, C., 1998, J. Microsc. **189**, 118–136.
Boyer, D., Tamarant, P., Maali, A., Lounis, B., Orrit, M., 2002, Science **297**, 1160–1163.
Brockman, J.M., Frutos, A.G., Corn, R.M., 1999, J. Am. Chem. Soc. **121**, 8044–8051.
Cao, Y.C., Jin, R., Nam, J.-M., Thaxton, C.S., Mirkin, C.A., 2003, J. Am. Chem. Soc. **125**, 14676–14677.
Cao, Y.-W.C., Jin, R., Mirkin, C.A., 2002, Science **297**, 1536–1540.
Chang, R.K., Furtak, R.E., 1982, Surface Enhanced Raman Scattering, Plenum, New York.
Chen, C.Y., Burstein, E., Lundquist, S., 1979, Solid State Commun. **32**, 63–66.
Chen, C.Y., Burstein, E., 1980, Phys. Rev. Lett. **45**, 1287–1291.
Chen, K., Liu, Y., Ameer, G., Backman, V., 2005, J. Biomed. Opt. **10**, 024005.
Cotton, T.M., Van Duyne, R.P., 1982, FEBS Lett. **147**, 81–84.
Culha, M., Stokes, D., Allain, L.R., Vo-Dinh, T., 2003, Anal. Chem. **75**, 6196–6201.
Dahlin, A., Zäch, M., Rindzevicius, T., Käll, M., Sutherland, D.S., Höök, F., 2005, J. Am. Chem. Soc. **127**, 5043–5048.
Dou, X., Takama, T., Yamaguchi, Y., Yamamoto, H., 1997, Anal. Chem. **69**, 1492–1495.
Dou, X., Takama, T., Yamaguchi, Y., Hirai, K., Yamamoto, H., Doi, S., Ozaki, Y., 1998a, Appl. Opt. **37**, 759–763.
Dou, X., Yamaguchi, Y., Yamamoto, H., Doi, S., Ozaki, Y., 1998b, J. Raman Spectrosc. **29**, 739–742.
Drachev, V.P., Thoreson, M.D., Khaliullin, E.N., Davisson, J., Shalaev, V.M., 2004, J. Phys. Chem. B **108**, 18046–18052.
Drachev, V.P., Nashine, V.C., Thoreson, M.D., Ben-Amotz, D., Davisson, V.J., Shalaev, V.M., 2005a, Langmuir **21**, 8368–8373.
Drachev, V.P., Thoreson, M.D., Nashine, V., Khaliullin, E.N., Ben-Amotz, D., Davisson, V.J., Shalaev, V.M., 2005b, J. Raman Spectrosc. **36**, 648–656.
Driskell, J.D., Kwarta, K.M., Lipert, R.J., Porter, M.D., 2005, Anal. Chem. **77**, 6147–6154.
Dubertret, B., Calame, M., Libchaber, A.J., 2001, Nat. Biotechnol. **19**, 365–370.
Dulkeith, E., Morteani, A.C., Niedereichholz, T., Klar, T.A., Feldmann, J., Levi, S.A., van Veggel, F.C.J.M., Reinhoudt, D.N., Möller, M., Gittins, D.I., 2002, Phys. Rev. Lett. **89**, 203002.
Dulkeith, E., Ringler, M., Klar, T.A., Feldmann, J., Munoz Javier, A., Parak, W.J., 2005, Nano Lett. **5**, 585–589.
Efrima, S., Bronk, B.V., 1998, J. Phys. Chem. B **102**, 5947–5950.

Elghanian, R., Storhoff, J.J., Mucic, R.C., Letsinger, R.L., Mirkin, C.A., 1997, Science **277**, 1078–1081.
El-Sayed, I.H., Huang, X., El-Sayed, M.A., 2005, Nano Lett. **5**, 829–834.
Englebienne, P., 1998, Analyst **123**, 1599–1603.
Englebienne, P., Van Hoonacker, A., Valsamis, J., 2000, Clin. Chem. **46**, 2000–2003.
Faraday, M., 1857, Philos. Trans. Roy. Soc. **147**, 145.
Faulk, W.P., 1971, Immunochemistry **8**, 1081–1083.
Ferraro, J.R., 1994, Introductory Raman Spectroscopy, Academic, New York.
Fischer, U.C., Zingsheim, H.P., 1981, J. Vacuum Sci. Technol. **19**, 881–885.
Fleischmann, M., Hendra, P.J., McQillian, A.J., 1974, Chem. Phys. Lett. **26**, 163–166.
Förster, T., 1948, Ann. Phys. **2**, 55–75.
Frey, N., Müller-Bardorff, M., Katus, H.A., 1998, Zeitschrift für Kardiologie **87**(Suppl. 2), 100–105.
Fromm, D.P., Sundaramurthy, A., Schuck, P.J., Kino, G., Moerner, W.E., 2004, Nano Lett. **4**, 957–961.
Gadenne, P., Gagnot, D., Masson, M., 1997, Physica A **241**, 161–165.
Gans, R., 1925, Ann. Phys. **76**, 29–38.
Garrell, R.L., Pemberton, J.E., 1993, Fundamentals and Applications of Surface Raman Spectroscopy, VCH Publishers, Deerfield Beach.
Geoghegan, W.D., Scillian, J.J., Ackermann, G.A., 1978, Immunol. Commun. **7**, 1–12.
Georganopoulou, D.G., Chang, L., Nam, J.-M., Thaxton, C.S., Mufson, E.J., Klein, W.L., Mirkin, C.A., 2005, Proc. Natl. Acad. Sci. U S A **102**, 2273–2276.
Goverde, B.C., Leuvering, J.H.W., Thal, P.J.H.M., Scherdes, J.C.M., Schuurs, A.H.W.M., 1982, Anal. Bioanal. Chem. (Fresenius Zeitschrift für Analytische Chemie) **311**, 361.
Graham, D., Mallinder, B.J., Whitcombe, D., Watson, N.D., Smith, W.E., 2002, Anal. Chem. **74**, 1069–1074.
Grubisha, D.S., Lipert, R.J., Park, H.-Y., Driskell, J., Porter, M.D., 2003, Anal. Chem. **75**, 5936–5943.
Haes, A.J., Van Duyne, R.P., 2002, J. Am. Chem. Soc. **124**, 10596–10604.
Haes, A.J., Hall, W.P., Chang, L., Klein, W.L., Van Duyne, R.P., 2004a, Nano Lett. **4**, 1029–1034.
Haes, A.J., Van Duyne, R.P., 2004, Anal. Bioanal. Chem. **379**, 920–930.
Haes, A.J., Zou, S., Schatz, G.C., Van Duyne, R.P., 2004b, J. Phys. Chem. B **108**, 109–116.
Haes, A.J., Zou, S., Schatz, G.C., Van Duyne, R.P., 2004c, J. Phys. Chem. B **108**, 6961–6968.
Haes, A.J., Chang, L., Klein, W.L., Van Duyne, R.P., 2005, J. Am. Chem. Soc. **127**, 2264–2271.
Hao, E., Schatz, G.C., 2004, J. Chem. Phys. **120**, 357–366.
Hawi, S.R., Rochanakij, S., Adar, F., Campbell, W.B., Nithipatikom, K., 1998, Anal. Biochem. **259**, 212–217.
Haynes, C.L., Van Duyne, R.P., 2003, J. Phys. Chem. B **107**, 7426–7433.
He, L., Musick, M.D., Nicewarner, S.R., Salinas, F.G., Benkovic, S.J., Natan, M.J., Keating, C.D., 2000, J. Am. Chem. Soc. **122**, 9071–9077.
Henglein, A., Meisel, D., 1998, J. Phys. Chem. B **102**, 8364–8366.
Hilger, A., Cüppers, N., Tenfelde, M., Kreibig, U., 2000, Eur. Phys. J. D **10**, 115–118.
Hillenbrand, R., Keilmann, F., 2001, Appl. Phys. B **73**, 239–243.
Hirsch, L.R., Jackson, J.B., Lee, A., Halas, N.J., West, J.L., 2003, Anal. Chem. **75**, 2377–2381.
Homola, J., Yee, S.S., Gauglitz, G., 1999, Sens. Actuators B **54**, 3–15.
Homola, J., 2003, Anal. Bioanal. Chem. **377**, 528–539.
Hüttmann, G., Birngruber, R., 1999, IEEE J. Select. Topics Quantum Electron. **5**, 954–962.
Inoue, M., Ohtaka, K., 1983, J. Phys. Soc. Jpn. **52**, 3853–3864.
International Human Genome Sequencing Consortium, 2001, Nature **409**, 860–921.

Isola, N.R., Stokes, D.L., Vo-Dinh, T., 1998, Anal. Chem. **70**, 1352–1356.
Jackson, J.B., Westcott, S.L., Hirsch, L.R., West, J.L., Halas, N.J., 2003, Appl. Phys. Lett. **82**, 257–259.
Jeanmarie, D.L., Van Duyne, R.P., 1977, J. Electroanal. Chem. **84**, 1–20.
Jin, R., Wu, G., Li, Z., Mirkin, C.A., Schatz, G.C., 2003, J. Am. Chem. Soc. **125**, 1643–1654.
Johnson, P.B., Christy, R.W., 1972, Phys. Rev. B **6**, 4370–4379.
Jonsson, U., Fagerstam, L., Ivarsson, B., Johnsson, B., Karlsson, R., Lundh, K., Lofas, S., Persson, B., Roos, H., Ronnberg, I., Sjolanders, S., Stenberg, E., Stahlberg, R., Uraniczky, C., Ostlin, H., Malmqvist, M., 1991, Biotechniques **11**, 620.
Kambhampati, P., Child, C.M., Foster, M.C., Campion, A., 1998, J. Chem. Phys. **108**, 5013–5026.
Karlsson, R., Michaelsson, A., Mattson, L., 1991, J. Immunol. Meth. **145**, 229–240.
Kim, Y., Johnson, R.C., Hupp, J.T., 2001, Nano Lett. **1**, 165–167.
Klar, T., Perner, M., Grosse, S., Spirkl, W., von Plessen, G., Feldmann, J., 1998, Phys. Rev. Lett. **80**, 4249–4252.
Klar, T.A., Jakobs, S., Dyba, M., Egner, A., Hell, S.W., 2000, Proc. Natl. Acad. Sci. U S A **97**, 8206–8210.
Kneipp, K., Wang, Y., Dasari, R.R., Feld, M.S., 1995, Spectrochim. Acta A Mol. Spectrosc. **51**, 481–487.
Kneipp, K., Wang, Y., Kneipp, H., Itzkan, I., Dasari, R.R., Feld, M.S., 1996, Phys. Rev. Lett. **76**, 2444–2447.
Kneipp, K., Yang, W., Kneipp, H., Perelman, L.T., Itzkan, I., Dasari, R.R., Feld, M.S., 1997, Phys. Rev. Lett. **78**, 1667–1670.
Kneipp, K., Kneipp, H., Kartha, V.B., Manoharan, R., Deinum, G., Itzkan, I., Dasari, R.R., Feld, M.S., 1998a, Phys. Rev. E **57**, R6281–R6284.
Kneipp, K., Kneipp, H., Manoharan, R., Hanlon, E.B., Itzkan, I., Dasari, R.R., Feld, M.S., 1998b, Appl. Spectrosc. **52**, 1493–1497.
Köhler, J.M., Csaki, A., Reichert, J., Möller, R., Straube, W., Fritsche, W., 2001, Sens. Actuators B **76**, 166–172.
Kreibig, U., Althoff, A., Pressmann, H., 1981, Surf. Sci. **106**, 308–317.
Kreibig, U., Vollmer, M., 1995, Optical Properties of Metal Clusters, Springer-Verlag, Berlin.
Kreibig, U., Gartz, M., Hilger, A., 1997, Ber. Bunsen-Ges. Phys. Chem. **101**, 1593.
Kretschmann, E., 1972, Untersuchungen zur Anregung und Streuung von Oberflächenplasmaschwingungen an Silberschichten. Doctoral Thesis., Universität Hamburg, Hamburg.
Kreuzer, M.P., Quidant, R., Badenes, G., Marco, M.-P., 2006, Biosens. Bioelectron. **21**, 1345–1349.
Landsberg, G., Mandelstam, L., 1928, Naturwissenschaften **16**, 557–558.
Lee, N.S., Hsieh, Y.Z., Paisley, R.F., Morris, M.D., 1988, Anal. Chem. **60**, 442–446.
Leuvering, J.H.W., Thal, P.J.H.M., van der Waart, M., Schuurs, A.H.W.M., 1980a, Anal. Bioanal. Chem. (Fresenius Zeitschrift für Analytische Chemie) **301**, 132.
Leuvering, J.H.W., Thal, P.J.H.M., van der Waart, M., Schuurs, A.H.W.M., 1980b, J. Immunoassay **1**, 77.
Leuvering, J.H.W., Thal, P.J.H.M., van der Waart, M., Schuurs, A.H.W.M., 1981, J. Immunol. Meth. **45**, 183–194.
Leuvering, J.H.W., Goverde, B.C., Thal, P.J.H.M., Schuurs, A.H.W.M., 1983, J. Immunol. Meth. **60**, 9–23.
Li, K.R., Stockman, M.I., Bergman, D.J., 2003, Phys. Rev. Lett. **91**, 227402.
Liedberg, B., Nylander, C., Lunström, I., 1983, Sens. Actuators **4**, 299–304.
Lindfors, K., Kalkbrenner, T., Stoller, P., Sandoghdar, V., 2004, Phys. Rev. Lett. **93**, 037401.
Link, S., El-Sayed, M.A., 2000, Int. Rev. Phys. Chem. **19**, 409.

Linnert, T., Mulvaney, P., Henglein, A., 1993, J. Phys. Chem. **97**, 679–682.
Liu, G.L., Doll, J.C., Lee, L.P., 2005, Opt. Express **13**, 8520–8525.
Löfas, S., Malmqvist, M., Rönnberg, I., Stenberg, E., Liedberg, B., Lundström, I., 1991, Sens. Actuators **5**, 79–84.
Lutz, M., 1995, Biospectroscopy **1**, 313–327.
Lyon, L.A., Music, M.D., Smith, P.C., Reiss, B.D., Pena, D.J., Natan, M.J., 1999, Sens. Actuators B **54**, 118–124.
Malinsky, M.D., Kelly, K.L., Schatz, G.C., Van Duyne, R.P., 2001, J. Am. Chem. Soc. **123**, 1471–1482.
Maxwell, D.J., Taylor, J.R., Nie, S., 2002, J. Am. Chem. Soc. **124**, 9606–9612.
McCall, S.L., Platzmann, P.M., Wolff, P.A., 1980, Phys. Lett. **77A**, 381–383.
McFarland, A.D., Van Duyne, R.P., 2003, Nano Lett. **3**, 1057–1062.
Michaels, A.M., Nirmal, M., Brus, L.E., 1999, J. Am. Chem. Soc. **121**, 9932–9939.
Mie, G., 1908, Ann. Phys. **3**, 25.
Mirkin, C.A., Letsinger, R.L., Mucic, R.C., Storhoff, J.J., 1996, Nature **382**, 607–609.
Mock, J.J., Barbic, M., Smith, D.R., Schultz, D.A., Schultz, S., 2002, J. Chem. Phys. **116**, 6755–6759.
Mock, J.J., Smith, D.R., Schultz, S., 2003, Nano Lett. **3**, 485–491.
Moeremans, M., Daneels, G., Van Dijck, A., Langanger, G., De Mey, J., 1984, J. Immunol. Meth. **74**, 353–360.
Moore, B.D., Stevenson, L., Watt, A., Flitsch, S., Turner, N.J., Cassidy, C., Graham, D., 2004, Nat. Biotechnol. **22**, 1133–1138.
Morjani, H., Riou, F.J., Nabiev, I., Lavelle, F., Manafit, M., 1993, Cancer Res. **53**, 4784–4790.
Moskovits, M., 1978, J. Chem. Phys. **69**, 4159–4161.
Moskovits, M., 1985, Rev. Mod. Phys. **57**, 783–826.
Mulvaney, P., 1996, Langmuir **12**, 788–800.
Murgida, D.H., Hildebrandt, P., 2001, J. Am. Chem. Soc. **123**, 4062–4068.
Nabiev, I., Chourpa, I., Manfait, M., 1994, J. Phys. Chem. **98**, 1344–1350.
Nam, J.-M., Stoeva, S.I., Mirkin, C.A., 2004, J. Am. Chem. Soc. **126**, 5932–5933.
Nath, N., Chilkoti, A., 2002, Anal. Chem. **74**, 504–509.
Nath, N., Chilkoti, A., 2004, Anal. Chem. **76**, 5370–5378.
Ni, J., Lipert, R.J., Dawson, B., Porter, M.D., 1999, Anal. Chem. **71**, 4903–4908.
Nie, S., Emory, S.R., 1997, Science **275**, 1102–1106.
Niki, K., Kawasaki, Y., Kimura, Y., Higuchi, Y., Yasuoka, N., 1987, Langmuir **3**, 982–986.
Oldenburg, S.J., Genick, C.C., Clark, K.A., Schultz, D.A., 2002, Anal. Biochem. **309**, 109–116.
O'Neal, D.P., Hirsch, L.R., Halas, N.J., Payne, J.D., West, J.L., 2004, Cancer Lett. **209**, 171–176.
Otto, A., 1984, Surface-enhanced Raman Scattering: 'Classical' and 'Chemical' Origins, Springer, Berlin.
Otto, A., 2001, Phys. Status Solidi (a) **188**, 1455–1470.
Palik, E.D., 1985, Handbook of Optical Constants of Solids, Academic Press, New York.
Park, S.-J., Taton, T.A., Mirkin, C.A., 2002, Science **295**, 1503–1506.
Picorel, R., Holt, R.E., Heald, R., Cotton, T.M., Seibert, M., 1991, J. Am. Chem. Soc. **113**, 2839–2843.
Pockrand, I., Swalen, J.D., Gordon, J.G., Philpott, M.R., 1978, Surf. Sci. **74**, 237–244.
Pockrand, I., Billmann, J., Otto, A., 1983, J. Chem. Phys. **78**, 6384–6390.
Pohl, D.W., Denk, W., Lanz, M., 1984, Appl. Phys. Lett. **44**, 651.
Powell, R.D., Halsey, C.M.R., Spector, D.L., Kaurin, S.L., McCann, J., Hainfeld, J.F., 1997, J. Histochem. Cytochem. **45**, 947–956.
Powell, R.D., Halsey, C.M.R., Hainfeld, J.F., 1998, Microsc. Res. Tech. **42**, 2–12.
Prikulis, J., Hanarp, P., Olofsson, L., Sutherland, D., Käll, M., 2004, Nano Lett. **4**, 1003–1007.

Quinten, M., Kreibig, U., Schönauer, D., Genzel, L., 1985, Surf. Sci. **156**, 741–750.
Raman, C.V., Krishnan, K.S., 1928, Nature **121**, 501–502.
Raschke, G., Kowarik, S., Franzl, T., Sönnichsen, C., Klar, T.A., Feldmann, J., Nichtl, A., Kürzinger, K., 2003, Nano Lett. **3**, 935–938.
Raschke, G., Brogl, S., Susha, A.S., Rogach, A.L., Klar, T.A., Feldmann, J., Fieres, B., Petkov, N., Bein, T., Nichtl, A., Kürzinger, K., 2004, Nano Lett. **4**, 1853–1857.
Rechberger, W., Hohenau, A., Leitner, A., Krenn, J.R., Lamprecht, B., Aussenegg, F.R., 2003, Opt. Commun. **220**, 137–141.
Reichert, J., Csaki, A., Köhler, J.M., Fritsche, W., 2000, Anal. Chem. **72**, 6025–6029.
Reinhard, B.M., Siu, M., Agarwal, H., Alivisatos, A.P., Liphardt, J., 2005, Nano Lett. **5**, 2246–2252.
Riboh, J.C., Haes, A.J., McFarland, A.D., Yonzon, C.R.Y., Van Duyne, R.P., 2003, J. Phys. Chem. B **107**, 1772–1780.
Rich, R.L., Myszka, D.G., 2000, Curr. Opin. Biotechnol. **11**, 54–61.
Rindzevicius, T., Alaverdyan, Y., Dahlin, A., Höök, F., Sutherland, D.S., Käll, M., 2005, Nano Lett. **5**, 2335–2339.
Rohr, T.E., Cotton, T., Fan, N., Tarcha, P.J., 1989, Anal. Biochem. **182**, 388–398.
Ruperez, A., Montes, R., Laserna, J.J., 1991, Vib. Spectrosc. **2**, 145–154.
Sarychev, A.K., Shalaev, V.M., 2000, Phys. Rep. **335**, 275–371.
Schrader, B., 1995, Infrared and Raman Spectroscopy: Methods and Applications, Wiley, Chichester.
Schultz, S., Smith, D.R., Mock, J.J., Schultz, D.A., 2000, Proc. Natl. Acad. Sci. U S A **97**, 996–1001.
Seibert, M., Picorel, R., Kim, J.-H., Cotton, T.M., 1992, Meth. Enzymol. **213**, 31–42.
Shafer-Peltier, K.E., Haynes, C.L., Glucksberg, M.R., Van Duyne, R.P., 2003, J. Am. Chem. Soc. **125**, 588–593.
Shalaev, V.M., Botet, R., Mercer, J., Stechel, E.B., 1996, Phys. Rev. B **54**, 8235–8242.
Shalaev, V.M., 2000, Nonlinear Optics of Random Media: Fractal Composites and Metal-Dielectric Films, Springer, Heidelberg, Berlin.
Sommerfeld, A., 1909, Ann. Phys. **28**, 665–736.
Sönnichsen, C., Geier, S., Hecker, N.E., von Plessen, G., Feldmann, J., Ditlbacher, H., Lamprecht, B., Krenn, J.R., Aussenegg, F.R., Chan, V.Z.-H., Spatz, J.P., Möller, M., 2000, Appl. Phys. Lett. **77**, 2949–2951.
Sönnichsen, C., Franzl, T., von Plessen, G., Feldmann, J., 2002a, New J Phys **4**, 93.
Sönnichsen, C., Franzl, T., Wilk, T., von Plessen, G., Feldmann, J., Wilson, O., Mulvaney, P., 2002b, Phys. Rev. Lett. **88**, 077402.
Sönnichsen, C., Reinhard, B.M., Liphardt, J., Alivisatos, A.P., 2005, Nat. Biotechnol. **23**, 741–745.
Souza, G.R., Miller, J.H., 2001, J. Am. Chem. Soc. **123**, 6734–6735.
Storhoff, J.J., Elghhainian, R., Mucic, R.C., Mirkin, C.A., Letsinger, R.L., 1998, J Am Chem Soc **120**, 1959–1964.
Storhoff, J.J., Lucas, A.D., Garimella, V., Bao, Y.P., Müller, U.R., 2004a, Nat. Biotechnol. **22**, 887.
Storhoff, J.J., Marla, S.S., Bao, P., Hagenow, S., Mehta, H., Lucas, A., Garimella, V., Patno, T., Buckingham, W., Cork, W., Müller, U.R., 2004b, Biosens. Bioelectron. **19**, 875–883.
Su, K.-H., Wei, Q.-H., Zhang, X., Mock, J.J., Smith, D.R., Schultz, S., 2003, Nano Lett. **3**, 1087–1090.
Sulk, R., Chan, C., Guicheteau, J., Gomez, C., Heyns, J.B.B., Corcoran, R., Carron, K., 1999, J. Raman Spectrosc. **30**, 853–859.
Taton, T.A., Mirkin, C.A., Letsinger, R.L., 2000, Science **289**, 1757–1760.
Taton, T.A., Lu, G., Mirkin, C.A., 2001, J. Am. Chem. Soc. **123**, 5164–5165.
Thanh, N.T.K., Rosenzweig, Z., 2002, Anal. Chem. **74**, 1624–1628.

Thomas, L.L., Kim, J.-H., Cotton, T.M., 1990, J. Am. Chem. Soc. **112**, 9378–9386.
Van Duyne, R.P., Hulteen, J.C., Treichel, D.A., 1993, J. Chem. Phys. **99**, 2101–2115.
Venter, J.C., et al., 2001, Science **291**, 1304–1351.
Vo-Dinh, T., Houck, K., Stokes, D.L., 1994, Anal. Chem. **66**, 3379–3383.
Vo-Dinh, T., Stokes, D.L., Griffin, G.D., Volkan, M., Kim, U.J., Simon, M.I., 1999, J. Raman Spectrosc. **30**, 785–793.
Volkan, M., Stokes, D.L., Vo-Dinh, T., 2000, Appl. Spectrosc. **54**, 1842–1848.
Wabuyele, M.B., Vo-Dinh, T., 2005, Anal. Chem. **77**, 7810–7815.
Weitz, D.A., Garoff, S., Gramila, T.J., 1982, Opt. Lett. **7**, 168–170.
Westphal, V., Hell, S.W., 2005, Phys. Rev. Lett. **94**, 143903.
Wokaun, A., Gordon, J.P., Liao, P.F., 1982, Phys. Rev. Lett. **48**, 957–960.
Wood, E., Sutton, C., Beezer, A.E., Creighton, J.A., Davis, A.F., Mitchell, J.C., 1997, Int. J. Pharma. **154**, 115–118.
Xu, H.X., Bjerneld, E.J., Käll, M., Börjesson, L., 1999, Phys. Rev. Lett. **83**, 4357–4360.
Yguerabide, J., Yguerabide, E.E., 1998a, Anal. Biochem. **262**, 157–176.
Yguerabide, J., Yguerabide, E.E., 1998b, Analytical Biochemistry **262**, 137–156.
Yguerabide, J., Yguerabide, E.E., 2001, J. Cell. Biochem. Suppl. **37**, 71–81.
Yonzon, C.R., Jeoung, E., Zou, S., Schatz, G.C., Mrksich, M., Van Duyne, R.P., 2004, J. Am. Chem. Soc. **126**, 12669–12676.
Zenneck, J., 1907, Ann. Phys. **23**, 846.
Zhou, H.S., Honma, I., Komiyama, H., Haus, J.W., 1994, Phys. Rev. B Condens. Matter **50**, 12052–12056.

Chapter 9

Thin metal-dielectric nanocomposites with a negative index of refraction

by

Alexander V. Kildishev, Thomas A. Klar*, Vladimir P. Drachev, Vladimir M. Shalaev

School of Electrical and Computer Engineering and Birck Nanotechnology Center, Purdue University, Indiana 47907

---

*On leave from Photonics and Optoelectronics Group, Physics Department and CeNS, Ludwig-Maximilians-Universität München, 80799 München, Germany

Nanophotonics with Surface Plasmons — *V.M. Shalaev & S. Kawata (Editors)*
Advances in Nano-Optics and Nano-Photonics — *© 2007 Published by Elsevier B.V.*
ISSN: 1871-0018 — DOI: 10.1016/S1871-0018(06)02009-7

Contents

§ 1. Introduction

The race in engineering metamaterials comprising a negative refractive index in the optical range has been fueled by the realization of negative index materials for GHz frequencies 6 years ago. Sheer miniaturization of the GHz resonant structures is one approach. Alternative designs make use of localized plasmon-resonant metal nanoparticles or nanoholes in metal films. Following this approach, a negative refractive index has been realized in the optical range very recently. These results are reviewed and summarized in this chapter. The chapter addresses the critical question on how to unambiguously retrieve the effective refractive index of a given thin nanostructured layer from data accessible to measurements and reveals difficulties in cascading individual layers in a bulk material. The major focus is made on overcoming absorptive losses. Numerical simulations show that a composite material comprising silver strips and a gain providing material can have a negative refractive index and 100% transmission, simultaneously. The chapter has a special focus on simple and intuitive ways of designing negative index metametarials (NIMs) where efficient magnetic resonators, which are used to provide the required negative permeability, are combined with bulk or semicontinuous metal elements to provide the necessary permittivity.

1.1. The index of refraction

The refractive index is the most fundamental parameter to describe the interaction of electromagnetic radiation with matter. It is a complex number $n = n' + \imath n''(\imath = \sqrt{-1})$ where n' has generally been considered to be positive. While the condition $n' < 0$ does not violate any fundamental physical law, materials with negative index have some unusual and counter-intuitive properties. For example, light, which is refracted at an interface between a positive and a negative index material, is bent in the "wrong" way with respect to the normal, group and phase velocities are antiparallel, wave and Pointing vectors are antiparallel, and the vectors $\vec{E}$, $\vec{H}$, and $\vec{k}$ form a left-handed system. Because of these properties, such materials are synonymously called "left handed" or negative-index materials. Theoretical work on negative-phase velocity dates back to Lamb (1904, in hydrodynamics) or Schuster (1904, in optics) and was considered in more detail by Mandel'shtam (1945) and Veselago (1968). A historical survey referring to these and other early works can be found in Holloway et al. (2003). Important discussions on terminology and later works is done by Caloz and Itoh (Caloz and Itoh, 2006).

In general, left-handed materials do not exist naturally, with some rare exceptions like bismuth that being placed in a waveguide shows $n' < 0$ at a wavelength of $\lambda \approx 60\,\mu$m (Podolskiy et al., 2005a). However, no naturally existing negative index material is known so far in the optical range of frequencies. Therefore, it is necessary to turn to man-made, artificial materials, which are composed in such a way that the averaged (effective) refractive index is less than zero, $n'_{\mathrm{eff}} < 0$. Such a material can be obtained using photonic crystals (PC) (Kosaka et al., 1998; Gralak et al., 2000; Notomi, 2000; Luo et al., 2002; Berrier et al., 2004). However in this case, the interior structure of the material is not subwavelength. Consequently, PCs do not show the full range of possible benefits of left-handed materials. For example, super-resolution, which has been predicted by Pendry (2000), is not achievable with photonic band gap materials because their periodicity is in the range of λ. A thin slab of a photonic crystal only restores small k-vector evanescent field components because the material can be considered as an effective medium only for long wavelengths, and large k-vector components are not restored (Luo et al., 2003; Smith et al., 2003; Lu et al., 2005). A truly effective refractive index $n'_{\mathrm{eff}} < 0$ can be achieved in metamaterials with structural dimensions far below the wavelength. Metamaterials for optical wavelengths must therefore be nano-crafted.

A possible – but not the only – approach to achieve a negative refractive index is to design a material where the (isotropic) permittivity $\varepsilon' = \varepsilon' + \imath\varepsilon''$ and the (isotropic) permeability $\mu = \mu' + \imath\mu''$ obey the equation

$$\varepsilon'|\mu| + \mu'|\varepsilon| < 0, \tag{1.1}$$

This leads to a negative real part of the refractive index $n = \sqrt{\varepsilon\mu}$ (Depine and Lakhtakia, 2004). Equation (1.1) is satisfied, if $\varepsilon' < 0$ and $\mu' < 0$. However, we note that this is not a necessary condition. There may be magnetically active media (i.e., $\mu \neq 1$) with a positive real part μ' for which eq. (1.1) is fulfilled and therefore show a negative n'.

To date, we only considered isotropic media where ε and μ are complex scalar numbers. It has been shown that in a waveguide with an anisotropic media, where ε and μ are tensors, a negative refractive index is feasible even if the material shows no magnetic response ($\mu = 1$). If, for example, $\varepsilon_{\perp} < 0$ and $\varepsilon_{\parallel} > 0$, then $n' < 0$ can be achieved (Podolskiy and Narimanov, 2005; Podolskiy et al., 2005a). Despite the fact that using anisotropic waveguide is a very promising approach, we will not focus on that topic here. This is mainly because so far a negative index for optical frequencies has only been achieved following the approach of magnetically active media.

This chapter is organized as follows: first, we recall how to achieve magnetic activity for GHz frequencies using metallic split-ring resonators (SRR) and how the SRRs have been successively scaled down to shift the magnetic resonance up to THz frequencies. When the optical range is approached, the finite skin depth of metals as well as localized plasmonic resonances must be considered in addition to the simple geometric scaling of metallic structures. Then, we discuss approaches to new design outlines, which are making active use of localized plasmonic effects. Metamaterials containing metal nanostructures as magnetically active components usually show low transmission due to reflection and absorption. We demonstrate an impedance-matched design to suppress reflection. To compensate for losses and obtain a fully transparent layer of NIM we add a gain material. A special focus is made on simple semi-analytical approaches to the simulation of optical NIMs. Finally, intuitive ways of designing NIMs are shown, where efficient magnetic resonators, which are used to provide the required negative permeability, are combined with bulk or semicontinuous metal elements to provide the necessary permittivity.

1.2. Downscaling split ring resonators

The first recipe of how to design a magnetically active material was suggested by Pendry in 1999 (Pendry et al., 1999). Two concentric split rings that face in opposite directions and are of subwavelength dimensions were predicted to give rise to $\mu' < 0$ (fig. 1(a)). One can regard this as an electronic circuit consisting of inductive and capacitive elements. The rings form the inductances and the two slits as well as the gap between the two rings can be considered as capacitors. A magnetic field, which is oriented perpendicular to the plane of drawing, induces an opposing magnetic field in the loop due to Lenz's law. This leads to a diamagnetic response and hence to a negative real part of the permeability. The capacitors (the two slits and the gap between the rings) are necessary to

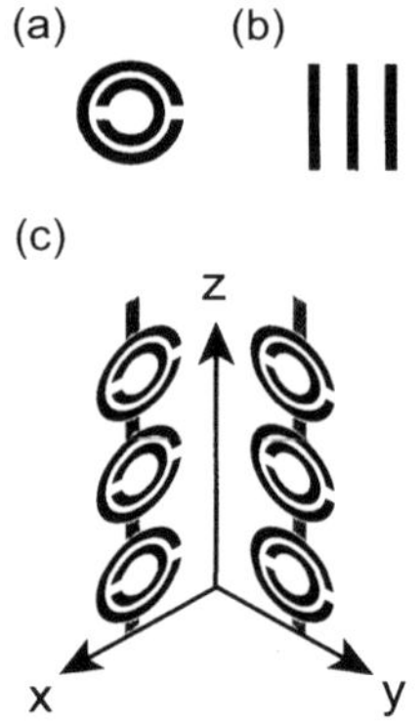

Fig. 1. (a) Magnetically resonant ($\mu' < 0$) metal structure: two counter-facing split rings of subwavelength dimensions (split ring resonator, SRR). (b) Electrically resonant ($\varepsilon' < 0$) metallic structure: metal rods. (c) A combination of both structures results in a negative index metamaterial ($n' < 0$).

assure that the wavelength of the resonance is larger than the dimensions of the SRR.

Soon after this theoretical prediction, Schultz and coworkers combined the SRRs with a material that shows negative electric response in the 10 GHz range and consists of metallic wires in order to reduce the charge-carrier density; and hence shift the plasmonic response from optical frequencies down to GHz frequencies (fig. 1(b)) (Pendry et al., 1996). The outcome was the first-ever metamaterial with simultaneously negative real parts of the permeability and permittivity (Smith et al., 2000) and consequently a negative refractive index of approximately 10 GHz (fig. 1(c)) (Shelby et al., 2001a, b). From then onwards the race to push left-handedness to higher frequencies was open. The GHz resonant SRRs had a diameter of several millimeters, but size reduction lead to a higher frequency response. The resonance frequency was pushed up to 1 THz using this scaling technique (Yen et al., 2004; Moser et al., 2005).

1.3. Metamaterials using localized plasmonic resonances

1.3.1. Metal nanorods

It was mentioned by Lagarkov and Sarychev (1996) that a pair of noble metal nanorods can show a large paramagnetic response, and it was first pointed out by Podolskiy et al. (2002) that such a pair of noble metal nanorods is also capable of a *diamagnetic response* at 1500 nm. In the

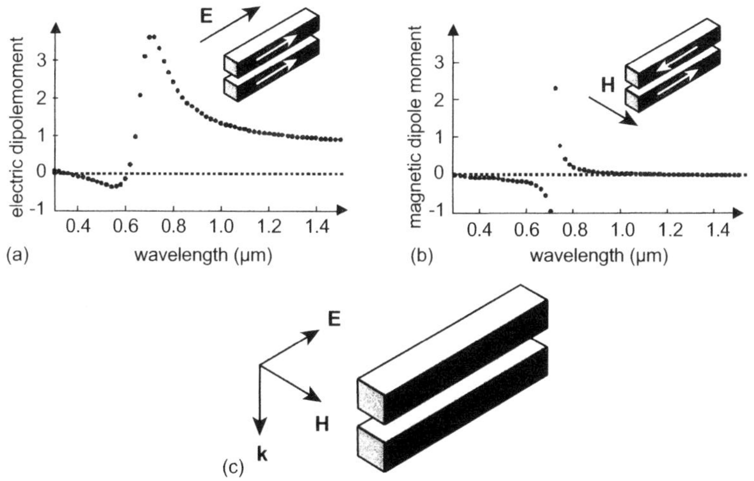

Fig. 2. Response of a pair of gold nanorods to radiation, simulated with coupled dipole approximation technique. (a) Electrical dipole moment: electric field oriented parallel to the axis of the rods. (b) Magnetic dipole moment: magnetic field oriented perpendicular to the plane of the rods. (c) A pair of rods illuminated from above with TM polarization. The pair of rods will have a double negative response to the field.

publication by Podolskiy et al. (2002), it was predicted for the first time that materials containing such pairs of rods can show a negative n' even for visible wavelengths. This issue has been discussed in more detail by Panina et al. (2002) and also by Podolskiy et al. (2003, 2005b). It is illustrated in fig. 2 how a pair of nanorods can show a negative response to an electromagnetic plane wave. Two gold rods are separated by a distance far less than the wavelength. The diameter of the cross-section of rods is also much less than the wavelength and length of the rods, but does not need to be in the range of half the wavelength. An alternating electric field parallel to both rods will induce parallel currents in both rods, which are in phase or out of phase with the original electric field, depending on whether the wavelength of the electric field is longer or shorter than the wavelength of the dipolar eigenresonance of the electrodynamically coupled rods. Figure 2(a) shows the induced electric dipole moment in case of the following specific dimensions as has been reported in Podolskiy et al. (2005b): a rod length of 162 nm, a diameter of 32 nm (assuming cylindrically shaped rods), and a distance of 80 nm.

Let us now consider the magnetic field, which shall be oriented perpendicular to the plane of the rods. This magnetic field will cause antiparallel currents in the two rods as shown in fig. 2(b). This can be considered as a dipolar magnetic mode. The magnetic response will be dia- or paramagnetic depending on whether the wavelength of the incoming magnetic field is shorter or longer than the dipolar magnetic eigenfrequency of the electrodynamically coupled rods (fig. 2(b), after Podolskiy et al., 2005b). In this description in terms of coupled plasmonic resonances the magnetic dipole resonance appears at the same wavelength as the electric quadrupole resonance. However, the latter does not contribute to the electromagnetic radiation in the direction given in fig. 2(c) (Podolskiy et al., 2003).

So far, the electromagnetic response has been discussed in terms of coupled plasmonic resonances. An alternative way of looking at it is that the antiparallel currents in the rods and the displacement currents at the ends of the two rods form a current loop or an inductance, while the gaps at the ends form two capacitors. The result is a resonant LC-circuit (Lagarkov and Sarychev, 1996; Engheta et al., 2005).

It is important that both resonances, the dipolar electric and the dipolar magnetic, are at similar wavelengths. This requires that the coupling between the two rods should not be too strong, otherwise the two resonances split further apart. It is seen in figs. 2(a) and 2(b) that there is a certain range of wavelengths (between 500 and 600 nm) where both, the induced electric and the induced magnetic dipole moments are opposing the incident fields. Hence, an electromagnetic plane wave impinging from above and with E and H oriented as shown in fig. 2(c) (TM polarization) will induce a double negative response.

To the best of our knowledge, the unambiguous measurement of a negative refractive index in the optical range (specifically, at the optical telecom wavelength of 1500 nm) was reported for the first time by Shalaev et al. (2005a, b). The metamaterial in which the negative refractive index was achieved is outlined in fig. 3. Pairs of nanorods were fabricated on a glass substrate using electron beam lithography. The actual structure of the gold nanorod doublets is shown in fig. 3(a). The nanorods are 50 nm thick, stacked on top of the glass substrate, and a 50 nm thick SiO_2 layer is used as a spacer. The upper rod is smaller in dimension than the lower. A scanning electron microscope (SEM) picture of a single pair and its dimension are shown in fig. 3(a). Pairs of nanorods are periodically repeated as depicted in fig. 3(b) and shown by a SEM micrograph in fig. 3(c). Figure 3(d) shows a unit cell of the periodic arrangement and gives more dimensions. A full description of the sample and its preparation is given in Shalaev et al. (2005a), Drachev et al. (2006), and Kildishev et al. (2006). An attempt to explain the reflection spectra of a similar sample containing pairs of gold nanoparticles has been made

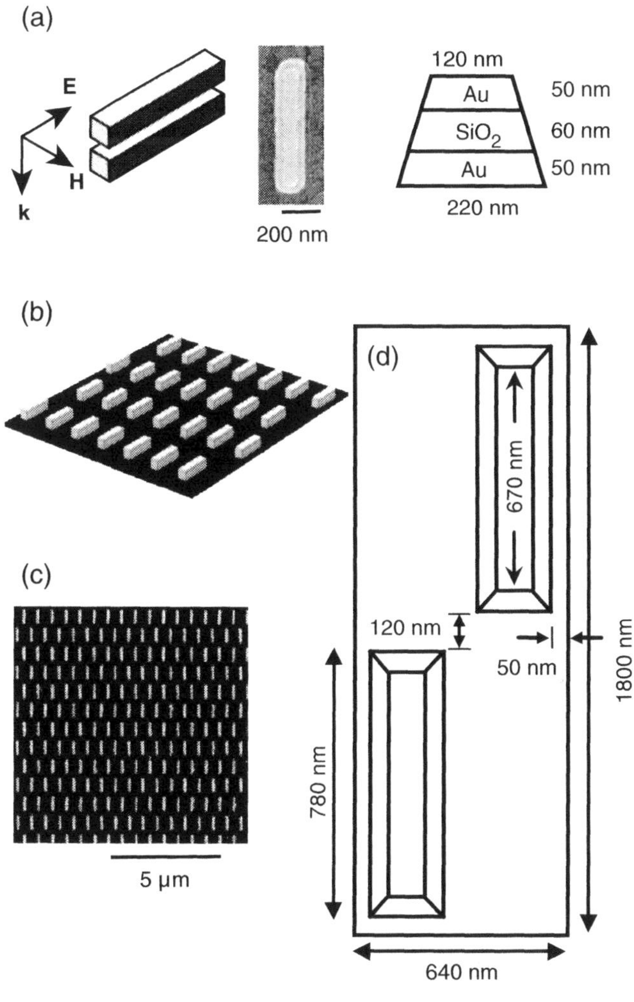

Fig. 3. (a) Left to right: scheme of nanorod pair and proper light polarization for negative index, SEM image, and dimensions. (b) Scheme of the arrangement of nanorod pairs. (c) SEM image of arranged nanorod pairs. (d) Dimensions of the arrangement (one unit cell is shown).

assuming a given dispersion of μ and ε (Grigorenko et al., 2005). However negative refractive index has been achieved in that work.[1]

[1] Although the authors claim a negative magnetic response achieved in their metamaterial in visible range, validation of their result with our 3D FDTD code and an independent study performed at Penn State University using a hybrid finite element - boundary integral (PFEBI) technique showed only a feeble always-positive magnetic response (to be discussed elsewhere).

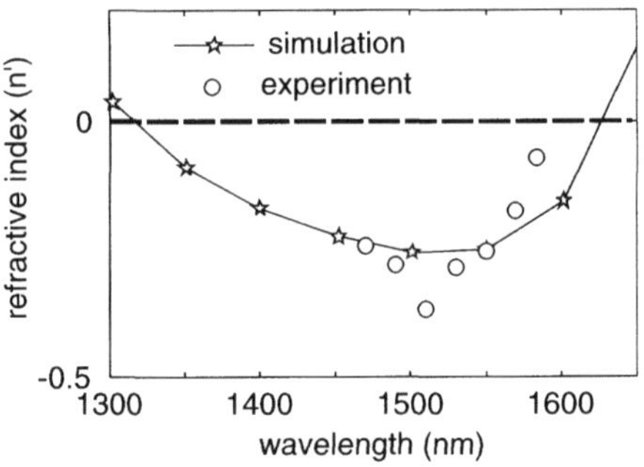

Fig. 4. Region of negative refraction for the real part of the refractive index of a layer of nanorod pairs shown in Fig. 3. Circles represent data, which are restored from experimentally determined transmission, reflection, and phase measurements. Stars represent FDTD simulation. A refractive index of $n' = -0.3 \pm 0.1$ was determined.

Figure 4 shows the results obtained in Shalaev et al. (2005a) for the real part of the refractive index of the metamaterial shown in fig. 3. The full circles show experimental results and the open triangles give the results as obtained from simulations using the finite difference method in time domain (FDTD). It is clearly seen that the real part of the refractive index becomes negative in the wavelength range from approximately 1400 to 1600 nm, which includes the important telecommunication band at 1500 nm.The figure gives a closer look to this frequency range. The experimental data prove that $n' = -0.3 \pm 0.1$ was obtained in Shalaev et al. (2005a).

It turns out to be not trivial to experimentally determine the exact value of the refractive index for a thin film. In the present case, the film of negative refraction was only 160 nm thick. Therefore, the straightforward method of determining n by applying Snell's law to the incoming and refracted beams cannot be used. A different method to unambiguously determine the refractive index requires the measurement of the transmission T, the reflectance R, and the absolute phases of the transmitted and reflected electric fields τ and ρ, respectively. If these four quantities are measured, the refractive index $n = n' + in''$ in a thin, passive ($n'' > 0$) film sandwiched between air (top) and a glass substrate (bottom) can be determined uniquely as has been discussed in Smith et al. (2002) and Kildishev et al. (2006) using transfer matrices

$$n = \frac{1}{k\Delta} \arccos \frac{1 - r^2 + n_s t^2}{[1 + n_s - (1 - n_s)r]t}, \tag{1.2}$$

where $k = 2\pi/\lambda$ is the wave vector of light in vacuum, Δ the thickness of the thin film, n_s the refractive index of the glass substrate, and r and t the complex reflection and transmission coefficients, respectively:

$$t = \sqrt{T}\,e^{i\tau}, \quad r = \sqrt{R}\,e^{i\rho}. \tag{1.3}$$

Figure 5(a) shows the transmission and reflection spectra of the NIM of fig. 3. In order to measure the absolute phase, the beam of a tunable semiconductor laser was split into two orthogonally polarized beams, where one beam passed through the NIM of thickness Δ while the other was used as a reference and passed only through the glass substrate at a spot not covered by the metamaterial (Drachev et al., 2006) (fig. 5(b)). The beams were recombined behind the glass substrate. The phase difference between the beam passing through the thin film and the reference

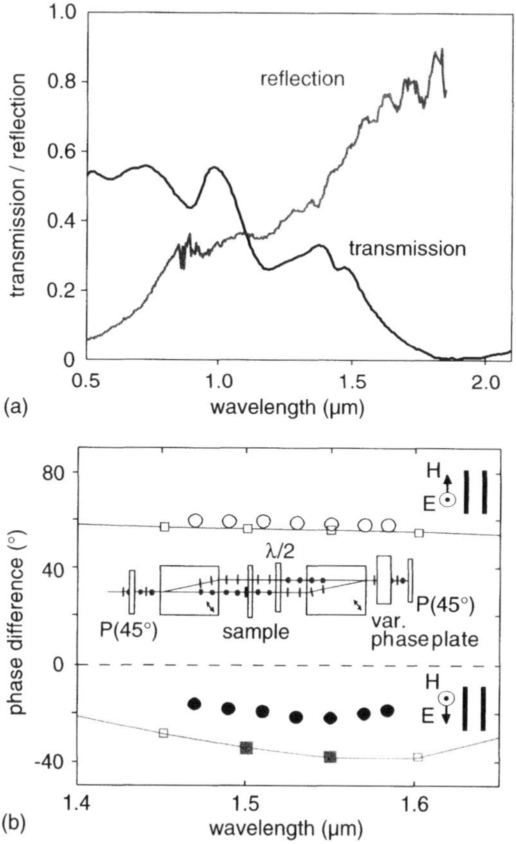

Fig. 5. (a) Measured transmission and reflection spectra of the sample shown in fig. 3(b). Setup for phase measurements. (b) Phase difference in the two light paths as shown in the inset. Circles are measured values, quadrangles and lines are from simulation. The light is delayed in case of TE polarisation (H-field parallel to rod pair, open symbols). In contrast, the phase is advanced in case of TM polarization.

beam propagating only through air of the same thickness Δ was determined using interferometry (inset in fig. 5(b)). The phase τ was delayed in the metamaterial by approximately 60° compared to air in case of TE polarization (electric field perpendicular to the plane of rods). In contrast, τ was advanced by approximately 20° in case of TM polarization (fig. 5(c), Shalaev et al., 2005a). The phase shifts in reflection ρ were obtained for both polarizations in a similar way. The advancement of τ for TM polarization was an indirect evidence of $n'<1$. However, to unambiguously prove that $n'<0$, the complete set (T,R,τ,ρ) must be obtained, so that n can be reconstructed using eq. (1.3) (Kildishev et al., 2006).

Nevertheless, one can use pure phase measurements to make an estimate for n' as has been pointed out in Kildishev et al. (2006). In the case of low reflection ($R \ll 1$), the following equation holds:

$$n' \approx \frac{\tau}{k\Delta}, \tag{1.4}$$

while in the limit of strong reflection ($R \approx 1$) the following equation holds:

$$n' \approx \frac{\tau - \rho - \frac{\pi}{2}}{k\Delta}. \tag{1.5}$$

These two formula indeed give an upper and lower bound to the correct value of n' according to eqs. (1.2) and (1.3) (see fig. (6)) (Kildishev et al., 2006).

1.3.2. Voids

It is an interesting approach for NIMs to take the inverse of a resonant structure (Falcone et al., 2004), e.g., a pair of voids as the inverse of a pair of nanorods (Zhang et al., 2005a, b, 2006). The basic idea is illustrated in fig. 7(a). Instead of a pair of metal nanoellipses separated by an oxide, which are similar to the pair of rods in fig. (2), two thin films of metal are separated by an oxide and mounted on a glass substrate. Then, an elliptically shaped void is etched in the films (fig. 7(a), right-hand side), thus forming the negative of the original-paired metal ellipse structure (fig. 7(a), left-hand side). Both samples should have similar resonance behavior if the orientation of the electric and magnetic fields are also interchanged. FDTD simulations have been performed to determine the refractive index of void metamaterials (Kildishev et al., 2006). The dimensions were chosen according to fig. 7(b) in the simulations in order to match the dimensions of the experimental sample reported in Zhang et al. (2006).

The simulations were carried out for both cases of polarizations: the electric field oriented along the long axis of the elliptical voids and perpendicular to it. It is seen that n' becomes negative in both cases; however,

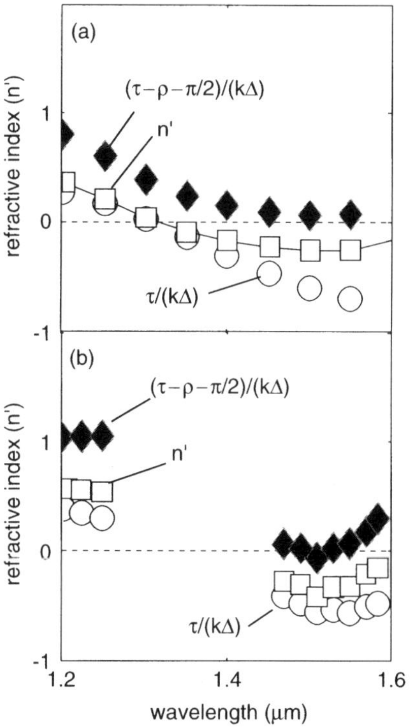

Fig. 6. Real part of the refractive index as determined by the exact formula (eq. (1.2), squares) or by phase-only assumptions according to eq. (2.5) (full diamonds) or eq. (1.4) (open circles). (a) Numerical simulations. (b) Experimental results.

the effect is more pronounced if the electric field is oriented along the short axis (fig. 7(c)). Furthermore, at approximately 1600 nm the real part of n is negative while the imaginary part is less than 1 indicating lower losses compared to the double-rod sample discussed before, where the imaginary part of the refractive index was 3 (Shalaev et al., 2005a). Experimental measurements with samples similar to those sketched in fig. 7(a), but with spherical voids instead of elliptical voids, confirmed a negative n' at a wavelength of 2 μm (Zhang et al., 2005b). The imaginary part n'' was large in that case, however it has been shown that further optimization can reduce n'' substantially (Zhang et al., 2006).

1.4. Pairs of metal strips for impedance-matched negative index metamaterials

Metamaterials using plasmon-resonant metal nanoparticles have two distinct problems, each of them reducing the overall transmission through the

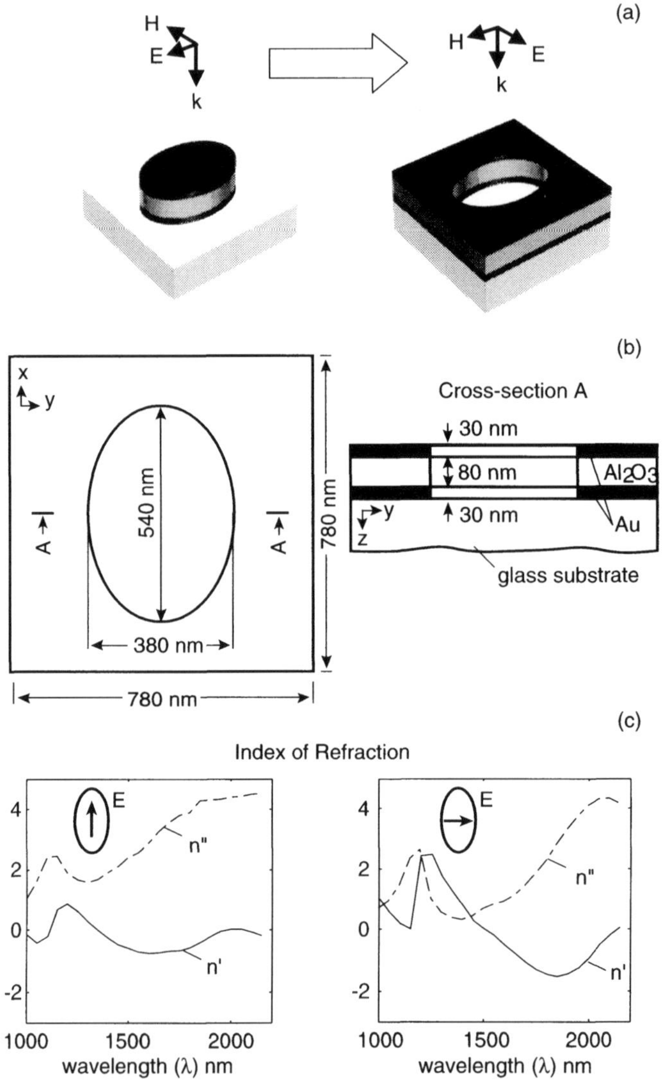

Fig. 7. (a) Left: nano-ellipse consisting of two 30 nm thick ellipses of gold separated by 80 nm of Al_2O_3. Right: An elementary cell of coupled elliptic voids. (b) Dimensions of the voids. The voids are repeated periodically in 2D. (c) Refractive index $n = n' + in''$ for light polarized parallel (left) or perpendicular (right) to the long axis of the voids as obtained from FDTD simulations.

metamaterial. The first one is absorptive losses (in terms of a large n''), because ohmic losses are generally large due to the excitation of localized plasmon resonances in the nanostructures. A possible solution to this problem will be discussed in the next section. In this section we will concentrate on the second issue, which is impedance matching. The impedance is given by $\eta^2 = (\eta' + i\eta'')^2 = \mu\varepsilon^{-1}$ and it is required that the impedances match at a boundary between two media in order to eliminate reflection. This condition is well known for microwaves and replaces Brewster's law for optical frequencies if $\mu \neq 1$ (Panina et al., 2002). Impedance is matched at a boundary between a NIM and air, if $\eta' \rightarrow 1$ and $\eta'' \rightarrow 0$ in the metamaterial.

In fig. 8(a) we introduce a metamaterial where the conditions $\eta \rightarrow 1 + 0\imath$, $n' < -1$, and $n'' < 1$ hold simultaneously for a visible wavelength. The structure consists of pairs of coupled silver strips. Both strips are 280 nm wide (x direction), 16 nm thick, and are infinitely long in the y direction. The two silver strips are separated in z direction by a 65 nm thick layer of Al_2O_3. The pairs of strips are periodically repeated in x direction with a period of 500 nm. We assume air above and below the layer of strips. In our finite element frequency-domain (FEMFD) simulations this layer of metamaterial is illuminated from above with plane waves at normal incidence (along z direction). The electric field is polarized in x direction. The magnetic field, which is parallel to the strips, induces antiparallel currents in the two silver strips. This leads to a magnetic response of the structure. We use FEMFD calculations to determine the spectra of the electrodynamic constants. Figure 8(b) shows the real parts of the permittivity (triangles) and the permeability (squares). It is seen that both are negative at wavelengths between 580 and 590 nm.

The spectra of the reflectance, transmission, absorption, refractive index, and impedance are displayed in fig. 9. It can be seen in fig. 9(a) that the transmission has a local maximum of 51% at 582 nm. This is because the reflection has a local minimum and the absorption is limited. Indeed, the impedance is matched quite well from 582 to 589 nm, i.e., $\eta' > 0.5$ and eventually reaching 1 at 586 nm, and simultaneously $|\eta''| < 0.5$ in the range 570–585 nm (fig. 9(c)). In total, this leads to a reflectance of less than 10% at 584 nm.

The absorption seems to have a local maximum at 586 nm, however it does not reproduce the spectrum of n''. This is mainly because the reflection at the interface between air and the metamaterial hinders the electromagnetic radiation from entering the metamaterial at longer wavelengths, and therefore, the effective absorption of radiation inside the metamaterial is low for longer wavelengths. Still, it accounts for almost 90% of the losses in the range of the "reflectance window" at 584 nm. In summary of this section, we have shown that a metamaterial consisting of pairs of silver strips as depicted in fig. 8(a) can form an impedance-matched negative index material for visible light. The transmission is limited to 50% almost solely due to absorption, while reflection losses play a minor role.

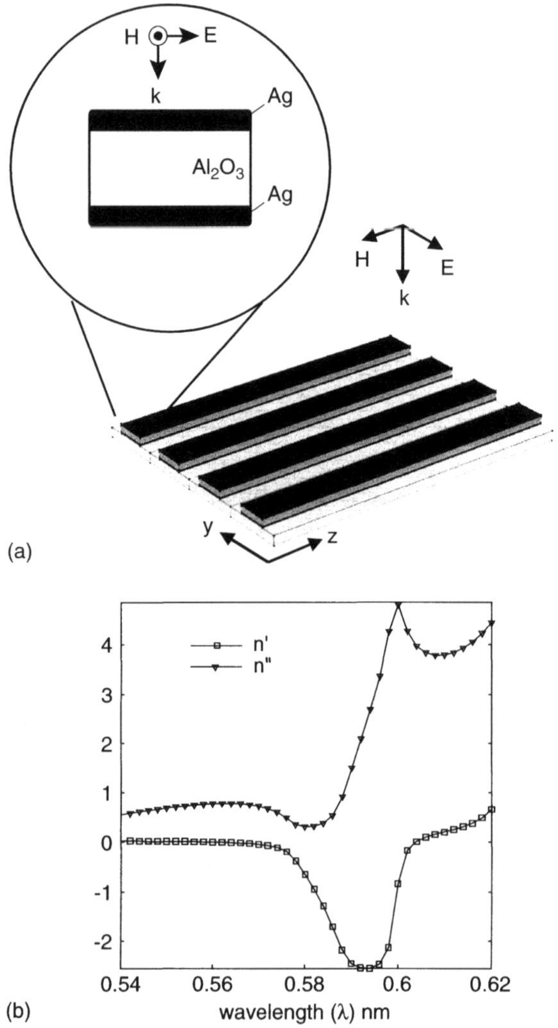

Fig. 8. (a) Double silver strips, separated by Al_2O_3. The strips are infinitely long in y direction and periodically repeated in x direction. The H-field is oriented in the y direction. Currents in the strips are antiparallel if the H-field is polarized in y direction. (b) Real and imaginary parts of the refractive index (n' and n'') simulated with FEMFD.

1.5. Gain, compensating for losses

It has been pointed out recently that energy can be transferred from a gain material to surface plasmon polaritons (Sudarkin and Demkovich, 1989; Tredicucci et al., 2000; Avrutsky, 2004; Nezhad et al., 2004; Seidel et al., 2005) or to plasmons in metal nanostructures (Bergman and

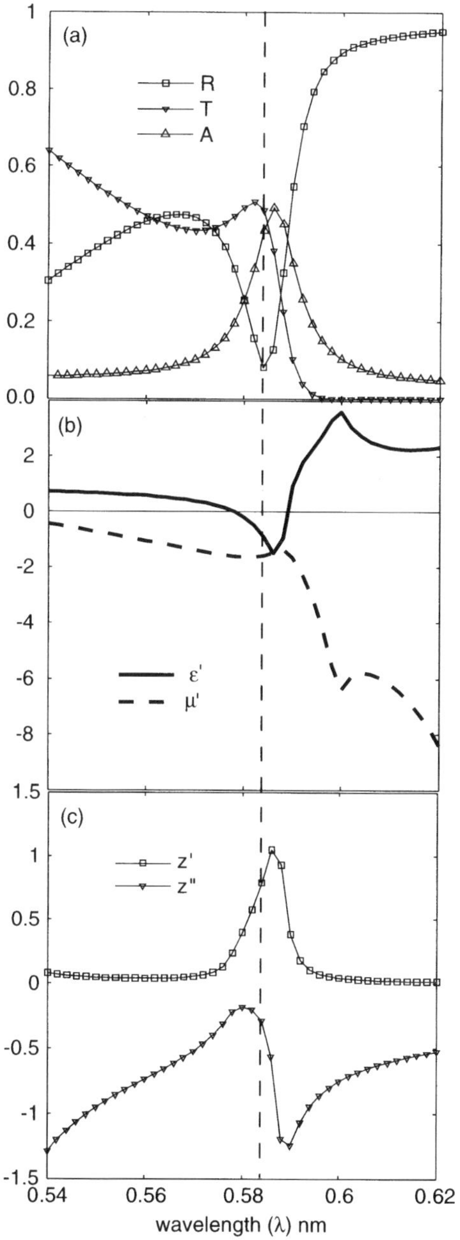

Fig. 9. Spectra of several optical constants of the structure shown in Fig. 8. (a) Reflection R, transmission T, and absorption A spectra. (b) Real parts of permeability (μ') and permittivity (ε'). (c) Real and imaginary part of the impedance. The vertical dashed line at 584 nm indicates a spectral region where the reflection is minimal, the transmission is high, and the real parts of permittivity and permeability are close to -1, while the real part of the impedance is close to $+1$, indicating impedance matching to air.

Stockman, 2003; Lawandy, 2004) using stimulated emission. Specifically, continuous thin films of metal have been used to confine lasing modes in quantum cascade lasers to the gain region and also to guide the lasing modes by surface plasmon modes (Tredicucci et al., 2000). Ramakrishna and Pendry suggested to staple gain materials such as semiconductor laser materials in between the negative index (or metal) layers of stacked near-field lenses (Ramakrishna and Pendry, 2003) in order to remove absorption and improve resolution. The requirement of a perfect near-field lens, where thin layers of positive and negative index materials are alternated, is that $\varepsilon_P = -\varepsilon_N$ and simultaneously $\mu_P = -\mu_N$, where the subscripts denote materials constants of positive (P) and negative (N) materials. This requirement naturally includes the conditions $\varepsilon''_P = -\varepsilon''_N$ and $\mu''_P = -\mu''_N$, i.e., the positive layers must provide gain in order to optimize the lens (Ramakrishna and Pendry, 2003).

In our discussion we would like to turn to the refractive index rather than the permittivity and the permeability, because the absorption (α) and gain (g) coefficients are more straightforwardly connected to the refractive index $n'' = \lambda(\alpha - g)/4\pi$. Further, instead of alternating negative and positive index materials we propose to "submerge" the negative index structures (e.g., containing metal nanorods) in gain media as shown in fig. 10. This could be achieved, for example, by spin-coating a solution of laser dye molecules or π-conjugated molecules on top of the negative index structures. Applying semiconductor nanocrystals (NCs) would be an alternative approach.

One might question whether the metal nanostructures nullify any attempt to amplify electromagnetic fields using gain materials in their close vicinity because gold nanoparticles are well known to quench fluorescence in an extremely efficient manner (Dulkeith et al., 2002; Imahori et al., 2004). In contrast, however, working solid state and organic

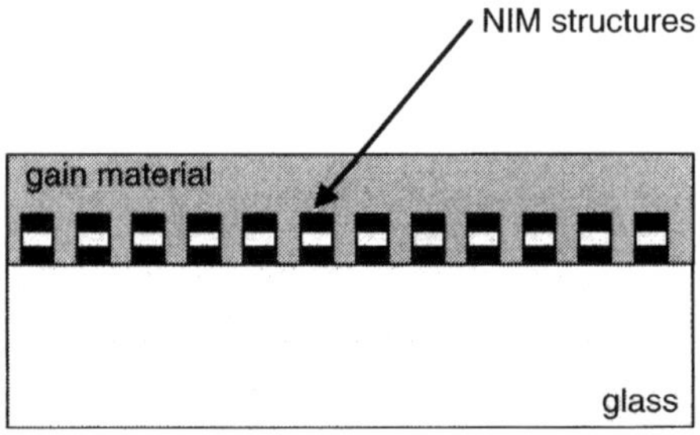

Fig. 10. Negative index metamaterial (e.g., double rods), filled with gain medium, e.g., a solid solution of dye molecules in a matrix.

semiconductor lasers show that sufficient gain can be provided so that in devices containing metal layers or metal nanoparticles the losses can be compensated. For instance, it has been shown that an optically pumped organic laser comprising a metal-nanoparticle distributed feedback (DFB) grating needs only a marginally increased pumping threshold (compared to organic lasers with metal-free DFB gratings) to be operative (Stehr et al., 2003). In the case of infrared quantum cascade lasers (QCL), a wave guiding metallic layer was shown to be beneficial for the laser power output (Tredicucci et al., 2000). This astonishing result is due to an increased overlap of the surface plasmon-guided mode profile with the gain region (the quantum cascade structure, in this case). This overlap offsets the increased losses (compared to a metal-free QCL) resulting from surface plasmon excitation. The net effect is an overall improved laser performance. We therefore conclude that it should indeed be feasible to use gain materials in order to compensate for the losses introduced by the resonant plasmonic metal nanoparticles in NIMs.

We want to give a specific example on the basis of the sample shown in figs. 8 and 9. For the moment, we assume that the metal strips are submerged in a 200 nm thick layer of gain material (fig. 11(a)). We further assume that the gain material and the metal strips do not influence each other. This is an assumption that certainly needs to be discussed, but for the moment we shall assume that the gain of the material is not influenced by the metal strips. At the wavelength of least reflectance (due to impedance matching, $\lambda = 584\,\text{nm}$), the strip material shows absorption of approximately 45% (fig. 9(a)). Applying Lambert–Beer's law and assuming that the absorptive loss should be fully compensated by the 400 nm thick gain layer, it turns out that a gain of $g = 3 \times 10^4\,\text{cm}^{-1}$ is required. Let us further assume that we use Rhodamine 6G dissolved in some optically inert polymer. Rhodamine 6G has a stimulated emission cross-section of $\sigma_{SE} = 3 \times 10^{-16}\,\text{cm}^2$ (Holzer et al., 2000) and therefore the concentration of excited dye molecules should be 170 mM. Alternatively, semiconductor nanocrystals such as CdSe NCs could be applied. It has been shown in Leatherdale et al. (2002) that the absorption cross-section per NC volume can be as large as $10^5\,\text{cm}^{-1}$. Because g and α are usually of similar magnitude, we conclude, that densely packed nanocrystal films can show gain in the order of $g \approx 10^5\,\text{cm}^{-1}$.

It is seen that the dye or nanocrystal concentrations need to be quite high to compensate for the losses. However, we have assumed in our rough estimation that the gain of the material in between the metal strips is not affected by the local fields in the vicinity of these metal strips. These fields can be quite high due to nanoplasmonic resonances. In fact, it has

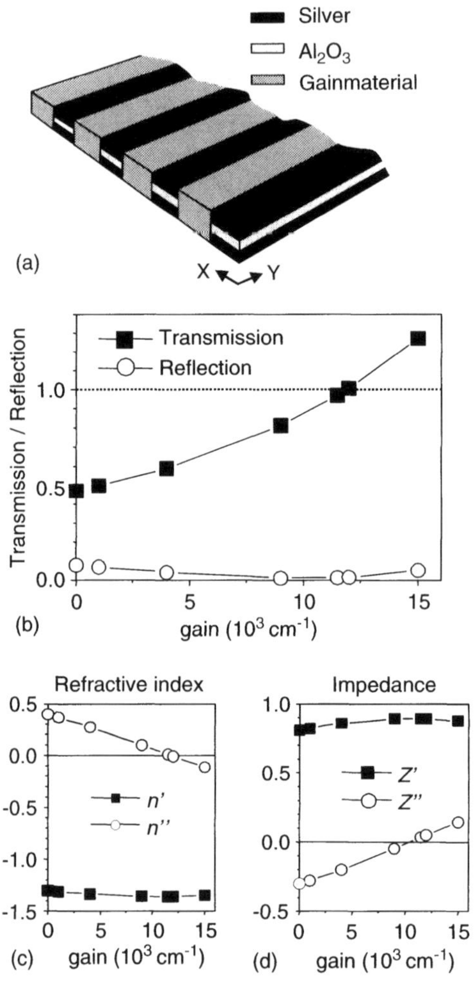

Fig. 11. (a) Same sample as in Fig. 8(a), but with gain providing material in between the double silver strips. Air is assumed above and below the layer, and the layer is irradiated with a plain wave (584 nm) from above, H-field polarized along the y direction. (b) Transmission and reflection as a function of the gain. At $g = 12{,}000\,\text{cm}^{-1}$, gain and losses cancel each other. Interestingly, the reflection shows also a minimum at $g = 12{,}000\,\text{cm}^{-1}$. (c, d) Refractive index and impedance as a function of gain. $n' \approx -1.35$ for all investigated gain levels.

been pointed out by Kim et al. (1999) and Lawandy (2004) that a gain medium and localized plasmonic resonance may lead to extremely high effective polarizabilities of the combined system. Therefore, the possibility may arise that each pair of gold nanorod as shown in fig. 2, or each strip as in fig. 8, shows a much larger response to an incoming electric field as the same metal structure in air.

In the example given above we have neglected that the gain material is in intimate contact with the silver strips. In order to get a better picture, we applied FEMFD simulations on the model shown in fig. 11(a). We took the same structure as shown in fig. 8, but now filled the gaps in between the double silver strips with a material that provides a fixed amount of gain between 0 and $15 \times 10^3\,\mathrm{cm}^{-1}$. Figures 11(b–d) show the transmittance (T), reflectance (R), refractive index (n' and n''), and impedance (η' and η'') as a function of gain (g). We found that at a gain of $12 \times 10^3\,\mathrm{cm}^{-1}$ the structure becomes transparent (fig. 11(b)), while the real part of the refractive index n' is almost unaffected by the gain material (fig. 11(c)). Furthermore, the impedance that has already been matched quite well without the gain medium (fig. 9) improves further when gain is applied, i.e., $\eta' \approx 1$ and $\eta'' \approx 0$ for $g = 12 \times 10^3\,\mathrm{cm}^{-1}$ (fig. 11(d)). The exact results for a gain of $g = 12 \times 10^3\,\mathrm{cm}^{-1}$ are $n' = -1.355$, $n'' = -0.008$, $\eta' = 0.89$, $\eta'' = 0.05$, $T = 100.5\%$, and $R = 1.6\%$.

Actually, if a critical magnitude of gain is surpassed, the polarizability and the field enhancement do not depend on nanoparticle shape or material any longer, but are solely limited by gain saturation in the gain medium (Lawandy, 2004). At present, we have not included gain saturation in our model. It could be envisioned that the gain material does not "simply" restore energy, which is lost due to absorption by the metal nanostructures, but it becomes an instrumental element of the NIM, e.g., heavily increasing the negative response of the pairs of nanorods. This will allow the design of NIMs of less overall metal content. The density of pairs of rods or the size of each pair may be reduced, while the overall effective negative response of the metamaterial remains strong. This exciting field certainly needs more consideration, which will be discussed elsewhere.

§ 2. Optical characteristics of cascaded NIMs

A possible approach to designing NIMs is a periodic array of elementary coupled metal-dielectric resonators. This work takes a closer look at approaches, which simultaneously provide fast calculation of the field inside a given metamaterial arranged of elementary periodic layers and calculation of its effective parameters. First, consider a simplified approach to defining equivalent optical properties of an elementary NIM layer (Kildishev et al., 2006). With this approach, an effective refractive index is ascribed to a layer of NIM, as if it were a layer of homogeneous medium. This assumption suggests that the periodic structure of NIM does not diffract the incident plane wave, and a classical direct problem of plane-wave propagation through a multilayer structure of homogeneous materials can be used in accordance with Born and Wolf (Born and Wolf, 1964). For a given monochromatic incident light it is then possible to measure complex reflectance and transmittance coefficients (r and t) and then unambiguously retrieve the refractive index of the NIM sample. This effective parameter can be conveniently retrieved from a characteristic matrix of a homogeneous film at normal incidence (Kildishev et al., 2006).

Provided that the characteristic matrix ($\mathbf{M}$) is decomposed using a diagonal matrix of eigenvalues ($\mathbf{A}$) and a matrix of eigenvectors ($\mathbf{V}$) as $\mathbf{M} = \mathbf{VAV}^{-1}$, the transmission through an elementary layer is given by

$$\mathbf{Q}_0 = \mathbf{MQ}, \tag{2.1}$$

where

$$\mathbf{Q} = \begin{pmatrix} H_2 \\ \eta_0^{-1} E_2 \end{pmatrix}, \quad \mathbf{Q}_0 = \begin{pmatrix} H_1 \\ \eta_0^{-1} E_1 \end{pmatrix},$$

$$\mathrm{V} = \begin{pmatrix} 1 & 1 \\ -\eta & \eta \end{pmatrix} = \begin{pmatrix} 1 & 0 \\ 0 & \eta \end{pmatrix} \begin{pmatrix} 1 & 1 \\ -1 & 1 \end{pmatrix}, \text{ and}$$

$$\mathbf{A} = \begin{pmatrix} \mathrm{e}^{\imath n k \Delta} & 0 \\ 0 & \mathrm{e}^{-\imath n k \Delta} \end{pmatrix}$$

E_1, and H_1 are the field values at the source-side interface and E_2, H_2 are the values at the output interface, Δ is a thickness of the film, $k = 2\pi/\lambda$, (λ is free-space wavelength), $\eta = \sqrt{\mu/\varepsilon}$ and $n = \sqrt{\mu\varepsilon}$ are the effective impedance and the refractive index of the film, respectively, and $\eta_0 = \sqrt{\mu_0/\varepsilon_0}$ is the impedance of free-space.

With this simple assumption it is thought that a cascaded bulk material can be arranged using a stack of q elementary layers with an effective transformation matrix $\mathbf{M}^q = \mathbf{VA}^q\mathbf{V}^{-1}$. In general, this straightforward approach assumes that the spatial harmonics of each layer interact only with the same harmonics of other layers in the stack. In essence, this loose assumption ignores any transformation of a given incident harmonic into the spatial harmonics of different order, which are either reflected or transmitted.

To illustrate this issue, consider another approach to obtaining effective parameters of a multilayer NIM arranged of thin infinite elementary layers with periodic distribution of elementary materials. Essentially, the enhanced method follows the recipe for a classical case of stratified media (e.g., Luneburg, 1964).

2.1. Bloch–Floquet waves in cascaded layers

A simpler 2D example is used here to illustrate the approach, just because derivations for the spatial harmonic analysis (SHA) in 2D are less difficult. Consider a single period (l) of an infinite interface of a free-space domain with the domain of a material characterized by a set of stepwise continuous permittivity values ($\varepsilon_{1,1}$, $\varepsilon_{2,1}$, ...), as shown in fig. 12(a). Introduce a local coordinate system, with the unit normal ($\hat{\mathbf{x}}$), the unit transverse vector ($\hat{\mathbf{z}}$) and the tangent unit vector ($\hat{\mathbf{y}}$). Provided that a TM

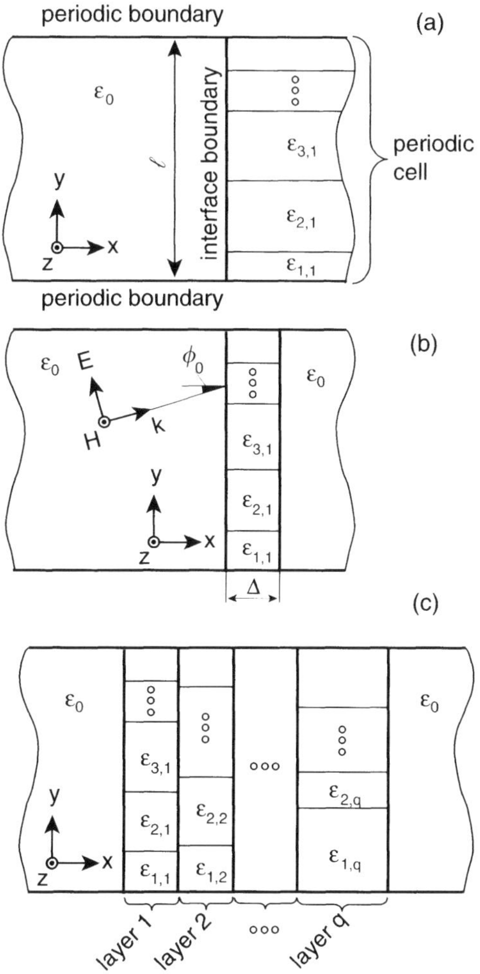

Fig. 12. (a) Interface of an elementary layer of NIM with free space. (b) An isolated elementary layer. (c) A cascaded multilayer NIM.

boundary-value problem is taken ($\vec{H} = \hat{\mathbf{z}}h$), then only the tangential components of the $\vec{H}$ and $\vec{E}$ field distributions at the interfaces are required in this case. Then, consider two scalar fields (h and d) as the distribution of the transverse magnetic field ($h = \hat{\mathbf{z}} \cdot \vec{H}$) and the distribution of the electric field ($d = \varepsilon\hat{\mathbf{y}} \cdot \vec{E}$). A monochromatic Maxwell equation $\hat{\mathbf{y}} \cdot (\nabla h \times \hat{\mathbf{z}}) = -\imath\omega d$ couples the fields

$$d = (\imath\omega)^{-1}h'. \tag{2.2}$$

Here the normal derivative of h is denoted as h', where $h' = \hat{\mathbf{x}} \cdot \nabla h$.

The core of any SHA approach is the transformation of the fields from a physical space to spatial spectral space using available proper functions (g^m). Provided that h and g^m are sets of discrete values obtained at a uniform grid on l, these sets are considered as two vectors (h and g^m). The following sum: $\Sigma_{p=-p_{\max}}^{p_{\max}} g_p^{v*} g_p^m$ is considered here as a scalar product of two vectors $(g^v, g^m)_l$, where p is a point of the grid on l arranged of $\widehat{p} = 2p_{\max} + 1$ points. Note that the proper functions g^m are orthonormal on l, i.e., $(g^v, g^m)_l = \delta(m - v)$.

Isolate, for example, the first elementary layer as shown in fig. 12(b). The magnetic field is defined by $h_1 = \Sigma_{m=-\infty}^{\infty} g_t^m c_{t,1}^m + g_r^m c_{r,1}^m$, where $g_t{}^m$ and $g_r{}^m$ are transmitted and reflected elementary fields of order m, respectively. In essence, the field h_1 is decomposed into elementary fields (the Bloch–Floquet waves), which are orthonormal on l. In a truncated approximation, $|m| < m_{\max}$, the vectors g^m can form a $\widehat{p} \times \widehat{m}$ matrix g, $\widehat{m} = 2m_{\max} + 1$, and the complex magnitudes of reflected and transmitted fields ($c_{r,1}^m$ and $c_{t,1}^m$) can be taken as the components of two different $\widehat{m}$-dimensional vectors $c_{r,1}$ and $c_{t,1}$. Then, the magnetic field in a matrix form is defined by $h_1 = g_{t,1}c_{t,1} + g_{r,1}c_{r,1}$.

The Bloch–Floquet theorem allows for the separation of variables, $g_t = v_1 u_1$ and $g_r = v_1 u_1^{-1}$, where u is a $\widehat{m} \times \widehat{m}$ matrix exponential, $u_1 = \exp \imath k_{x,1} kx$, of a proper values matrix, $k_{x,1}^m$; and v_1 is a $\widehat{p} \times \widehat{m}$ matrix constructed of orthogonal vectors v_1^m. For the free-space case, indices in u, v, and $k_x{}^m$ are dropped, and the proper functions u are defined through $k_x^m = [1 - (k_y^m)^2]^{1/2}$, where $k_y^m = m\lambda l^{-1} + \sin\phi_0$ and ϕ_0 is the angle of incidence (shown in fig. 12(b)). The wavefront $v^m = \widehat{p}^{\,-1/2} \exp \imath k_y^m ky$ is just an orthonormal Fourier component of the m-th order.

2.2. *Eigenvalue problem*

To obtain both $k_{x,1}$ and v_1 for a given elementary layer, where the permittivity of elementary materials (a piecewise continuous function ε_1) is periodic in y direction, but constant in x direction, it is necessary to attain an eigenvalue problem formulation. Since in this case (Luneburg, 1964),

$$k^2\varepsilon_1 h_1 + \nabla^2 h_1 - f_1 \partial_y h_1 = 0, \tag{2.3}$$

where f_1 is the logarithmic derivative, $f_1 = \varepsilon_1' \varepsilon_1^{-1}$, ε_1 and ε_1^{-1} are $\widehat{p} \times \widehat{p}$ diagonal matrices, and k^2 a scalar.

Then, the above equation can be rewritten using $h_1 = v_1(u_1 c_{t,1} - u_1^{-1} c_{r,1})$. Next, introducing a_1 as a $\widehat{m} \times \widehat{m}$ matrix mapping an orthogonal basis v_1 into the free-space basis v ($v_1 = va_1$), substituting $\imath(v^{-1} f_1 v) k_y$ with $\imath(k_y v^{-1}\varepsilon_1 v - v^{-1}\varepsilon_1 v k_y)\imath v^{-1}\varepsilon_1^{-1} v k_y$ and using γ_1 for $v^{-1}\varepsilon_1 v$ and i for the $\widehat{m} \times \widehat{m}$ identity matrix, eq. (2.3) is further simplified as

$$a_1 k_{x,1}^2 a_1^{-1} = \gamma_1 \left(\mathrm{i} - k_y \gamma_1^{-1} k_y\right). \tag{2.4}$$

The transform a_1 is required because in contrast with the free-space case, each wavefront v_1^m in an elementary inhomogeneous layer is not a single Fourier component anymore; but as a "physical function"[2] it still can be expressed as a superposition of the Fourier components. Note that eq. (2.4) is written in an eigenvalue form since $k_{x,1}^2$ is a diagonal matrix. The equation can be solved either numerically or analytically for both $k_{x,1}^2$ and a_1, provided that γ_1 and k_y are known.

2.3. *Mixed boundary-value problem*

Transverse magnetic field continuity together with the conservation of the tangential electric field gives the standard boundary conditions (BC), $h_0 = h_1$, $\varepsilon_1 h_0' = h_1'$, where the pairs h, h' represent the magnetic field and its normal derivative just before and after the interface; ε_1 is the value of permittivity. First, to simplify the notations (and further programming) a matrix nomenclature is defined as

$$\mathbf{n}(m_1, m_2) = \begin{pmatrix} m_1 & m_2 \\ m_2 & m_1 \end{pmatrix}, \quad \mathbf{d}(m_1, m_2) = \begin{pmatrix} m_1 & \mathrm{o} \\ \mathrm{o} & m_2 \end{pmatrix}, \quad \mathbf{i} = \begin{pmatrix} \mathrm{i} & -\mathrm{i} \\ \mathrm{i} & \mathrm{i} \end{pmatrix},$$
$$\mathbf{s}(v_1, v_2) = \begin{pmatrix} v_1 \\ v_2 \end{pmatrix}, \tag{2.5}$$

where **d**, **n**, and **i** are partitioned square matrices, o, i are two $\widehat{m} \times \widehat{m}$ matrices (the null matrix and the identity matrix, respectively), and m_1, m_2 are general square matrices of the same size; s is a stacked vector made of two equal vectors (v_1, v_2) with $\widehat{m}$ components.

[2] That is, a piecewise continuous function with limited variation on l.

After using $h_0 = v\,(uc_{t,0} - u^{-1}c_{r,0})$, $h_1 = va_1(u_1c_{t,1} - u_1^{-1}c_{r,1})$, and taking the normal derivatives, a spectral form of the BC is

$$c_0 = s_1c_1. \tag{2.6}$$

Here $c_0 = \mathbf{d}(u, u^{-1})\mathrm{s}(c_{t,0}, c_{r,0})$, $c_1 = \mathbf{d}(u_1, u_1^{-1})\mathrm{s}(c_{t,1}, c_{r,1})$, and $s_1 = \mathbf{i}^{-1}\mathbf{d}(a_1, k_x^{-1}\gamma_1^{-1}a_1k_{x,1})\mathbf{i}$.

At the second interface (as shown in fig. 12(b)), the equation for the elementary layer is given by

$$b_1c_1 = s_1^{-1}c_2. \tag{2.7}$$

Here $b_1 = \mathbf{d}(\beta_1, \beta_1^{-1})$, the matrix exponential $\beta_1 = \exp \imath k_{x,1}\delta_1$ adjusts the phases for a scaled thickness of layer ($\delta_1 = 2\pi\Delta_1/\lambda$).

Combining eqs. (2.6) and (2.7) gives the following form:

$$c_0 = s_1b_1^{-1}s_1^{-1}c_2. \tag{2.8}$$

Since k_x is given as a common matrix for all layers, a possible alternative is employing the following normalization $\mathbf{c} = \mathbf{i}c$, where the upper and lower partitions of $\mathbf{c}$ correspond to a magnetic component and a normalized electric component, respectively. These Fourier components are both continuous across any interlayer interface and form the basis for wave matching. From now on the arguments in $\mathbf{d}(\tilde{a}^{-1}, \tilde{k}_x^{-1}\tilde{a}^{-1}\tilde{\gamma})$ are dropped and a subscript is added to denote a layer number starting from the source side; thus, eq. (2.8) is simplified to

$$\mathbf{c}_0 = \mathbf{d}_1\mathbf{b}_1^{-1}\mathbf{d}_1^{-1}\mathbf{c}_2, \tag{2.9}$$

where the linear operators $\mathbf{d}_1 = \mathbf{d}(a_1, k_x^{-1}\gamma_1^{-1}a_1k_{x,1})$ and $\mathbf{b}_1 = \mathbf{i}b_1\mathbf{i}^{-1}$ are individual for each layer with a given distribution of elementary materials (γ_1), and defined matrices of the proper values $k_{x,1}$, and the proper vectors a_1.

If the trivial case of a uniform slab of permittivity $\bar{\varepsilon}_1$ is taken, then $a_1 = \mathrm{i}$, $\gamma_1 = \bar{\varepsilon}_1\mathrm{i}$ and a generalized analog of eq. (2.1) is

$$\begin{pmatrix} \tilde{h}_0 \\ \tilde{h}'_0 \end{pmatrix} = \begin{pmatrix} \mathrm{i} & \mathrm{o} \\ \mathrm{o} & \eta_1 \end{pmatrix}\begin{pmatrix} \mathrm{i} & -\mathrm{i} \\ \mathrm{i} & \mathrm{i} \end{pmatrix}\begin{pmatrix} \beta_1^{-1} & \mathrm{o} \\ \mathrm{o} & \beta_1 \end{pmatrix}\begin{pmatrix} \mathrm{i} & -\mathrm{i} \\ \mathrm{i} & \mathrm{i} \end{pmatrix}^{-1} \begin{pmatrix} \mathrm{i} & \mathrm{o} \\ \mathrm{o} & \eta_1^{-1} \end{pmatrix}\begin{pmatrix} \tilde{h}_2 \\ \tilde{h}'_2 \end{pmatrix}, \tag{2.10}$$

where the tangential fields are the correspondent matrices of the Fourier transforms, $\tilde{h} = v^{-1}h$, $\tilde{h}' = (\imath kv)^{-1}h'$; $\eta_1 = \mathrm{diag}(\bar{\varepsilon}_1^{-1}k_x^m)$ (from eq. (2.4), $k_{x,1}^2 = \varepsilon_1\mathrm{i} - k_y^2$, and $k_{x,1}^m = [\bar{\varepsilon}_1 - (k_y^m)^2]^{1/2}$). Then for example, validation of eq. (2.10) for a plane wave at normal incidence gives a familiar result, shown earlier in eq. (1.1). Note that in general, the following identity holds:

$$\begin{pmatrix} \tilde{h}_0 \\ \tilde{h}'_0 \end{pmatrix} = \begin{pmatrix} a_1 & \mathrm{o} \\ \mathrm{o} & k_x^{-1}\gamma_1^{-1}a_1 k_{x,1} \end{pmatrix} \begin{pmatrix} \cos k_{x,1}\delta_1 & -\imath \sin k_{x,1}\delta_1 \\ -\imath \sin k_{x,1}\delta_1 & \cos k_{x,1}\delta_1 \end{pmatrix} \begin{pmatrix} a_1^{-1} & \mathrm{o} \\ \mathrm{o} & k_{x,1}^{-1}a_1^{-1}\gamma_1 k_x \end{pmatrix} \begin{pmatrix} \tilde{h}_2 \\ \tilde{h}'_2 \end{pmatrix}, \quad (2.11)$$

where $\cos k_{x,1}\delta_1$ and $\sin k_{x,1}\delta_1$ are arranged of the adequate matrix exponentials.

2.4. A simple validation test

A simplified single-layer model for validating the simulation method is shown in fig. 13(a). The sample structure is intentionally made of a very thin metallic grating (with 10 nm thickness). The grating is arranged of 400 nm gold strips separated by narrow strips of silica, the period of the structure is 480 nm. A large aspect ratio of the metallic strips and large electric resonance at about 1.2 μm are among the main challenges of the test model. To obtain a good set of the reference data the structure was simulated using a commercial software package with fifth-order finite elements. The validity of the FEM solution was verified by using the same model with different levels of additional meshing refinement and an adaptive solver. The results were stable upon the use of 41,000 degrees of freedom (field variables), where the bulk of the resources had been spent for the free-space buffer, nonreflecting layers, and adequate meshing at the corners.

In contrast to FEM, the spatial harmonic analysis appeared much more efficient. The problem was stabilized after the use of 11 eigenvalues with the calculation time of about 100 times less (4 s) versus the FEM solver with the same number of wavelength points and the same hardware. It should be noted that the amount of simulation time using SHA is approximately proportional to the total number of elementary layers and scales approximately as the square (or cube) of the total number of eigenvalues in 2D (or 3D) problems, while the performance of FEM solvers is decreasing very moderately with the increase in the total number of stacked layers.

Both models appeared to be quite sensitive to the material properties of the metal. Although in both cases, the interpolated complex refractive

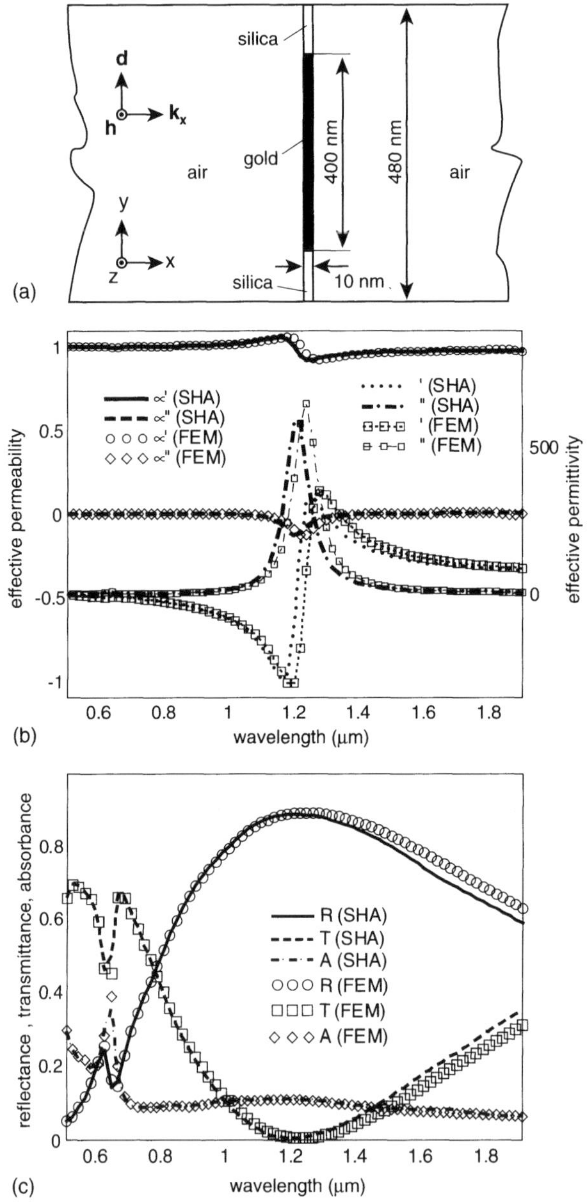

Fig. 13. (a) Sketch of a resonant elementary layer. (b) Effective permeability and permittivity obtained for the elementary layer using FEM and SHA (using 11 eigenvalues). (c) Comparison of the reflection, transmission, and absorption spectra obtained in simulations using FEM and SHA (11 eigenvalues).

index was based on the experimental table of Johnson and Christy (1972). In our opinion, the overall differences in the effective characteristics shown in figs. 13(b) and 13(c) are most likely due to the differences in the interpolation approach used in the FEM solver.

Above a simple validation of the modeling approach, the test model of fig. 13(a) reveals all the typical features of the periodic structures with localized plasmonic resonances, e.g., at the same wavelength the electric resonance is always accompanied by a satellite magnetic antiresonance and vice versa.

2.5. Cascading the elementary layers

Using eq. (2.8) cascading a subset of q elementary layers (depicted in fig. 12(b)) gives

$$c_0 = w_q c_q. \tag{2.12}$$

Here $w_q = \prod_{v=1}^{q} \tilde{w}_v$, $\tilde{w}_v = s_v b_v^{-1} s_v^{-1}$ is an elementary transform due to v-th layer and q the total number of layers in the subset stack.

2.6. Reflection and transmission coefficients

The major workload in the above method falls on the calculation of the proper values and vectors ($k_{x,v}$ and a_v) for each elementary layer. Once the values are obtained, the characteristic matrices of each layer are arranged as $s_v b_v^{-1} s_v^{-1}$.

Introducing the transformations $r_v c_{t,v} = c_{r,v}$, $t_v c_{t,v} = c_{t,q}$ (with the initial values given by $t_q = \mathrm{i}$ and $r_q = \mathrm{o}$), and then from eq. (2.12) the matrices of spatial spectral reflectance and transmittance are defined as

$$\begin{aligned} r_{v-1} &= \left(w_v^{1,0} + w_v^{1,1}\rho_v\right)\left(w_v^{0,0} + w_v^{0,1} r_v\right)^{-1}, \\ t_{v-1} &= t_v\left(w_v^{0,0} + w_v^{0,1} r_v\right)^{-1}, \end{aligned} \tag{2.13}$$

where the characteristic matrix is partitioned as

$$\tilde{w} = \left(\begin{array}{c|c} w^{0,0} & w^{0,1} \\ \hline w^{1,0} & w^{1,1} \end{array}\right).$$

The partitions are calculated using s_v, which is a matrix with symmetrical partitions, $s_v = \mathbf{n}(s_v^0, s_v^1)$ with $s_v^0 = \frac{1}{2}(k_x^{-1}\gamma_v^{-1} a_v k_{x,v} + a_v)$ and $s_v^1 = \frac{1}{2}(k_x^{-1}\gamma_v^{-1} a_v k_{x,v} - a_v)$ and a similar matrix $s_v^{-1} = \mathbf{n}(\tilde{s}_v^0, \tilde{s}_v^1)$ with

$\tilde{s}_\nu^0 = \frac{1}{2}(k_{x,\nu}^{-1} a_\nu^{-1} \gamma_\nu k_x + a_\nu^{-1})$ and $\tilde{s}_\nu^1 = \frac{1}{2}(k_{x,\nu}^{-1} a_\nu^{-1} \gamma_\nu k_x - a_\nu^{-1})$, then

$$w_\nu^{0,0} = s_\nu^0 \beta_\nu^{-1} \tilde{s}_\nu^0 + s_\nu^1 \beta_\nu \tilde{s}_\nu^1, \quad w_\nu^{0,1} = s_\nu^0 \beta_\nu^{-1} \tilde{s}_\nu^1 + s_\nu^1 \beta_\nu \tilde{s}_\nu^0,$$
$$w_\nu^{1,0} = s_\nu^1 \beta_\nu^{-1} \tilde{s}_\nu^0 + s_\nu^0 \beta_\nu \tilde{s}_\nu^1, \quad w_\nu^{1,1} = s_\nu^1 \beta_\nu^{-1} \tilde{s}_\nu^1 + s_\nu^0 \beta_\nu \tilde{s}_\nu^0. \tag{2.14}$$

Thus, for example, a single-layer structure is calculated as follows: $t_0 = (w_1^{0,0})^{-1}$, $r_0 = w_1^{1,0}(w_1^{0,0})^{-1}$; then, the transmitted and reflected Bloch–Floquet waves are $c_{t,1} = t_0 c_{t,0}$ and $c_{r,0} = r_0 c_{t,0}$.

2.7. *Discussions*

It also follows from the analysis of eqs. (2.8)–(2.14) that:

(i) None of asymmetric multilayer composites can be effectively described either by the simplified homogenization approach (eq. (2.1)) or through its generalized analog (eq. (2.10)). (A multilayer composite is asymmetric, if it contains an odd number of elementary layers, and the layers are not mirror-symmetric relative to the central layer; all structures with even number of distinct layers are always asymmetric.)

(ii) Effective optical parameters (including an effective negative refractive index) obtained in a single symmetric subset of elementary layers may not guarantee the same effective parameters in a bulk material arranged of identical subsets, not just because of absorptive losses, but also due to new interactions of near-field waves introduced by the use of cascading.

To illustrate (i) consider a classical example of a subset structure with two homogeneous lossless layers (Born and Wolf, 2002, p. 72). The characteristic matrix of two layers with thicknesses Δ_1 and Δ_2, and indices n_1 and η_1, is computed using eq. (2.10) as $\mathbf{w}_2 = \tilde{\mathbf{w}}_1 \tilde{\mathbf{w}}_2$, with $\tilde{\mathbf{w}}_1 = \mathbf{d}_1 \mathbf{b}_1^{-1} \mathbf{d}_1^{-1}$ and $\tilde{\mathbf{w}}_2 = \mathbf{d}_2 \mathbf{b}_2^{-1} \mathbf{d}_2^{-1}$. The effective characteristic matrix ($\mathbf{w}_{\text{eff}}$) of an equivalent single layer, which is defined as

$$\mathbf{w}_{\text{eff}} = \left(\begin{array}{c|c} \mathbf{w}_{\text{eff}}^{0,0} & \mathbf{w}_{\text{eff}}^{0,1} \\ \hline \mathbf{w}_{\text{eff}}^{1,0} & \mathbf{w}_{\text{eff}}^{1,1} \end{array}\right) = \begin{pmatrix} \cos n_{\text{eff}} k_x \delta & -\imath \eta_{\text{eff}} \sin n_{\text{eff}} k_x \delta \\ -\imath \eta_{\text{eff}}^{-1} \sin n_{\text{eff}} k_x \delta & \cos n_{\text{eff}} k_x \delta \end{pmatrix}, \tag{2.15}$$

(using a scaled thickness $\delta = k(\Delta_1 + \Delta_2)$ and the effective index n_{eff}), should be equal to the characteristic matrix of the double layer $\mathbf{w}_2$. To be equivalent to $\mathbf{w}_{\text{eff}}$, the product $\mathbf{w}_2 = \tilde{\mathbf{w}}_1 \tilde{\mathbf{w}}_2$, must have identical diagonal partitions since $\mathbf{w}_{\text{eff}}^{0,0} = \mathbf{w}_{\text{eff}}^{1,1}$ in eq. (2.15). This is true only if the product commutes, i.e., $\tilde{\mathbf{w}}_1\tilde{\mathbf{w}}_2 = \tilde{\mathbf{w}}_2\tilde{\mathbf{w}}_1$ leaving the only trivial case of $n_1 = n_2$ possible. Therefore, even a simple stack of two distinct lossless films cannot be adequately modeled by a single effective layer. Physically, the condition $\tilde{\mathbf{w}}_1\tilde{\mathbf{w}}_2 = \tilde{\mathbf{w}}_2\tilde{\mathbf{w}}_1$ means that the effective parameters of a multilayer NIM should not depend on the side that is chosen for illumination, i.e., its structure should be symmetric.

Note that although $\tilde{\mathbf{w}}_1\tilde{\mathbf{w}}_2\tilde{\mathbf{w}}_3 = \tilde{\mathbf{w}}_3\tilde{\mathbf{w}}_2\tilde{\mathbf{w}}_1$ is always true for any triple-layered structure, since the first and the last layers are equal ($\tilde{\mathbf{w}}_1 = \tilde{\mathbf{w}}_3$), the homogenization of $\mathbf{w}_3 = \mathbf{d}_1(\mathbf{b}_1^{-1}\mathbf{d}_{12}\mathbf{b}_2^{-1}\mathbf{d}_{12}^{-1}\mathbf{b}_1^{-1})\mathbf{d}_1^{-1}$ is not very simple even for the structure with homogeneous elementary layers.

Now to exemplify (ii) consider a cascaded structure arranged of identical symmetric substructures. A three-layer substructure is of natural choice here, since the majority of known structures falling in this category (as shown in figs. 2, 3, 7, and 8) are $w_3 = s_1 b_1^{-1} s_{12} b_2^{-1} s_{12}^{-1} b_1^{-1} s_1^{-1}$, where $s_{12} = s_1^{-1} s_2 = \mathbf{i}^{-1}\mathbf{d}_1^{-1}\mathbf{d}_2\mathbf{i}$. The diagonally partitioned matrix $\mathbf{d}_1^{-1}\mathbf{d}_2$ is responsible for interactions between the layers. Cascading p triple-layer substructures suggests taking the p-th power of the characteristic matrix w_3. Although the result is straightforward since $(w_3)^p = s_1(b_1^{-1} s_{12} b_2^{-1} s_{12}^{-1} b_1^{-1})^p s_1^{-1}$, it is clear that new interactions of near-field waves introduced by cascading will change the effective properties of the cascaded structure in comparison to those of the initial three-layer substructure, unless it is possible to write w_3 as $w_3 = s_{\mathrm{eff}} b_{\mathrm{eff}} s_{\mathrm{eff}}^{-1}$, where b_{eff} is a diagonal matrix of effective eigenvalues and s_{eff} a matrix of effective eigenvectors.

§ 3. Combining magnetic resonators with semicontinuous films

The above method built on SHA has been usefully applied to the practical problem of optimizing nanostrip magnetic resonators in combination with homogeneous and later with semicontinuous metal films (Chettiar et al., 2006). A unit cell of the geometry is shown in fig. 14(a) along with the orientation of the incident field. It consists of two main parts, a pair of nanostrips, which act as a magnetic resonator giving negative permeability, and two metal films adding negative permittivity. Note that resonant conditions are required to give negative permeability. In contrast, negative permittivity is given by the bulk metal of the metal films. The silica layer in the left represents the substrate over which the structure is fabricated. The thin layer of silica on the right is to protect the top metal layer from oxidation or other possible degradations.

It is preferable to use noble metals (gold and silver) because of losses. But the metal films create a problem when designing an NIM at the

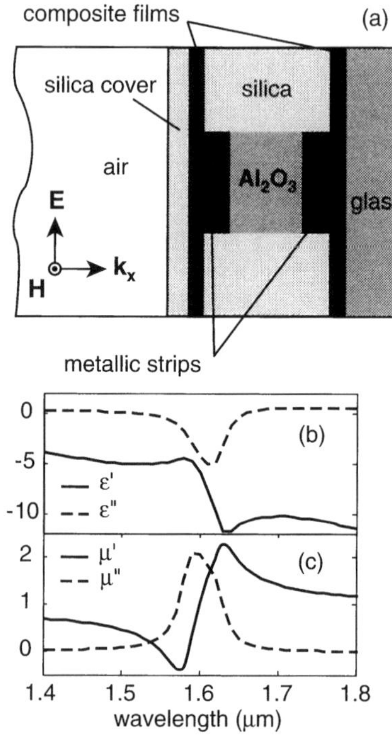

Fig. 14. (a) Schematic representation of the NIM structure. (b) Effective permittivity for a representative structure. (c) Effective permeability.

telecommunication wavelength of 1.5 μm. Alas, at a wavelength of 1.5 μm both gold and silver have highly negative permittivity, while the nanostrip magnetic resonator cannot provide a permeability that is comparably negative. This results in a huge impedance mismatch causing low transmittance. The problem can be circumvented by using a thinner metal film. But the metal film thickness cannot be reduced arbitrarily as the minimum thickness depends critically on the material properties and fabrication limitations.

But instead of using bulk metal, the films can be formed of a mixture of silica and metal. These materials are known as semicontinuous metal films. Such a film can be fabricated using very basic techniques. The permittivity of semicontinuous metal films is described by the effective medium theory (EMT). According to EMT, semicontinuous metal films have a permittivity that is much less negative as compared to bulk metal films. Hence, a semicontinuous metal film is thick enough to be fabricated easily while having a permittivity similar in magnitude as the permeability created by magnetic resonators. This is critical for impedance matching.

The structures were simulated using SHA. Selected results were verified with the results obtained using a commercial FEM solver. The effective material properties were extracted from the simulated reflection and transmission data. Figures 14(b) and 14(c) show the effective permittivity and permeability, respectively, for sample geometry. For this case the semicontinuous metal films had a metal filling fraction of 65% and the metal was silver. We note that for some wavelengths both permittivity and permeability have negative real parts.

Figure 15(a) shows the effective refractive index and we note that the real part of the refractive index is negative between 1500 and 1650 nm. Figure 15(b) shows the transmittance and reflectance spectra. The transmittance spectrum has a maximum value of 27% at 1570 nm. The refractive index has a value of $-1.85+0.93i$ at the wavelength of

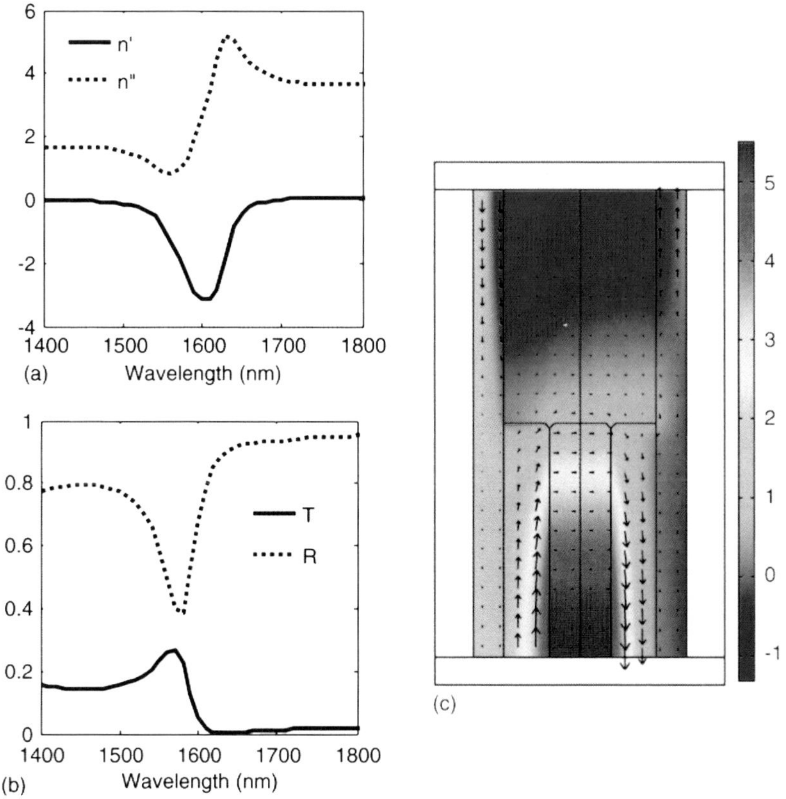

Fig. 15. (a) Effective refractive index for a representative structure. (b) Transmittance and reflectance spectra. (c) Arrows of electric field vectors plotted on top of the gray scale map of the normalized magnetic field.

maximum transmittance. A related structure was simulated in a commercial FEM solver to obtain the field plots, which would provide us with a much better understanding of the nature of the resonance. The structure was simulated at a wavelength of 1570 nm and the resulting quiver plot of the displacement vector field is shown in figure 15(c). For saving time, only one half of the structure was simulated along with proper application of mirror boundary conditions. From the arrow plot it is clear that there is a strong magnetic resonance due to the two coupled silver strips, which is instrumental in providing the negative permeability.

3.1. *Sensitivity of the design*

The basis structure designed to operate at 1.5 μm is shown in fig. 16(a). The structure is a combination of a periodic metal-dielectric composite (PMD) and a semicontinuous silver film (SSF). The PMD is arranged of infinitely long silver strips separated by an alumina spacer. To prevent aging of the film, an additional layer of silica is added on top of the upper SSF. (The effect of the protective layer has been investigated and appears to be of limited importance.)

The structure demonstrates both negative permittivity (fig. 16(b)) and permeability (fig. 16(c)) about the target wavelength of 1.5 μm. To explore sensitivity of the design to geometrical parameters a number of geometrical characteristics have been changed. For example, fig. 16 shows permittivity (fig. 16(b)), permeability (fig. 16(c)), and refractive index (fig. 16(d)) spectra changing according to the variation of the width W of the PMD shown in fig. 16(a). For example, the spectra in figs. 16(b–e) are depicted for the variation of W ranging from 260 to 380 nm with 20 nm stepping. Simulations demonstrate stable negative refractive index near 1.5 μm (illustrated in the inset, fig. 16(e)).

3.2. *Conclusion*

Recently, metamaterials have been designed that show a negative real part of the refractive index at the telecom wavelength of 1500 nm or 200 THz. Keeping in mind that it took only 5 years to come from 10 GHz to 200 THz, we have no doubt that a negative refractive index metamaterial will be soon available also for the visible range. We have shown in numerical simulations that the following two key remedies are now available to overcome major obstacles that currently limit the

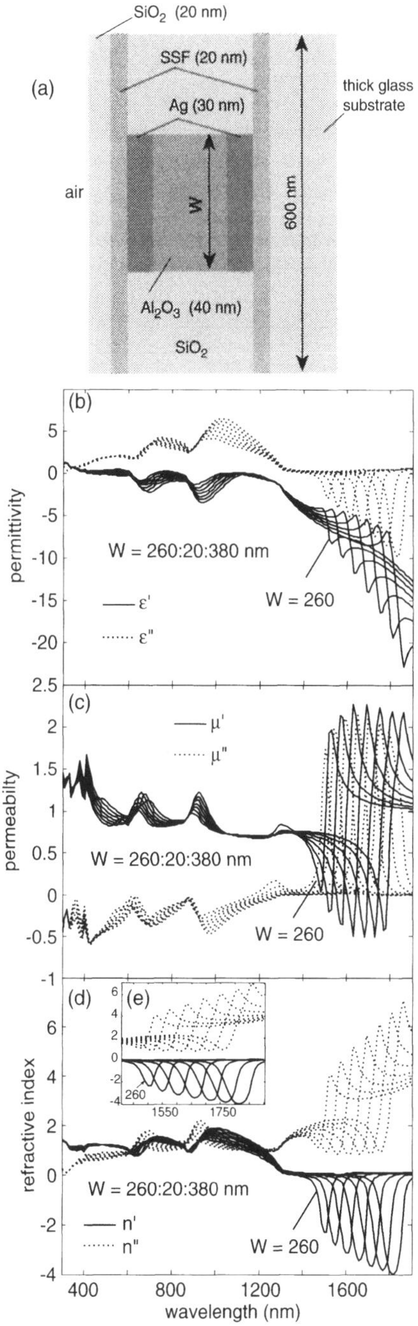

Fig. 16. NIM with semicontinuous films. (a) Basis structure. (b) Permittivity spectra. (c) Permeability spectra. (d) Refractive index spectra for different widths of the silver strips. (e) Zoomed region of negative refraction.

development of optical NIM: (1) impedance-matching designs are capable of suppressing high reflectance, and (2) gain materials embedded in metallic nanostructures can fully compensate for absorptive losses while still retaining the negative refractive index.

The instrumental idea of the new design of fig. 16(a) is based on combining narrow-band resonant PMDs and broadband nonresonant films. While a PMD provides negative effective permeability, an adequate negative effective permittivity is achieved due to nonresonant SSFs.

The approach has been accomplished in the following steps addressing major difficulties of the design process:

- the basis periodic structure (PMD) has electric and magnetic responses at different wavelengths. A remedy is to use different substructures in order to obtain negative magnetic and electric responses at the same wavelength;
- any basis periodic resonant structure exhibits a complementary antiresonance. While for example, a normal magnetic resonance of μ' shown in fig. 9(b) is paired with the anomalous electric resonance (antiresonance) at the target wavelength of 584 nm, a normal resonance of ε' (fig. 13(b)) of an electrically resonant periodic structure of fig. 13a comes with the anomalous magnetic resonance (μ' in fig. 13(b)) at about 1.2 μm. These anti-resonances could decrease (and at times nullify) the effect of magnetic resonance provided that an isolated element of PMD is both magnetically and electrically resonant at the same target wavelength. Therefore, it is essential to use a substructure that has no electrical antiresonance. Silver films are used to add negative permittivity since silver provides negative permittivity naturally (at nonresonant condition);
- the interaction between the silver film and the magnetic substructure gives rise to additional resonances. Therefore, the film is fused with the magnetic substructure to minimize those additional resonances;
- a homogeneous silver film provides an intense negative permittivity and the permeability provided by the magnetic substructure (PMD) cannot match it. As a result, in impedance mismatch causes excess reflection. That is why; semicontinuous metal films with controlled effective permittivity are used instead of continuous metal films.

The design starts with the initial optimization of a given PMD to obtain an optimal negative magnetic response for the structure. Once the negative magnetic response is preliminary optimized, metal films are added to provide an adequate negative electric response. The combined multilayered structure requires additional optimization due to the interaction of the elementary layers of PMDs with metallic strips.

Acknowledgment

We would like to acknowledge fruitful collaboration with V. A. Podolskiy, A. K. Sarychev, W. Cai, U. K. Chettiar, and H.-K. Yuan.

References

Avrutsky, I., 2004, Phys. Rev. B **70**, 155416.
Bergman, D.J., Stockman, M.I., 2003, Phys. Rev. Lett. **90**, 027402.
Berrier, A., Mulot, M., Swillo, M., Qiu, M., Thylen, L., Talneau, A., Anand, S., 2004, Phys. Rev. Lett. **93**, 073902.
Born, M., Wolf, E., 1964, Principles of Optics, Cambridge University Press, Cambridge.
Caloz, C., Itoh, T., 2006, Electromagnetic Metamaterials, John Wiley, Hoboken.
Chettiar, U.K., Kildishev, A.V., Klar, T.A., Yuan, H.-K., Cai, W., Sarychev, A.K., Drachev, V.P., Shalaev, V.M., 2006, From low-loss to lossless optical negtive index materials, CLEO/QELS Conference, Long Beach, CA.
Depine, R.A., Lakhtakia, A., 2004, Microw. Opt. Technol. Lett. **41**, 315–316.
Drachev, V.P., Cai, W., Chettiar, U., Yuan, H.K., Sarychev, A.K., Kildishev, A.V., Klimeck, G., Shalaev, V.M., 2006, Laser Phys. Lett. **3**, 49–55.
Dulkeith, E., Morteani, A.C., Niedereichholz, T., Klar, T.A., Feldmann, J., Levi, S.A., van Veggel, F.C.J.M., Reinhoudt, D.N., Möller, M., Gittins, D.I., 2002, Phys. Rev. Lett. **89**, 203002.
Engheta, N., Salandrino, A., Alu, A., 2005, Phys. Rev. Lett. **95**, 095504.
Falcone, F., Lopetegi, T., G. Laso, M.A., Baena, J.D., Bonache, J., Beruete, M., Marques, R., Martin, F., Sorolla, M., 2004, Phys. Rev. Lett. **93**, 197401.
Gralak, B., Enoch, S., Tayeb, G., 2000, J. Opt. Soc. Am. A **17**, 1012–1020.
Grigorenko, A.N., Geim, A.K., Gleeson, H.F., Zhang, Y., Firsov, A.A., Khrushchev, I.Y., Petrovic, J., 2005, Nature **438**, 335–338.
Holloway, C.L., Kuester, E.F., Baker-Jarvis, J., Kabos, P., 2003, IEEE Trans. Antennas Propagation **51**, 2596–2603.
Holzer, W., Gratz, H., Schmitt, T., Penzkofer, A., Costela, A., Garcia-Moreno, I., Sastre, R., Duarte, F.J., 2000, Chem. Phys. **256**, 125–136.
Imahori, H., Kashiwagi, Y., Endo, Y., Hanada, T., Nishimura, Y., Yamazaki, I., Araki, Y., Ito, O., Fukuzumi, S., 2004, Langmuir **20**, 73–81.
Johnson, P.B., Christy, R.W., 1972, Phys. Rev. B **6**, 4370–4379.
Kildishev, A.V., Cai, W., Chettiar, U.K., Yuan, H.K., Sarychev, A.K., Drachev, V.P., Shalaev, V.M., 2006, J. Opt. Soc. Am. B **23**, 423–433.
Kim, W., Safonov, V.P., Shalaev, V.M., Armstrong, R.L., 1999, Phys. Rev. Lett. **82**, 4811–4814.
Kosaka, H., Kawashima, T., Tomita, A., Notomi, M., Tamamura, T., Suto, T., Kawakami, S., 1998, Phys. Rev. B **58**, 10096–10099.
Lagarkov, A.N., Sarychev, A.K., 1996, Phys. Rev. B **53**, 6318–6336.
Lamb, H., 1904, Proc. Lond. Math. Soc. **1**, 473–479.
Lawandy, N.M., 2004, Appl. Phys. Lett. **85**, 5040–5042.

Leatherdale, C.A., Woo, W.K., Mikulec, F.V., Bawendi, M.G., 2002, J. Phys. Chem. B **106**, 7619–7622.
Lu, Z., Murakowski, J.A., Schuetz, C.A., Shi, S., Schneider, G.J., Prather, D.W., 2005, Phys. Rev. Lett. **95**, 153901.
Luneburg, R.K., 1964, Mathematical Theory of Optics, UCLA Press, Berkeley.
Luo, C., Johnson, S.G., Joannopoulos, J.D., Pendry, J.B., 2002, Phys. Rev. B **65**, 201104.
Luo, C., Johnson, S.G., Joannopoulos, J.D., Pendry, J.B., 2003, Phys. Rev. B **68**, 045115.
Mandel'shtam, L.I., 1945, Zh. Eksp. Teor. Fiz. **15**, 475–478.
Moser, H.O., Casse, B.D.F., Wilhelmi, O., Saw, B.T., 2005, Phys. Rev. Lett. **94**, 063901.
Nezhad, M.P., Tetz, K., Fainman, Y., 2004, Opt. Expr. **12**, 4072–4079.
Notomi, M., 2000, Phys. Rev. B **62**, 10696–10705.
Panina, L.V., Grigorenko, A.N., Makhnovskiy, D.P., 2002, Phys. Rev. B **66**, 155411.
Pendry, J.B., 2000, Phys. Rev. Lett. **85**, 3966–3969.
Pendry, J.B., Holden, A.J., Stewart, W.J., Youngs, I., 1996, Phys. Rev. Lett. **76**, 4773–4776.
Pendry, J.B., Holden, A.J., Robbins, D.J., Stewart, W.J., 1999, IEEE Trans. Microw. Theory Tech. **47**, 2075–2084.
Podolskiy, V.A., Alekseev, L., Narimanov, E.E., 2005a, J. Mod. Opt. **52**, 2343–2349.
Podolskiy, V.A., Narimanov, E.E., 2005, Phys. Rev. B **71**, 201101(R).
Podolskiy, V.A., Sarychev, A.K., Narimanov, E.E., Shalaev, V.M., 2005b, J. Opt. A **7**, S32–S37.
Podolskiy, V.A., Sarychev, A.K., Shalaev, V.M., 2002, J. Nonlinear Opt. Phys. Mater. **11**, 65–74.
Podolskiy, V.A., Sarychev, A.K., Shalaev, V.M., 2003, Opt. Expr. **11**, 735–745.
Ramakrishna, S.A., Pendry, J.B., 2003, Phys. Rev. B **67**, 201101.
Schuster, A., 1904, An Introduction to the Theory of Optics, Edward Arnold, London.
Seidel, J., Grafström, S., Eng, L., 2005, Phys. Rev. Lett. **94**, 177401.
Shalaev, V.M., Cai, W., Chettiar, U.K., Yuan, H.K., Sarychev, A.K., Drachev, V.P., Kildishev, A.V., 2005a, http://www.arxiv.org/abs/physics/0504091, Apr. 13, 2005.
Shalaev, V.M., Cai, W., Chettiar, U.K., Yuan, H.K., Sarychev, A.K., Drachev, V.P., Kildishev, A.V., 2005b, Opt. Lett. **30**, 3356–3358.
Shelby, R.A., Smith, D.R., Nemat-Nasser, S.C., Schultz, S., 2001a, Appl. Phys. Lett. **78**, 489–491.
Shelby, R.A., Smith, D.R., Schultz, S., 2001b, Science **292**, 77–79.
Smith, D.R., Padilla, W.J., Vier, D.C., Nemat-Nasser, S.C., Schultz, S., 2000, Phys. Rev. Lett. **84**, 4184–4187.
Smith, D.R., Schultz, S., Markos, P., Soukoulis, C.M., 2002, Phys. Rev. B **65**, 195104.
Smith, D.R., Schurig, D., Rosenbluth, M., Schultz, S., Ramakrishna, S.A., Pendry, J.B., 2003, Appl. Phys. Lett. **82**, 1506–1508.
Stehr, J., Crewett, J., Schindler, F., Sperling, R., von Plessen, G., Lemmer, U., Lupton, J.M., Klar, T.A., Feldmann, J., Holleitner, A.W., Forster, M., Scherf, U., 2003, Adv. Mater. **15**, 1726.
Sudarkin, A.N., Demkovich, P.A., 1989, Soviet Physics, Tech. Phys. **34**, 764–766.
Tredicucci, A., Gmachl, C., Capasso, F., Hutchinson, A.L., Sivco, D.L., Cho, A.Y., 2000, Appl. Phys. Lett. **76**, 2164–2166.
Veselago, V.G., 1968, Soviet Physics Uspekhi **10**, 509–514.
Yen, T.J., Padilla, W.J., Fang, N., Vier, D.C., Smith, D.R., Pendry, J.B., Basov, D.N., Zhang, X., 2004, Science **303**, 1494–1496.
Zhang, S., Fan, W., Malloy, K.J., Brueck, S.R.J., Panoiu, N.C., Osgood, R.M., 2005a, Opt. Expr. **13**, 4922–4930.
Zhang, S., Fan, W., Malloy, K.J., Brueck, S.R.J., Panoiu, N.C., Osgood, R.M., 2006, J. Opt. Soc. Am. B **23**, 434–438.
Zhang, S., Fan, W., Panoiu, N.C., Malloy, K.J., Osgood, R.M., Brueck, S.R.J., 2005b, Phys. Rev. Lett. **95**, 137404.

Author Index

H

I

J

K

L

M

O

P

Q

R

T

Z

Subject Index